WORLD HEALTH ORGANIZATION

INTERNATIONAL AGENCY FOR RESEARCH ON CANCER

IARC MONOGRAPHS
ON THE
EVALUATION OF CARCINOGENIC RISKS TO HUMANS

Solar and Ultraviolet Radiation

VOLUME 55

This publication represents the views and expert opinions
of an IARC Working Group on the
Evaluation of Carcinogenic Risks to Humans,
which met in Lyon,

11–18 February 1992

1992

IARC MONOGRAPHS

In 1969, the International Agency for Research on Cancer (IARC) initiated a programme on the evaluation of the carcinogenic risk of chemicals to humans involving the production of critically evaluated monographs on individual chemicals. In 1980 and 1986, the programme was expanded to include the evaluation of the carcinogenic risk associated with exposures to complex mixtures and other agents.

The objective of the programme is to elaborate and publish in the form of monographs critical reviews of data on carcinogenicity for agents to which humans are known to be exposed, and on specific exposure situations, to evaluate these data in terms of human risk with the help of international working groups of experts in chemical carcinogenesis and related fields; and to indicate where additional research efforts are needed.

This project is supported by PHS Grant No. 2-UO1 CA33193-10 awarded by the US National Cancer Institute, Department of Health and Human Services. Additional support has been provided since 1986 by the Commission of the European Communities.

©International Agency for Research on Cancer 1992

ISBN 92 832 1255 X

ISSN 0250-9555

All rights reserved. Application for rights of reproduction or translation, in part or *in toto*, should be made to the International Agency for Research on Cancer.

Distributed for the International Agency for Research on Cancer
by the Secretariat of the World Health Organization

PRINTED IN THE UNITED KINGDOM

CONTENTS

NOTE TO THE READER 11
LIST OF PARTICIPANTS 13
PREAMBLE
    Background 19
    Objective and Scope 19
    Selection of Topics for Monographs 20
    Data for Monographs 21
    The Working Group 21
    Working Procedures 21
    Exposure Data 22
    Evidence for Carcinogenicity in Humans 23
    Studies of Cancer in Experimental Animals 27
    Other Relevant Data 29
    Summary of Data Reported 30
    Evaluation 32
    References 36
GENERAL REMARKS 39
SOLAR AND ULTRAVIOLET RADIATION
1. Exposure data 43
   1.1 Nomenclature 43
      1.1.1 Optical radiation 43
      1.1.2 Quantities and units 45
      1.1.3 Units of biologically effective ultraviolet radiation 46
   1.2 Methods for measuring ultraviolet radiation 47
      1.2.1 Spectroradiometry 47
      1.2.2 Wavelength-independent (thermal) detectors 48
      1.2.3 Wavelength-dependent detectors 48
   1.3 Sources and exposures 49
      1.3.1 Solar ultraviolet radiation 50
         (a) Measurements of terrestrial solar radiation 54
         (b) Personal exposures 57
      1.3.2 Exposure to artificial sources of ultraviolet radiation .. 58
         (a) Sources 58
            (i) Incandescent sources 58
            (ii) Gas discharge lamps 59

             (iii) Arc lamps 59
             (iv) Fluorescent lamps 59
             (v) Metal halide lamps 59
             (vi) Electrodeless lamps 59
        (b) Human exposure 60
             (i) Cosmetic use 60
             (ii) Medical and dental applications 63
             (iii) Occupational exposures 66
             (iv) General lighting 70
        (c) Regulations and guidelines 70
             (i) Cosmetic use 70
             (ii) Occupational exposure 71
2. Studies of cancer in humans 73
   2.1 Solar radiation 73
      2.1.1 Nonmelanocytic skin cancer 73
          (a) Case reports 73
              (i) Studies of xeroderma pigmentosum patients 73
              (ii) Studies of transplant recipients 73
          (b) Descriptive studies 74
              (i) Host factors 74
              (ii) Anatomical distribution 74
              (iii) Geographical variation 75
              (iv) Migration 76
              (v) Occupation 76
          (c) Cross-sectional studies 77
          (d) Case–control studies 83
          (e) Cohort studies 86
          (f) Collation of results 91
      2.1.2 Cancer of the lip 93
          (a) Descriptive studies 93
              (i) Geographical variation 93
              (ii) Occupation 94
          (b) Case–control studies 94
      2.1.3 Malignant melanoma of the skin 95
          (a) Case reports 95
          (b) Descriptive studies 95
              (i) Sex distribution 95
              (ii) Age distribution 95
              (iii) Anatomical distribution 96
              (iv) Ethnic origin 96
              (v) Geographical variation 96
              (vi) Migration 99
              (vii) Socioeconomic status and occupation 99

			(c) Case–control studies	100

- (c) Case–control studies 100
    - (i) Australia 100
    - (ii) Europe 102
    - (iii) North America 106
- (d) Collation of results 113
    - (i) Total sun exposure: potential exposure by place of residence 113
    - (ii) Biological response to total sun exposure 113
    - (iii) Total sun exposure assessed by questionnaire 115
    - (iv) Short periods of residence implying high potential exposure 115
    - (v) Occupational exposure 115
    - (vi) Intermittent exposure 115
    - (vii) Sunburn 122
- 2.1.4 Malignant melanoma of the eye 122
    - (a) Case reports 122
    - (b) Descriptive studies 125
        - (i) Ethnic origin 125
        - (ii) Place of birth and residence 125
        - (iii) Occupation 127
        - (iv) History of skin cancer 127
    - (c) Case–control studies 127
- 2.1.5 Other cancers 130
- 2.2 Artificial sources of ultraviolet radiation 130
    - 2.2.1 Nonmelanocytic skin cancer 130
    - 2.2.2 Malignant melanoma of the skin 130
    - 2.2.3 Malignant melanoma of the eye 134
- 2.3 Premalignant conditions 134
    - 2.3.1 Basal-cell naevus syndrome 134
    - 2.3.2 Dysplastic naevus syndrome 134
- 2.4 Molecular genetics of human skin cancers 135
    - 2.4.1 *ras* Gene mutations 135
    - 2.4.2 p53 Gene mutations 135

3. Studies of cancer in animals 139
    - 3.1 Experimental conventions 139
        - 3.1.1 Species studied 139
        - 3.1.2 Wavelength ranges 139
        - 3.1.3 Measured doses 140
        - 3.1.4 Protocols 140
    - 3.2 Broad-spectrum radiation 141
        - 3.2.1 Sunlight 141
        - 3.2.2 Solar-simulated radiation 142
        - 3.2.3 Sources emitting UVC, UVB and UVA radiation 142
    - 3.3 Sources emitting mainly UVB radiation 144
        - 3.3.1 Mouse 144

		3.3.2 Rat	146
		3.3.3 Hamster	146
		3.3.4 Guinea-pig	146
		3.3.5 Fish	146
		3.3.6 Opossum	146
	3.4	Sources emitting mainly UVC radiation	147
		3.4.1 Mouse	147
		3.4.2 Rat	148
	3.5	Sources emitting mainly UVA radiation	148
	3.6	Interaction of wavelengths	150
		3.6.1 Interaction of exposures given on the same day	150
		3.6.2 Long-term interactions	151
	3.7	Additional experimental observations	151
		3.7.1 Tumour types	151
		3.7.2 Dose and effect	153
		3.7.3 Dose delivery	154
		3.7.4 Action spectra	154
		3.7.5 Pigmentation	155
	3.8	Administration with known chemical carcinogens	155
		3.8.1 Administration with polycyclic aromatic hydrocarbons	156
		(a) 3,4-Benzo[a]pyrene	156
		(b) 7,12-Dimethylbenz[a]anthracene	156
		3.8.2 Administration with other agents with promoting activity	157
		(a) Croton oil	157
		(b) 12-O-Tetradecanoylphorbol 13-acetate	158
		(c) Benzoyl peroxide	158
		(d) Methyl ethyl ketone peroxide	159
	3.9	Interaction with immunosuppressive agents	160
	3.10	Molecular genetics of animal skin tumours induced by ultraviolet radiation	161
4.	Other relevant data		163
	4.1	Transmission and absorption in biological tissues	163
		4.1.1 Epidermis	163
		(a) Humans	163
		(b) Experimental systems	164
		(c) Epidermal chromophores	165
		(d) Enhancement of epidermal penetration of ultraviolet radiation	166
		4.1.2 Eye	166
		(a) Humans	166
		(b) Experimental systems	166
	4.2	Adverse effects (other than cancer)	167
		4.2.1 Epidermis	167
		(a) Humans	167
		(i) Erythema and pigmentation (sunburn and suntanning)	167

CONTENTS

		(ii) Pigmented naevi	169
		(iii) Ultrastructural changes	170
		(iv) Keratosis	172
		(v) Photosensitivity disorders	172
	(b)	Experimental systems	173
	(c)	Comparison of humans and animals	174
4.2.2	Immune response		175
	(a)	Humans	175
		(i) Contact hypersensitivity (allergy)	175
		(ii) Lymphocytes	176
		(iii) Infectious diseases	177
		(iv) Photosensitive diseases	177
	(b)	Experimental systems	177
		(i) Contact hypersensitivity	177
		(ii) Delayed hypersensitivity to injected antigens	179
		(iii) Immunology of ultraviolet-induced skin cancer	180
		(iv) Transplantation immunity	180
		(v) Infectious diseases	181
		(vi) Human lymphocytes *in vitro*	182
	(c)	Comparison of humans and animals	182
4.2.3	Eye		183
	(a)	Humans	183
		(i) Anterior eye (cornea, conjunctiva)	183
		(ii) Lens	183
		(iii) Posterior eye	183
	(b)	Experimental systems	184
		(i) Anterior eye	184
		(ii) Lens	184
		(iii) Posterior eye	184
	(c)	Comparison of humans and animals	184
4.3 Photoproduct formation			185
4.3.1	DNA photoproducts		185
	(a)	Cyclobutane-type pyrimidine dimers	185
	(b)	Pyrimidine–pyrimidone (6-4) photoproducts	186
	(c)	Thymine glycols	187
	(d)	Cytosine damage	188
	(e)	Purine damage	188
	(f)	DNA strand breaks	188
	(g)	DNA–protein cross-links	189
4.3.2	Other chromophores and targets		189
	(a)	Chromophores	189
	(b)	Membranes	190

- 4.4 Human excision repair disorders 191
  - 4.4.1 Xeroderma pigmentosum 191
  - 4.4.2 Trichothiodystrophy 192
  - 4.4.3 Cockayne's syndrome 193
  - 4.4.4 Role of immunosuppression 193
- 4.5 Genetic and related effects 194
  - 4.5.1 Humans 194
    - (a) Epidermis 195
      - (i) Broad-spectrum ultraviolet radiation, including solar simulation 195
      - (ii) UVA radiation 196
      - (iii) UVB radiation 196
      - (iv) UVC radiation 197
    - (b) Lymphocytes 198
      - (i) Broad-spectrum ultraviolet radiation 198
      - (ii) UVA radiation 199
      - (iii) UVB radiation 199
  - 4.5.2 Experimental systems 199
    - (a) DNA damage 199
    - (b) Mutagenicity 200
    - (c) Chromosomal effects 202
    - (d) Transformation 203
    - (e) Effects of cellular and viral gene expression 204

5. Summary of data reported and evaluation 217
   - 5.1 Exposure data 217
   - 5.2 Human carcinogenicity data 218
     - 5.2.1 Solar radiation 218
       - (a) Nonmelanocytic skin cancer 218
       - (b) Cancer of the lip 219
       - (c) Malignant melanoma of the skin 219
       - (d) Melanoma of the eye 220
       - (e) Other cancers 220
     - 5.2.2 Artificial sources of ultraviolet radiation 220
     - 5.2.3 Molecular genetics of human skin cancers 221
   - 5.3 Carcinogenicity in experimental animals 221
   - 5.4 Other relevant data 222
     - 5.4.1 Transmission and absorption 222
     - 5.4.2 Effects on the skin 222
     - 5.4.3 Effects on the immune response 222
     - 5.4.4 DNA photoproducts 223
     - 5.4.5 Genetic and related effects 223
   - 5.5 Evaluation 227

6. References 229

CONTENTS

SUMMARY OF FINAL EVALUATIONS 281

GLOSSARY OF TERMS 283

Appendix 1. Topical sunscreens 285
1. General 285
2. Protective effects 286
   2.1 Against DNA damage 286
   2.2 Against acute and chronic actinic damage 286
   2.3 Against immunological alterations 286
   2.4 Against tumour formation 286
3. Adverse effects 287
   3.1 Acute toxicity 287
   3.2 Chronic toxicity 287
   3.3 Reduced vitamin D synthesis 287
4. References 288

CUMULATIVE INDEX TO THE *MONOGRAPHS* SERIES 291

NOTE TO THE READER

The term 'carcinogenic risk' in the *IARC Monographs* series is taken to mean the probability that exposure to an agent will lead to cancer in humans.

Inclusion of an agent in the *Monographs* does not imply that it is a carcinogen, only that the published data have been examined. Equally, the fact that an agent has not yet been evaluated in a monograph does not mean that it is not carcinogenic.

The evaluations of carcinogenic risk are made by international working groups of independent scientists and are qualitative in nature. No recommendation is given for regulation or legislation.

Anyone who is aware of published data that may alter the evaluation of the carcinogenic risk of an agent to humans is encouraged to make this information available to the Unit of Carcinogen Identification and Evaluation, International Agency for Research on Cancer, 150 cours Albert Thomas, 69372 Lyon Cedex 08, France, in order that the agent may be considered for re-evaluation by a future Working Group.

Although every effort is made to prepare the monographs as accurately as possible, mistakes may occur. Readers are requested to communicate any errors to the Unit of Carcinogen Identification and Evaluation, so that corrections can be reported in future volumes.

IARC WORKING GROUP ON THE EVALUATION OF CARCINOGENIC RISKS TO HUMANS
VOLUME 55: SOLAR AND ULTRAVIOLET RADIATION

Lyon, 11–18 February 1992

LIST OF PARTICIPANTS

Members[1]

C. Arlett, MRC Cell Mutation Unit, University of Brighton, Falmer, Brighton BN1 9RR, United Kingdom

B. Bridges, MRC Cell Mutation Unit, University of Brighton, Falmer, Brighton BN1 9RR, United Kingdom (*Chairman*)

A. Brøgger, Department of Genetics, Institute for Cancer Research, Montebello, 0310 Oslo 3, Norway

B.L. Diffey, Regional Medical Physics Department, Dryburn Hospital, Durham DH1 5TW, United Kingdom

J.M. Elwood, Hugh Adam Cancer Epidemiology Unit, University of Otago Medical School, PO Box 913, Dunedin, New Zealand

E.A. Emmett, Worksafe Australia, National Occupational Health and Safety Commission, 92 Parramatta Road, Camperdown, NSW 2050, Australia

D. English, NHMRC Research Unit, The Queen Elizabeth II Medical Centre, University of Western Australia, Nedlands, WA 6009, Australia

P.D. Forbes, Temple University, Biohazards Control Office, Environmental Health and Safety Offices, 3307 North Broad Street, Philadelphia, PA 19140, USA

R.P. Gallagher, British Columbia Cancer Agency, 600 West 10th Avenue, Vancouver, BC V52 4E6, Canada

J.W. Grisham, Department of Pathology, University of North Carolina, Brinkhous-Bullitt Building CB# 7525, Chapel Hill, NC 27599, USA

---

[1]Unable to attend: J. Marshall, Department of Ophthalmology, Block 8, UNDS, St Thomas's Hospital, London SE1 7EH, United Kingdom

K. Kraemer, National Cancer Institute, Division of Cancer Etiology, Laboratory of Molecular Carcinogenesis, Building 37, Room 3D-06, Bethesda, MD 20892, USA

J.C. van der Leun, Institute of Dermatology, University Hospital Utrecht, Heidelberglaan 100, 3584 CX Utrecht, The Netherlands

T. Mack, University of Southern California, School of Medicine, Department of Preventive Medicine, Parkview Medical Building, 1420 San Pablo Street, Los Angeles, CA 90033-9987, USA (*Vice-Chairman*)

W.L. Morison, Johns Hopkins University, Baltimore, MD 21205, USA

S. Olin, ILSI Risk Science Institute, 1126 Sixteenth Street NW, Washington DC 20036, USA

A. Østerlind, Danish Cancer Registry, Rosenvaengets Hovedvej 35, Box 839, 2100 Copenhagen, Denmark

D.H. Sliney, Laser Branch, US Army Environmental Hygiene Agency, Aberdeen Proving Ground, MD 21010-5422, USA

F. Stenbäck, Department of Pathology, University of Oulu, Kajaanintie 52 D, 90220 Oulu 22, Finland

R.M. Tyrrell, Swiss Institute for Experimental Cancer Research, 1066 Epalinges-sur-Lausanne, Switzerland

A.R. Young, Photobiology Department, St John's Institute of Dermatology, St Thomas's Hospital, London SE1 7EH, United Kingdom

Representative of the US Food and Drug Administration

J.Z. Beer, US Food and Drug Administration, Center for Devices and Radiological Health, 12709 Twinbrook Parkway, Rockville, MD 20852, USA

Observers

R.C. Burton, Department of Surgery, John Hunter Hospital, Locked Bag #1, Newcastle Mail Centre, Newcastle, NSW 2310, Australia

C.J. Portier, National Institute of Environmental Health Sciences, PO Box 12233, Research Triangle Park, NC 27709, USA

Secretariat

B. Armstrong, Deputy Director
H. Bartsch, Unit of Environmental and Host Factors
P. Boffetta, Unit of Analytical Epidemiology
J.R.P. Cabral, Unit of Mechanisms of Carcinogenesis
E. Cardis, Director's Office
M. Friesen, Unit of Environmental and Host Factors
M.-J. Ghess, Unit of Carcinogen Identification and Evaluation
J. Hall, Unit of Mechanisms of Carcinogenesis
E. Heseltine, Lajarthe, St Léon-sur-Vézère, France
A. Kricker, Unit of Descriptive Epidemiology

PARTICIPANTS

V. Krutovskikh, Unit of Multistage Carcinogenesis
D. McGregor, Unit of Carcinogen Identification and Evaluation
D. Mietton, Unit of Carcinogen Identification and Evaluation
H. Møller, Unit of Carcinogen Identification and Evaluation
R. Montesano, Unit of Mechanisms of Carcinogenesis
T. Nakazawa, Unit of Multistage Carcinogenesis
I. O'Neill, Unit of Environmental and Host Factors
M. Parkin, Unit of Descriptive Epidemiology
C. Partensky, Unit of Carcinogen Identification and Evaluation
I. Peterschmitt, Unit of Carcinogen Identification and Evaluation, Geneva, Switzerland
D. Shuker, Unit of Environmental and Host Factors
L. Tomatis, Director
H. Vainio, Unit of Carcinogen Identification and Evaluation
J. Wilbourn, Unit of Carcinogen Identification and Evaluation
H. Yamasaki, Unit of Multistage Carcinogenesis

Secretarial assistance

J. Cazeaux
M. Lézère
S. Reynaud

PREAMBLE

IARC MONOGRAPHS PROGRAMME ON THE EVALUATION OF CARCINOGENIC RISKS TO HUMANS[1]

PREAMBLE

1. BACKGROUND

In 1969, the International Agency for Research on Cancer (IARC) initiated a programme to evaluate the carcinogenic risk of chemicals to humans and to produce monographs on individual chemicals. The *Monographs* programme has since been expanded to include consideration of exposures to complex mixtures of chemicals (which occur, for example, in some occupations and as a result of human habits) and of exposures to other agents, such as radiation and viruses. With Supplement 6 (IARC, 1987a), the title of the series was modified from *IARC Monographs on the Evaluation of the Carcinogenic Risk of Chemicals to Humans* to *IARC Monographs on the Evaluation of Carcinogenic Risks to Humans*, in order to reflect the widened scope of the programme.

The criteria established in 1971 to evaluate carcinogenic risk to humans were adopted by the working groups whose deliberations resulted in the first 16 volumes of the *IARC Monographs* series. Those criteria were subsequently updated by further ad-hoc working groups (IARC, 1977, 1978, 1979, 1982, 1983, 1987b, 1988, 1991a; Vainio *et al.*, 1992).

2. OBJECTIVE AND SCOPE

The objective of the programme is to prepare, with the help of international working groups of experts, and to publish in the form of monographs, critical reviews and evaluations of evidence on the carcinogenicity of a wide range of human exposures. The *Monographs* may also indicate where additional research efforts are needed.

The *Monographs* represent the first step in carcinogenic risk assessment, which involves examination of all relevant information in order to assess the strength of the available evidence that certain exposures could alter the incidence of cancer in humans. The second step is quantitative risk estimation. Detailed, quantitative evaluations of epidemiological data may be made in the *Monographs*, but without extrapolation beyond the range of the data

---

[1]This project is supported by PHS Grant No. 2 UO1 CA33193-10 awarded by the US National Cancer Institute, Department of Health and Human Services. Since 1986, the programme has also been supported by the Commission of the European Communities.

available. Quantitative extrapolation from experimental data to the human situation is not undertaken.

The term 'carcinogen' is used in these monographs to denote an exposure that is capable of increasing the incidence of malignant neoplasms; the induction of benign neoplasms may in some circumstances (see p. 28) contribute to the judgement that the exposure is carcinogenic. The terms 'neoplasm' and 'tumour' are used interchangeably.

Some epidemiological and experimental studies indicate that different agents may act at different stages in the carcinogenic process, and several different mechanisms may be involved. The aim of the *Monographs* has been, from their inception, to evaluate evidence of carcinogenicity at any stage in the carcinogenesis process, independently of the underlying mechanisms. Information on mechanisms may, however, be used in making the overall evaluation (IARC, 1991a; Vainio *et al.*, 1992; see also pp. 33-34).

The *Monographs* may assist national and international authorities in making risk assessments and in formulating decisions concerning any necessary preventive measures. The evaluations of IARC working groups are scientific, qualitative judgements about the evidence for or against carcinogenicity provided by the available data. These evaluations represent only one part of the body of information on which regulatory measures may be based. Other components of regulatory decisions may vary from one situation to another and from country to country, responding to different socioeconomic and national priorities. **Therefore, no recommendation is given with regard to regulation or legislation, which are the responsibility of individual governments and/or other international organizations.**

The *IARC Monographs* are recognized as an authoritative source of information on the carcinogenicity of a wide range of human exposures. A users' survey, made in 1988, indicated that the *Monographs* are consulted by various agencies in 57 countries. Each volume is generally printed in 4000 copies for distribution to governments, regulatory bodies and interested scientists. The *Monographs* are also available *via* the Distribution and Sales Service of the World Health Organization.

3. SELECTION OF TOPICS FOR MONOGRAPHS

Topics are selected on the basis of two main criteria: (a) there is evidence of human exposure, and (b) there is some evidence or suspicion of carcinogenicity. The term 'agent' is used to include individual chemical compounds, groups of related chemical compounds, physical agents (such as radiation) and biological factors (such as viruses). Exposures to mixtures of agents may occur in occupational exposures and as a result of personal and cultural habits (like smoking and dietary practices). Chemical analogues and compounds with biological or physical characteristics similar to those of suspected carcinogens may also be considered, even in the absence of data on a possible carcinogenic effect in humans or experimental animals.

The scientific literature is surveyed for published data relevant to an assessment of carcinogenicity. The IARC surveys of chemicals being tested for carcinogenicity (IARC, 1973-1990) and directories of on-going research in cancer epidemiology (IARC, 1976-1991) often indicate those exposures that may be scheduled for future meetings. Ad-hoc working groups convened by IARC in 1984, 1989 and 1991 gave recommendations as to which agents should be evaluated in the *IARC Monographs* series (IARC, 1984, 1989, 1991b).

As significant new data on subjects on which monographs have already been prepared become available, re-evaluations are made at subsequent meetings, and revised monographs are published.

4. DATA FOR MONOGRAPHS

The *Monographs* do not necessarily cite all the literature concerning the subject of an evaluation. Only those data considered by the Working Group to be relevant to making the evaluation are included.

With regard to biological and epidemiological data, only reports that have been published or accepted for publication in the openly available scientific literature are reviewed by the working groups. In certain instances, government agency reports that have undergone peer review and are widely available are considered. Exceptions may be made on an ad-hoc basis to include unpublished reports that are in their final form and publicly available, if their inclusion is considered pertinent to making a final evaluation (see pp. 32 *et seq.*). In the sections on chemical and physical properties, on analysis, on production and use and on occurrence, unpublished sources of information may be used.

5. THE WORKING GROUP

Reviews and evaluations are formulated by a working group of experts. The tasks of the group are: (i) to ascertain that all appropriate data have been collected; (ii) to select the data relevant for the evaluation on the basis of scientific merit; (iii) to prepare accurate summaries of the data to enable the reader to follow the reasoning of the Working Group; (iv) to evaluate the results of experimental and epidemiological studies on cancer; (v) to evaluate data relevant to the understanding of mechanism of action; and (vi) to make an overall evaluation of the carcinogenicity of the exposure to humans.

Working Group participants who contributed to the considerations and evaluations within a particular volume are listed, with their addresses, at the beginning of each publication. Each participant who is a member of a working group serves as an individual scientist and not as a representative of any organization, government or industry. In addition, nominees of national and international agencies and industrial associations may be invited as observers.

6. WORKING PROCEDURES

Approximately one year in advance of a meeting of a working group, the topics of the monographs are announced and participants are selected by IARC staff in consultation with other experts. Subsequently, relevant biological and epidemiological data are collected by IARC from recognized sources of information on carcinogenesis, including data storage and retrieval systems such as BIOSIS, Chemical Abstracts, CANCERLIT, MEDLINE and TOXLINE—including EMIC and ETIC for data on genetic and related effects and teratogenicity, respectively.

For chemicals and some complex mixtures, the major collection of data and the preparation of first drafts of the sections on chemical and physical properties, on analysis, on production and use and on occurrence are carried out under a separate contract funded by the US National Cancer Institute. Representatives from industrial associations may assist in the preparation of sections on production and use. Information on production and trade is obtained from governmental and trade publications and, in some cases, by direct contact with industries. Separate production data on some agents may not be available because their publication could disclose confidential information. Information on uses may be obtained from published sources but is often complemented by direct contact with manufacturers. Efforts are made to supplement this information with data from other national and international sources.

Six months before the meeting, the material obtained is sent to meeting participants, or is used by IARC staff, to prepare sections for the first drafts of monographs. The first drafts are compiled by IARC staff and sent, prior to the meeting, to all participants of the Working Group for review.

The Working Group meets in Lyon for seven to eight days to discuss and finalize the texts of the monographs and to formulate the evaluations. After the meeting, the master copy of each monograph is verified by consulting the original literature, edited and prepared for publication. The aim is to publish monographs within nine months of the Working Group meeting.

The available studies are summarized by the Working Group, with particular regard to the qualitative aspects discussed below. In general, numerical findings are indicated as they appear in the original report; units are converted when necessary for easier comparison. The Working Group may conduct additional analyses of the published data and use them in their assessment of the evidence; the results of such supplementary analyses are given in square brackets. When an important aspect of a study, directly impinging on its interpretation, should be brought to the attention of the reader, a comment is given in square brackets.

7. EXPOSURE DATA

Sections that indicate the extent of past and present human exposure, the sources of exposure, the people most likely to be exposed and the factors that contribute to the exposure are included at the beginning of each monograph.

Most monographs on individual chemicals, groups of chemicals or complex mixtures include sections on chemical and physical data, on analysis, on production and use and on occurrence. In monographs on, for example, physical agents, biological factors, occupational exposures and cultural habits, other sections may be included, such as: historical perspectives, description of an industry or habit, chemistry of the complex mixture or taxonomy.

For chemical exposures, the Chemical Abstracts Services Registry Number, the latest Chemical Abstracts Primary Name and the IUPAC Systematic Name are recorded; other synonyms are given, but the list is not necessarily comprehensive. For biological agents, taxonomy and structure are described, and the degree of variability is given, when applicable.

Information on chemical and physical properties and, in particular, data relevant to identification, occurrence and biological activity are included. For biological agents, mode of replication, life cycle, target cells, persistence and latency, host response and description of nonmalignant disease caused by them are given. A description of technical products of chemicals includes trades names, relevant specifications and available information on composition and impurities. Some of the trade names given may be those of mixtures in which the agent being evaluated is only one of the ingredients.

The purpose of the section on analysis is to give the reader an overview of current methods, with emphasis on those widely used for regulatory purposes. Methods for monitoring human exposure are also given, when available. No critical evaluation or recommendation of any of the methods is meant or implied. The IARC publishes a series of volumes, *Environmental Carcinogens: Methods of Analysis and Exposure Measurement* (IARC, 1978–91), that describe validated methods for analysing a wide variety of chemicals and mixtures. For biological agents, methods of detection and exposure assessment are described, including their sensitivity, specificity and reproducibility.

The dates of first synthesis and of first commercial production of a chemical or mixture are provided; for agents which do not occur naturally, this information may allow a reasonable estimate to be made of the date before which no human exposure to the agent could have occurred. The dates of first reported occurrence of an exposure are also provided. In addition, methods of synthesis used in past and present commercial production and different methods of production which may give rise to different impurities are described.

Data on production, international trade and uses are obtained for representative regions, which usually include Europe, Japan and the USA. It should not, however, be inferred that those areas or nations are necessarily the sole or major sources or users of the agent. Some identified uses may not be current or major applications, and the coverage is not necessarily comprehensive. In the case of drugs, mention of their therapeutic uses does not necessarily represent current practice nor does it imply judgement as to their therapeutic efficacy.

Information on the occurrence of an agent or mixture in the environment is obtained from data derived from the monitoring and surveillance of levels in occupational environments, air, water, soil, foods and animal and human tissues. When available, data on the generation, persistence and bioaccumulation of the agent are also included. In the case of mixtures, industries, occupations or processes, information is given about all agents present. For processes, industries and occupations, a historical description is also given, noting variations in chemical composition, physical properties and levels of occupational exposure with time. For biological agents, the epidemiology of infection is described.

Statements concerning regulations and guidelines (e.g., pesticide registrations, maximal levels permitted in foods, occupational exposure limits) are included for some countries as indications of potential exposures, but they may not reflect the most recent situation, since such limits are continuously reviewed and modified. The absence of information on regulatory status for a country should not be taken to imply that that country does not have regulations with regard to the exposure. For biological agents, legislation and control, including vaccines and therapy, are described.

8. EVIDENCE FOR CARCINOGENICITY IN HUMANS

(a) *Types of studies considered*

Three types of epidemiological studies of cancer contribute to the assessment of carcinogenicity in humans—cohort studies, case–control studies and correlation studies. Rarely, results from randomized trials may be available. Case reports of cancer in humans may also be reviewed.

Cohort and case–control studies relate individual exposures under study to the occurrence of cancer in individuals and provide an estimate of relative risk (ratio of incidence in those exposed to incidence in those not exposed) as the main measure of association.

In correlation studies, the units of investigation are usually whole populations (e.g., in particular geographical areas or at particular times), and cancer frequency is related to a summary measure of the exposure of the population to the agent, mixture or exposure circumstance under study. Because individual exposure is not documented, however, a causal relationship is less easy to infer from correlation studies than from cohort and case–control studies. Case reports generally arise from a suspicion, based on clinical experience, that the concurrence of two events—that is, a particular exposure and occurrence of a cancer—has happened rather more frequently than would be expected by chance. Case reports usually lack complete ascertainment of cases in any population, definition or enumeration of the population at risk and estimation of the expected number of cases in the absence of exposure. The uncertainties surrounding interpretation of case reports and correlation studies make them inadequate, except in rare instances, to form the sole basis for inferring a causal relationship. When taken together with case–control and cohort studies, however, relevant case reports or correlation studies may add materially to the judgement that a causal relationship is present.

Epidemiological studies of benign neoplasms, presumed preneoplastic lesions and other end-points thought to be relevant to cancer are also reviewed by working groups. They may, in some instances, strengthen inferences drawn from studies of cancer itself.

(b) Quality of studies considered

The *Monographs* are not intended to summarize all published studies. Those that are judged to be inadequate or irrelevant to the evaluation are generally omitted. They may be mentioned briefly, particularly when the information is considered to be a useful supplement to that in other reports or when they provide the only data available. Their inclusion does not imply acceptance of the adequacy of the study design or of the analysis and interpretation of the results, and limitations are clearly outlined in square brackets at the end of the study description.

It is necessary to take into account the possible roles of bias, confounding and chance in the interpretation of epidemiological studies. By 'bias' is meant the operation of factors in study design or execution that lead erroneously to a stronger or weaker association than in fact exists between disease and an agent, mixture or exposure circumstance. By 'confounding' is meant a situation in which the relationship with disease is made to appear stronger or to appear weaker than it truly is as a result of an association between the apparent causal factor and another factor that is associated with either an increase or decrease in the incidence of the disease. In evaluating the extent to which these factors have been minimized in an individual study, working groups consider a number of aspects of design and analysis as described in the report of the study. Most of these considerations apply equally to case–control, cohort and correlation studies. Lack of clarity of any of these aspects in the reporting of a study can decrease its credibility and the weight given to it in the final evaluation of the exposure.

Firstly, the study population, disease (or diseases) and exposure should have been well defined by the authors. Cases of disease in the study population should have been identified in a way that was independent of the exposure of interest, and exposure should have been assessed in a way that was not related to disease status.

Secondly, the authors should have taken account in the study design and analysis of other variables that can influence the risk of disease and may have been related to the exposure of interest. Potential confounding by such variables should have been dealt with either in the design of the study, such as by matching, or in the analysis, by statistical adjustment. In cohort studies, comparisons with local rates of disease may be more appropriate than those with national rates. Internal comparisons of disease frequency among individuals at different levels of exposure should also have been made in the study.

Thirdly, the authors should have reported the basic data on which the conclusions are founded, even if sophisticated statistical analyses were employed. At the very least, they should have given the numbers of exposed and unexposed cases and controls in a case–control study and the numbers of cases observed and expected in a cohort study. Further tabulations by time since exposure began and other temporal factors are also important. In a cohort study, data on all cancer sites and all causes of death should have been given, to reveal the possibility of reporting bias. In a case–control study, the effects of investigated factors other than the exposure of interest should have been reported.

Finally, the statistical methods used to obtain estimates of relative risk, absolute rates of cancer, confidence intervals and significance tests, and to adjust for confounding should have been clearly stated by the authors. The methods used should preferably have been the generally accepted techniques that have been refined since the mid-1970s. These methods have been reviewed for case–control studies (Breslow & Day, 1980) and for cohort studies (Breslow & Day, 1987).

(c) Inferences about mechanism of action

Detailed analyses of both relative and absolute risks in relation to temporal variables, such as age at first exposure, time since first exposure, duration of exposure, cumulative exposure and time since exposure ceased, are reviewed and summarized when available. The analysis of temporal relationships can be useful in formulating models of carcinogenesis. In particular, such analyses may suggest whether a carcinogen acts early or late in the process of carcinogenesis, although at best they allow only indirect inferences about the mechanism of action. Special attention is given to measurements of biological markers of carcinogen exposure or action, such as DNA or protein adducts, as well as markers of early steps in the carcinogenic process, such as proto-oncogene mutation, when these are incorporated into epidemiological studies focused on cancer incidence or mortality. Such measurements may allow inferences to be made about putative mechanisms of action (IARC, 1991a; Vainio et al., 1992).

(d) Criteria for causality

After the quality of individual epidemiological studies of cancer has been summarized and assessed, a judgement is made concerning the strength of evidence that the agent, mixture or exposure circumstance in question is carcinogenic for humans. In making their judgement, the Working Group considers several criteria for causality. A strong association (i.e., a large relative risk) is more likely to indicate causality than a weak association, although it is recognized that relative risks of small magnitude do not imply lack of causality and may be important if the disease is common. Associations that are replicated in several studies of the same design or using different epidemiological approaches or under different circumstances of exposure are more likely to represent a causal relationship than isolated observations from single studies. If there are inconsistent results among investigations, possible reasons are sought (such as differences in amount of exposure), and results of studies judged to be of high quality are given more weight than those from studies judged to be methodologically less sound. When suspicion of carcinogenicity arises largely from a single study, these data are not combined with those from later studies in any subsequent reassessment of the strength of the evidence.

If the risk of the disease in question increases with the amount of exposure, this is considered to be a strong indication of causality, although absence of a graded response is not necessarily evidence against a causal relationship. Demonstration of a decline in risk after cessation of or reduction in exposure in individuals or in whole populations also supports a causal interpretation of the findings.

Although a carcinogen may act upon more than one target, the specificity of an association (i.e., an increased occurrence of cancer at one anatomical site or of one morphological type) adds plausibility to a causal relationship, particularly when excess cancer occurrence is limited to one morphological type within the same organ.

Although rarely available, results from randomized trials showing different rates among exposed and unexposed individuals provide particularly strong evidence for causality.

When several epidemiological studies show little or no indication of an association between an exposure and cancer, the judgement may be made that, in the aggregate, they show evidence of lack of carcinogenicity. Such a judgement requires first of all that the studies giving rise to it meet, to a sufficient degree, the standards of design and analysis described above. Specifically, the possibility that bias, confounding or misclassification of exposure or outcome could explain the observed results should be considered and excluded with reasonable certainty. In addition, all studies that are judged to be methodologically sound should be consistent with a relative risk of unity for any observed level of exposure and, when considered together, should provide a pooled estimate of relative risk which is at or near unity and has a narrow confidence interval, due to sufficient population size. Moreover, no individual study nor the pooled results of all the studies should show any consistent tendency for relative risk of cancer to increase with increasing level of exposure. It is important to note that evidence of lack of carcinogenicity obtained in this way from several epidemiological studies can apply only to the type(s) of cancer studied and to dose levels and intervals between first exposure and observation of disease that are the same as or less than those observed in all the studies. Experience with human cancer indicates that, in some cases, the period from first exposure to the development of clinical cancer is seldom less than 20 years; latent periods substantially shorter than 30 years cannot provide evidence for lack of carcinogenicity.

9. STUDIES OF CANCER IN EXPERIMENTAL ANIMALS

For several agents (e.g., aflatoxins, 4-aminobiphenyl, bis(chloromethyl)ether, diethylstilboestrol, melphalan, 8-methoxypsoralen (methoxsalen) plus ultraviolet radiation, mustard gas and vinyl chloride), evidence of carcinogenicity in experimental animals preceded evidence obtained from epidemiological studies or case reports. Information compiled from the first 41 volumes of the *IARC Monographs* (Wilbourn *et al.*, 1986) shows that, of the 44 agents and mixtures for which there is *sufficient* or *limited evidence* of carcinogenicity to humans (see p. 32), all 37 that have been tested adequately produce cancer in at least one animal species. Although this association cannot establish that all agents and mixtures that cause cancer in experimental animals also cause cancer in humans, nevertheless, **in the absence of adequate data on humans, it is biologically plausible and prudent to regard agents and mixtures for which there is sufficient evidence (see p. 33) of carcinogenicity in experimental animals as if they presented a carcinogenic risk to humans.** The possibility that a given agent may cause cancer through a species-specific mechanism which does not operate in humans (see p. 34) should also be taken into consideration.

The nature and extent of impurities or contaminants present in the chemical or mixture being evaluated are given when available. Animal strain, sex, numbers per group, age at start of treatment and survival are reported.

Other types of studies summarized include: experiments in which the agent or mixture was administered in conjunction with known carcinogens or factors that modify carcinogenic effects; studies in which the end-point was not cancer but a defined precancerous lesion; and experiments on the carcinogenicity of known metabolites and derivatives.

For experimental studies of mixtures, consideration is given to the possibility of changes in the physicochemical properties of the test substance during collection, storage, extraction, concentration and delivery. Chemical and toxicological interactions of the components of mixtures may result in nonlinear dose–response relationships.

An assessment is made as to the relevance to human exposure of samples tested in experimental systems, which may involve consideration of: (i) physical and chemical characteristics, (ii) constituent substances that indicate the presence of a class of substances, (iii) the results of tests for genetic and related effects, including genetic activity profiles, DNA adduct profiles, proto-oncogene mutation and expression and suppressor gene inactivation. The relevance of results obtained with viral strains analogous to that being evaluated in the monograph must also be considered.

(a) Qualitative aspects

An assessment of carcinogenicity involves several considerations of qualitative importance, including (i) the experimental conditions under which the test was performed, including route and schedule of exposure, species, strain, sex, age, duration of follow-up; (ii) the consistency of the results, for example, across species and target organ(s); (iii) the spectrum of neoplastic response, from preneoplastic lesions and benign tumours to malignant neoplasms; and (iv) the possible role of modifying factors.

As mentioned earlier (p. 21), the *Monographs* are not intended to summarize all published studies. Those studies in experimental animals that are inadequate (e.g., too short a duration, too few animals, poor survival; see below) or are judged irrelevant to the evaluation are generally omitted. Guidelines for conducting adequate long-term carcinogenicity experiments have been outlined (e.g., Montesano et al., 1986).

Considerations of importance to the Working Group in the interpretation and evaluation of a particular study include: (i) how clearly the agent was defined and, in the case of mixtures, how adequately the sample characterization was reported; (ii) whether the dose was adequately monitored, particularly in inhalation experiments; (iii) whether the doses and duration of treatment were appropriate and whether the survival of treated animals was similar to that of controls; (iv) whether there were adequate numbers of animals per group; (v) whether animals of both sexes were used; (vi) whether animals were allocated randomly to groups; (vii) whether the duration of observation was adequate; and (viii) whether the data were adequately reported. If available, recent data on the incidence of specific tumours in historical controls, as well as in concurrent controls, should be taken into account in the evaluation of tumour response.

When benign tumours occur together with and originate from the same cell type in an organ or tissue as malignant tumours in a particular study and appear to represent a stage in the progression to malignancy, it may be valid to combine them in assessing tumour incidence (Huff et al., 1989). The occurrence of lesions presumed to be preneoplastic may in certain instances aid in assessing the biological plausibility of any neoplastic response observed. If an agent or mixture induces only benign neoplasms that appear to be end-points that do not readily undergo transition to malignancy, it should nevertheless be suspected of being a carcinogen and it requires further investigation.

(b) *Quantitative aspects*

The probability that tumours will occur may depend on the species, sex, strain and age of the animal, the dose of the carcinogen and the route and length of exposure. Evidence of an increased incidence of neoplasms with increased level of exposure strengthens the inference of a causal association between the exposure and the development of neoplasms.

The form of the dose–response relationship can vary widely, depending on the particular agent under study and the target organ. Since many chemicals require metabolic activation before being converted into their reactive intermediates, both metabolic and pharmacokinetic aspects are important in determining the dose–response pattern. Saturation of steps such as absorption, activation, inactivation and elimination may produce nonlinearity in the dose–response relationship, as could saturation of processes such as DNA repair (Hoel et al., 1983; Gart et al., 1986).

(c) *Statistical analysis of long-term experiments in animals*

Factors considered by the Working Group include the adequacy of the information given for each treatment group: (i) the number of animals studied and the number examined histologically, (ii) the number of animals with a given tumour type and (iii) length of survival. The statistical methods used should be clearly stated and should be the generally accepted techniques refined for this purpose (Peto et al., 1980; Gart et al., 1986). When there is no difference in survival between control and treatment groups, the Working Group usually compares the proportions of animals developing each tumour type in each of the groups. Otherwise, consideration is given as to whether or not appropriate adjustments have been made for differences in survival. These adjustments can include: comparisons of the proportions of tumour-bearing animals among the effective number of animals (alive at the time the first tumour is discovered), in the case where most differences in survival occur before tumours appear; life-table methods, when tumours are visible or when they may be considered 'fatal' because mortality rapidly follows tumour development; and the Mantel-Haenszel test or logistic regression, when occult tumours do not affect the animals' risk of dying but are 'incidental' findings at autopsy.

In practice, classifying tumours as fatal or incidental may be difficult. Several survival-adjusted methods have been developed that do not require this distinction (Gart et al., 1986), although they have not been fully evaluated.

10. OTHER RELEVANT DATA

(a) *Absorption, distribution, metabolism and excretion*

Concise information is given on absorption, distribution (including placental transfer) and excretion in both humans and experimental animals. Kinetic factors that may affect the dose–response relationship, such as saturation of uptake, protein binding, metabolic activation, detoxification and DNA repair processes, are mentioned. Studies that indicate the metabolic fate of the agent in humans and in experimental animals are summarized briefly, and comparisons of data from humans and animals are made when possible. Comparative information on the relationship between exposure and the dose that reaches the target site may be of particular importance for extrapolation between species.

(b) *Toxic effects*

Data are given on acute and chronic toxic effects (other than cancer), such as organ toxicity, increased cell proliferation, immunotoxicity and endocrine effects. The presence and toxicological significance of cellular receptors is described.

(c) *Reproductive and developmental effects*

Effects on reproduction, teratogenicity, fetotoxicity and embryotoxicity are also summarized briefly.

(d) *Genetic and related effects*

Tests of genetic and related effects are described in view of the relevance of gene mutation and chromosomal damage to carcinogenesis (Vainio et al., 1992).

The adequacy of the reporting of sample characterization is considered and, where necessary, commented upon; with regard to complex mixtures, such comments are similar to those described for animal carcinogenicity tests on p. 28. The available data are interpreted critically by phylogenetic group according to the end-points detected, which may include DNA damage, gene mutation, sister chromatid exchange, micronucleus formation, chromosomal aberrations, aneuploidy and cell transformation. The concentrations employed are given, and mention is made of whether use of an exogenous metabolic system affected the test result. These data are given as listings of test systems, data and references; bar graphs (activity profiles) and corresponding summary tables with detailed information on the preparation of the profiles (Waters et al., 1987) are given in appendices.

Positive results in tests using prokaryotes, lower eukaryotes, plants, insects and cultured mammalian cells suggest that genetic and related effects could occur in mammals. Results from such tests may also give information about the types of genetic effect produced and about the involvement of metabolic activation. Some end-points described are clearly genetic in nature (e.g., gene mutations and chromosomal aberrations), while others are to a greater or lesser degree associated with genetic effects (e.g., unscheduled DNA synthesis). In-vitro tests for tumour-promoting activity and for cell transformation may be sensitive to changes that are not necessarily the result of genetic alterations but that may have specific relevance to the process of carcinogenesis. A critical appraisal of these tests has been published (Montesano *et al.*, 1986).

Genetic or other activity manifest in experimental mammals and humans is regarded as being of greater relevance than that in other organisms. The demonstration that an agent or mixture can induce gene and chromosomal mutations in whole mammals indicates that it may have carcinogenic activity, although this activity may not be detectably expressed in any or all species. Relative potency in tests for mutagenicity and related effects is not a reliable indicator of carcinogenic potency. Negative results in tests for mutagenicity in selected tissues from animals treated *in vivo* provide less weight, partly because they do not exclude the possibility of an effect in tissues other than those examined. Moreover, negative results in short-term tests with genetic end-points cannot be considered to provide evidence to rule out carcinogenicity of agents or mixtures that act through other mechanisms (e.g., receptor-mediated effects, cellular toxicity with regenerative proliferation, peroxisome proliferation) (Vainio *et al.*, 1992). Factors that may lead to misleading results in short-term tests have been discussed in detail elsewhere (Montesano *et al.*, 1986).

When available, data relevant to mechanisms of carcinogenesis that do not involve structural changes at the level of the gene are also described.

The adequacy of epidemiological studies of reproductive outcome and genetic and related effects in humans is evaluated by the same criteria as are applied to epidemiological studies of cancer.

(e) Structure–activity considerations

This section describes structure–activity relationships that may be relevant to an evaluation of the carcinogenicity of an agent.

11. SUMMARY OF DATA REPORTED

In this section, the relevant epidemiological and experimental data are summarized. Only reports, other than in abstract form, that meet the criteria outlined on p. 21 are considered for evaluating carcinogenicity. Inadequate studies are generally not summarized: such studies are usually identified by a square-bracketed comment in the preceding text.

(a) Exposures

Human exposure is summarized on the basis of elements such as production, use, occurrence in the environment and determinations in human tissues and body fluids. Quantitative data are given when available.

(b) Carcinogenicity in humans

Results of epidemiological studies that are considered to be pertinent to an assessment of human carcinogenicity are summarized. When relevant, case reports and correlation studies are also summarized.

(c) Carcinogenicity in experimental animals

Data relevant to an evaluation of carcinogenicity in animals are summarized. For each animal species and route of administration, it is stated whether an increased incidence of neoplasms or preneoplastic lesions was observed, and the tumour sites are indicated. If the agent or mixture produced tumours after prenatal exposure or in single-dose experiments, this is also indicated. Negative findings are also summarized. Dose–response and other quantitative data may be given when available.

(d) Other data relevant to an evaluation of carcinogenicity and its mechanisms

Data on biological effects in humans that are of particular relevance are summarized. These may include toxicological, kinetic and metabolic considerations and evidence of DNA binding, persistence of DNA lesions or genetic damage in exposed humans. Toxicological information, such as that on cytotoxicity and regeneration, receptor binding and hormonal and immunological effects, and data on kinetics and metabolism in experimental animals are given when considered relevant to the possible mechanism of the carcinogenic action of the agent. The results of tests for genetic and related effects are summarized for whole mammals, cultured mammalian cells and nonmammalian systems.

When available, comparisons of such data for humans and for animals, and particularly animals that have developed cancer, are described.

Structure–activity relationships are mentioned when relevant.

For the agent, mixture or exposure circumstance being evaluated, the available data on end-points or other phenomena relevant to mechanisms of carcinogenesis from studies in humans, experimental animals and tissue and cell test systems are summarized within one or more of the following descriptive dimensions:

(i) Evidence of genotoxicity (i.e., structural changes at the level of the gene): for example, structure–activity considerations, adduct formation, mutagenicity (effect on specific genes), chromosomal mutation/aneuploidy

(ii) Evidence of effects on the expression of relevant genes (i.e., functional changes at the intracellular level): for example, alterations to the structure or quantity of the product of a proto-oncogene or tumour suppressor gene, alterations to metabolic activation/inactivation/DNA repair

(iii) Evidence of relevant effects on cell behaviour (i.e., morphological or behavioural changes at the cellular or tissue level): for example, induction of mitogenesis, compensatory cell proliferation, preneoplasia and hyperplasia, survival of premalignant or malignant cells (immortalization, immunosuppression), effects on metastatic potential

(iv) Evidence from dose and time relationships of carcinogenic effects and interactions between agents: for example, early/late stage, as inferred from epidemiological studies; initiation/promotion/progression/malignant conversion, as defined in animal carcinogenicity experiments; toxicokinetics

These dimensions are not mutually exclusive, and an agent may fall within more than one of them. Thus, for example, the action of an agent on the expression of relevant genes could be summarized under both the first and second dimension, even if it were known with reasonable certainty that those effects resulted from genotoxicity.

12. EVALUATION

Evaluations of the strength of the evidence for carcinogenicity arising from human and experimental animal data are made, using standard terms.

It is recognized that the criteria for these evaluations, described below, cannot encompass all of the factors that may be relevant to an evaluation of carcinogenicity. In considering all of the relevant data, the Working Group may assign the agent, mixture or exposure circumstance to a higher or lower category than a strict interpretation of these criteria would indicate.

(a) *Degrees of evidence for carcinogenicity in humans and in experimental animals and supporting evidence*

These categories refer only to the strength of the evidence that an exposure is carcinogenic and not to the extent of its carcinogenic activity (potency) nor to the mechanisms involved. A classification may change as new information becomes available.

An evaluation of degree of evidence, whether for a single agent or a mixture, is limited to the materials tested, as defined physically, chemically or biologically. When the agents evaluated are considered by the Working Group to be sufficiently closely related, they may be grouped together for the purpose of a single evaluation of degree of evidence.

(i) *Carcinogenicity in humans*

The applicability of an evaluation of the carcinogenicity of a mixture, process, occupation or industry on the basis of evidence from epidemiological studies depends on the variability over time and place of the mixtures, processes, occupations and industries. The Working Group seeks to identify the specific exposure, process or activity which is considered most likely to be responsible for any excess risk. The evaluation is focused as narrowly as the available data on exposure and other aspects permit.

The evidence relevant to carcinogenicity from studies in humans is classified into one of the following categories:

Sufficient evidence of carcinogenicity: The Working Group considers that a causal relationship has been established between exposure to the agent, mixture or exposure circumstance and human cancer. That is, a positive relationship has been observed between the exposure and cancer in studies in which chance, bias and confounding could be ruled out with reasonable confidence.

Limited evidence of carcinogenicity: A positive association has been observed between exposure to the agent, mixture or exposure circumstance and cancer for which a causal interpretation is considered by the Working Group to be credible, but chance, bias or confounding could not be ruled out with reasonable confidence.

Inadequate evidence of carcinogenicity: The available studies are of insufficient quality, consistency or statistical power to permit a conclusion regarding the presence or absence of a causal association, or no data on cancer in humans are available.

Evidence suggesting lack of carcinogenicity: There are several adequate studies covering the full range of levels of exposure that human beings are known to encounter, which are mutually consistent in not showing a positive association between exposure to the agent, mixture or exposure circumstance and any studied cancer at any observed level of exposure. A conclusion of 'evidence suggesting lack of carcinogenicity' is inevitably limited to the cancer sites, conditions and levels of exposure and length of observation covered by the available studies. In addition, the possibility of a very small risk at the levels of exposure studied can never be excluded.

In some instances, the above categories may be used to classify the degree of evidence related to carcinogenicity in specific organs or tissues.

(ii) *Carcinogenicity in experimental animals*

The evidence relevant to carcinogenicity in experimental animals is classified into one of the following categories:

Sufficient evidence of carcinogenicity: The Working Group considers that a causal relationship has been established between the agent or mixture and an increased incidence of malignant neoplasms or of an appropriate combination of benign and malignant neoplasms in (a) two or more species of animals or (b) in two or more independent studies in one species carried out at different times or in different laboratories or under different protocols.

Exceptionally, a single study in one species might be considered to provide sufficient evidence of carcinogenicity when malignant neoplasms occur to an unusual degree with regard to incidence, site, type of tumour or age at onset.

Limited evidence of carcinogenicity: The data suggest a carcinogenic effect but are limited for making a definitive evaluation because, e.g., (a) the evidence of carcinogenicity is restricted to a single experiment; or (b) there are unresolved questions regarding the adequacy of the design, conduct or interpretation of the study; or (c) the agent or mixture increases the incidence only of benign neoplasms or lesions of uncertain neoplastic potential, or of certain neoplasms which may occur spontaneously in high incidences in certain strains.

Inadequate evidence of carcinogenicity: The studies cannot be interpreted as showing either the presence or absence of a carcinogenic effect because of major qualitative or quantitative limitations, or no data on cancer in experimental animals are available.

Evidence suggesting lack of carcinogenicity: Adequate studies involving at least two species are available which show that, within the limits of the tests used, the agent or mixture is not carcinogenic. A conclusion of evidence suggesting lack of carcinogenicity is inevitably limited to the species, tumour sites and levels of exposure studied.

(b) *Other data relevant to an evaluation of carcinogenicity*

Other evidence judged to be relevant to an evaluation of carcinogenicity and of sufficient importance to affect the overall evaluation is then described. This may include data on preneoplastic lesions, tumour pathology, genetic and related effects, structure–activity relationships, metabolism and pharmacokinetics, and physicochemical parameters.

Data relevant to mechanisms of the carcinogenic action are also evaluated. The strength of the evidence that any carcinogenic effect observed is due to a particular mechanism is assessed, using terms such as weak, moderate or strong. Then, the Working Group assesses if that particular mechanism is likely to be operative in humans. The strongest indications that a particular mechanism operates in humans come from data on humans or biological specimens obtained from exposed humans. The data may be considered to be especially relevant if they show that the agent in question has caused changes in exposed humans that are on the causal pathway to carcinogenesis. Such data may, however, never become available, because it is at least conceivable that certain compounds may be kept from human use solely on the basis of evidence of their toxicity and/or carcinogenicity in experimental systems.

For complex exposures, including occupational and industrial exposures, chemical composition and the potential contribution of carcinogens known to be present are considered by the Working Group in its overall evaluation of human carcinogenicity. The Working Group also determines the extent to which the materials tested in experimental systems are related to those to which humans are exposed.

(c) *Overall evaluation*

Finally, the body of evidence is considered as a whole, in order to reach an overall evaluation of the carcinogenicity to humans of an agent, mixture or circumstance of exposure.

An evaluation may be made for a group of chemical compounds that have been evaluated by the Working Group. In addition, when supporting data indicate that other, related compounds for which there is no direct evidence of capacity to induce cancer in humans or in animals may also be carcinogenic, a statement describing the rationale for this conclusion is added to the evaluation narrative; an additional evaluation may be made for this broader group of compounds if the strength of the evidence warrants it.

The agent, mixture or exposure circumstance is described according to the wording of one of the following categories, and the designated group is given. The categorization of an agent, mixture or exposure circumstance is a matter of scientific judgement, reflecting the strength of the evidence derived from studies in humans and in experimental animals and from other relevant data.

Group 1 — The agent (mixture) is carcinogenic to humans.
The exposure circumstance entails exposures that are carcinogenic to humans.

This category is used when there is *sufficient evidence* of carcinogenicity in humans. Exceptionally, an agent (mixture) may be placed in this category when evidence in humans is less than sufficient but there is *sufficient evidence* of carcinogenicity in experimental animals and strong evidence in exposed humans that the agent (mixture) acts through a relevant mechanism of carcinogenicity.

Group 2

This category includes agents, mixtures and exposure circumstances for which, at one extreme, the degree of evidence of carcinogenicity in humans is almost sufficient, as well as those for which, at the other extreme, there are no human data but for which there is evidence of carcinogenicity in experimental animals. Agents, mixtures and exposure circumstances are assigned to either group 2A (probably carcinogenic to humans) or group 2B (possibly carcinogenic to humans) on the basis of epidemiological and experimental evidence of carcinogenicity and other relevant data.

Group 2A—The agent (mixture) is probably carcinogenic to humans.
The exposure circumstance entails exposures that are probably carcinogenic to humans.

This category is used when there is *limited evidence* of carcinogenicity in humans and *sufficient evidence* of carcinogenicity in experimental animals. In some cases, an agent (mixture) may be classified in this category when there is *inadequate evidence* of carcinogenicity in humans and *sufficient evidence* of carcinogenicity in experimental animals and strong evidence that the carcinogenesis is mediated by a mechanism that also operates in humans. Exceptionally, an agent, mixture or exposure circumstance may be classified in this category solely on the basis of *limited evidence* of carcinogenicity in humans.

Group 2B—The agent (mixture) is possibly carcinogenic to humans.
The exposure circumstance entails exposures that are possibly carcinogenic to humans.

This category is used for agents, mixtures and exposure circumstances for which there is *limited evidence* of carcinogenicity in humans and less than *sufficient evidence* of carcinogenicity in experimental animals. It may also be used when there is *inadequate evidence* of carcinogenicity in humans but there is *sufficient evidence* of carcinogenicity in experimental animals. In some instances, an agent, mixture or exposure circumstance for which there is *inadequate evidence* of carcinogenicity in humans but *limited evidence* of carcinogenicity in experimental animals together with supporting evidence from other relevant data may be placed in this group.

Group 3—The agent (mixture or exposure circumstance) is not classifiable as to its carcinogenicity to humans.

This category is used most commonly for agents, mixtures and exposure circumstances for which the evidence of carcinogenicity is inadequate in humans and inadequate or limited in experimental animals.

Exceptionally, agents (mixtures) for which the evidence of carcinogenicity is inadequate in humans but sufficient in experimental animals may be placed in this category when there is strong evidence that the mechanism of carcinogenicity in experimental animals does not operate in humans.

Agents, mixtures and exposure circumstances that do not fall into any other group are also placed in this category.

Group 4—The agent (mixture) is probably not carcinogenic to humans.

This category is used for agents or mixtures for which there is *evidence suggesting lack of carcinogenicity* in humans and in experimental animals. In some instances, agents or mixtures for which there is *inadequate evidence* of carcinogenicity in humans but *evidence suggesting lack of carcinogenicity* in experimental animals, consistently and strongly supported by a broad range of other relevant data, may be classified in this group.

References

Breslow, N.E. & Day, N.E. (1980) *Statistical Methods in Cancer Research*, Vol. 1, *The Analysis of Case-control Studies* (IARC Scientific Publications No. 32), Lyon, IARC

Breslow, N.E. & Day, N.E. (1987) *Statistical Methods in Cancer Research*, Vol. 2, *The Design and Analysis of Cohort Studies* (IARC Scientific Publications No. 82), Lyon, IARC

Gart, J.J., Krewski, D., Lee, P.N., Tarone, R.E. & Wahrendorf, J. (1986) *Statistical Methods in Cancer Research*, Vol. 3, *The Design and Analysis of Long-term Animal Experiments* (IARC Scientific Publications No. 79), Lyon, IARC

Hoel, D.G., Kaplan, N.L. & Anderson, M.W. (1983) Implication of nonlinear kinetics on risk estimation in carcinogenesis. *Science*, *219*, 1032–1037

Huff, J.E., Eustis, S.L. & Haseman, J.K. (1989) Occurrence and relevance of chemically induced benign neoplasms in long-term carcinogenicity studies. *Cancer Metastasis Rev.*, *8*, 1–21

IARC (1973–1990) *Information Bulletin on the Survey of Chemicals Being Tested for Carcinogenicity/Directory of Agents Being Tested for Carcinogenicity*, Numbers 1–14, Lyon

    Number 1 (1973)   52 pages
    Number 2 (1973)   77 pages
    Number 3 (1974)   67 pages
    Number 4 (1974)   97 pages
    Number 5 (1975)   88 pages
    Number 6 (1976)   360 pages
    Number 7 (1978)   460 pages
    Number 8 (1979)   604 pages
    Number 9 (1981)   294 pages
    Number 10 (1983)   326 pages
    Number 11 (1984)   370 pages
    Number 12 (1986)   385 pages
    Number 13 (1988)   404 pages
    Number 14 (1990)   369 pages

IARC (1976–1991)

Directory of On-going Research in Cancer Epidemiology 1976. Edited by C.S. Muir & G. Wagner, Lyon

Directory of On-going Research in Cancer Epidemiology 1977 (IARC Scientific Publications No. 17). Edited by C.S. Muir & G. Wagner, Lyon

Directory of On-going Research in Cancer Epidemiology 1978 (IARC Scientific Publications No. 26). Edited by C.S. Muir & G. Wagner, Lyon

Directory of On-going Research in Cancer Epidemiology 1979 (IARC Scientific Publications No. 28). Edited by C.S. Muir & G. Wagner, Lyon

Directory of On-going Research in Cancer Epidemiology 1980 (IARC Scientific Publications No. 35). Edited by C.S. Muir & G. Wagner, Lyon

Directory of On-going Research in Cancer Epidemiology 1981 (IARC Scientific Publications No. 38). Edited by C.S. Muir & G. Wagner, Lyon

Directory of On-going Research in Cancer Epidemiology 1982 (IARC Scientific Publications No. 46). Edited by C.S. Muir & G. Wagner, Lyon

Directory of On-going Research in Cancer Epidemiology 1983 (IARC Scientific Publications No. 50). Edited by C.S. Muir & G. Wagner, Lyon

Directory of On-going Research in Cancer Epidemiology 1984 (IARC Scientific Publications No. 62). Edited by C.S. Muir & G. Wagner, Lyon

Directory of On-going Research in Cancer Epidemiology 1985 (IARC Scientific Publications No. 69). Edited by C.S. Muir & G. Wagner, Lyon

Directory of On-going Research in Cancer Epidemiology 1986 (IARC Scientific Publications No. 80). Edited by C.S. Muir & G. Wagner, Lyon

Directory of On-going Research in Cancer Epidemiology 1987 (IARC Scientific Publications No. 86). Edited by D.M. Parkin & J. Wahrendorf, Lyon

Directory of On-going Research in Cancer Epidemiology 1988 (IARC Scientific Publications No. 93). Edited by M. Coleman & J. Wahrendorf, Lyon

Directory of On-going Research in Cancer Epidemiology 1989/90 (IARC Scientific Publications No. 101). Edited by M. Coleman & J. Wahrendorf, Lyon

Directory of On-going Research in Cancer Epidemiology 1991 (IARC Scientific Publications No. 110). Edited by M. Coleman & J. Wahrendorf, Lyon

IARC (1977) *IARC Monographs Programme on the Evaluation of the Carcinogenic Risk of Chemicals to Humans. Preamble* (IARC intern. tech. Rep. No. 77/002), Lyon

IARC (1978) *Chemicals with* Sufficient Evidence *of Carcinogenicity in Experimental Animals*—IARC Monographs *Volumes 1-17* (IARC intern. tech. Rep. No. 78/003), Lyon

IARC (1978-1991) *Environmental Carcinogens. Methods of Analysis and Exposure Measurement*:

    Vol. 1. *Analysis of Volatile Nitrosamines in Food* (IARC Scientific Publications No. 18). Edited by R. Preussmann, M. Castegnaro, E.A. Walker & A.E. Wasserman (1978)

    Vol. 2. *Methods for the Measurement of Vinyl Chloride in Poly(vinyl chloride), Air, Water and Foodstuffs* (IARC Scientific Publications No. 22). Edited by D.C.M. Squirrell & W. Thain (1978)

    Vol. 3. *Analysis of Polycyclic Aromatic Hydrocarbons in Environmental Samples* (IARC Scientific Publications No. 29). Edited by M. Castegnaro, P. Bogovski, H. Kunte & E.A. Walker (1979)

    Vol. 4. *Some Aromatic Amines and Azo Dyes in the General and Industrial Environment* (IARC Scientific Publications No. 40). Edited by L. Fishbein, M. Castegnaro, I.K. O'Neill & H. Bartsch (1981)

    Vol. 5. *Some Mycotoxins* (IARC Scientific Publications No. 44). Edited by L. Stoloff, M. Castegnaro, P. Scott, I.K. O'Neill & H. Bartsch (1983)

    Vol. 6. N-*Nitroso Compounds* (IARC Scientific Publications No. 45). Edited by R. Preussmann, I.K. O'Neill, G. Eisenbrand, B. Spiegelhalder & H. Bartsch (1983)

    Vol. 7. *Some Volatile Halogenated Hydrocarbons* (IARC Scientific Publications No. 68). Edited by L. Fishbein & I.K. O'Neill (1985)

    Vol. 8. *Some Metals: As, Be, Cd, Cr, Ni, Pb, Se, Zn* (IARC Scientific Publications No. 71). Edited by I.K. O'Neill, P. Schuller & L. Fishbein (1986)

    Vol. 9. *Passive Smoking* (IARC Scientific Publications No. 81). Edited by I.K. O'Neill, K.D. Brunnemann, B. Dodet & D. Hoffmann (1987)

    Vol. 10. *Benzene and Alkylated Benzenes* (IARC Scientific Publications No. 85). Edited by L. Fishbein & I.K. O'Neill (1988)

    Vol. 11. *Polychlorinated Dioxins and Dibenzofurans* (IARC Scientific Publications No. 108). Edited by C. Rappe, H.R. Buser, B. Dodet & I.K. O'Neill (1991)

IARC (1979) *Criteria to Select Chemicals for* IARC Monographs (IARC intern. tech. Rep. No. 79/003), Lyon

IARC (1982) *IARC Monographs on the Evaluation of the Carcinogenic Risk of Chemicals to Humans, Supplement 4, Chemicals, Industrial Processes and Industries Associated with Cancer in Humans (IARC Monographs, Volumes 1 to 29)*, Lyon

IARC (1983) *Approaches to Classifying Chemical Carcinogens According to Mechanism of Action* (IARC intern. tech. Rep. No. 83/001), Lyon

IARC (1984) *Chemicals and Exposures to Complex Mixtures Recommended for Evaluation in* IARC *Monographs and Chemicals and Complex Mixtures Recommended for Long-term Carcinogenicity Testing* (IARC intern. tech. Rep. No. 84/002), Lyon

IARC (1987a) *IARC Monographs on the Evaluation of Carcinogenic Risks to Humans, Supplement 6, Genetic and Related Effects: An Updating of Selected* IARC *Monographs from Volumes 1 to 42*, Lyon

IARC (1987b) *IARC Monographs on the Evaluation of Carcinogenic Risks to Humans, Supplement 7, Overall Evaluations of Carcinogenicity: An Updating of* IARC *Monographs Volumes 1 to 42*, Lyon

IARC (1988) *Report of an IARC Working Group to Review the Approaches and Processes Used to Evaluate the Carcinogenicity of Mixtures and Groups of Chemicals* (IARC intern. tech. Rep. No. 88/002), Lyon

IARC (1989) *Chemicals, Groups of Chemicals, Mixtures and Exposure Circumstances to be Evaluated in Future IARC Monographs, Report of an ad hoc Working Group* (IARC intern. tech. Rep. No. 89/004), Lyon

IARC (1991a) *A Consensus Report of an* IARC Monographs *Working Group on the Use of Mechanims of Carcinogenesis in Risk Identification* (IARC intern. tech. Rep. No. 91/002), Lyon

IARC (1991b) *Report of an Ad-hoc* IARC Monographs *Advisory Group on Viruses and Other Biological Agents Such as Parasites* (IARC intern. tech. Rep. No. 91/001), Lyon

Montesano, R., Bartsch, H., Vainio, H., Wilbourn, J. & Yamasaki, H., eds (1986) *Long-term and Short-term Assays for Carcinogenesis—A Critical Appraisal* (IARC Scientific Publications No. 83), Lyon, IARC

Peto, R., Pike, M.C., Day, N.E., Gray, R.G., Lee, P.N., Parish, S., Peto, J., Richards, S. & Wahrendorf, J. (1980) Guidelines for simple, sensitive significance tests for carcinogenic effects in long-term animal experiments. In: *IARC Monographs on the Evaluation of the Carcinogenic Risk of Chemicals to Humans, Supplement 2, Long-term and Short-term Screening Assays for Carcinogens: A Critical Appraisal*, Lyon, pp. 311-426

Vainio, H., Magee, P., McGregor, D. & McMichael, A., eds (1992) *Mechanisms of Carcinogenesis in Risk Identification* (IARC Scientific Publications No. 116), Lyon, IARC

Waters, M.D., Stack, H.F., Brady, A.L., Lohman, P.H.M., Haroun, L. & Vainio, H. (1987) Apendix 1. Activity profiles for genetic and related tests. In: *IARC Monographs on the Evaluation of Carcinogenic Risks to Humans, Suppl. 6, Genetic and Related Effects: An Updating of Selected IARC Monographs from Volumes 1 to 42*, Lyon, IARC, pp. 687-696

Wilbourn, J., Haroun, L., Heseltine, E., Kaldor, J., Partensky, C. & Vainio, H. (1986) Response of experimental animals to human carcinogens: an analysis based upon the IARC Monographs Programme. *Carcinogenesis*, 7, 1853-1863

GENERAL REMARKS

This fifty-fifth volume of *IARC Monographs* contains evaluations of carcinogenic risks associated with human exposure to solar and ultraviolet (UV) radiation from medical and cosmetic devices, general illumination and industrial sources. Ultraviolet radiation (UVR) was considered previously (IARC, 1986) in a volume in which furocoumarins were evaluated. Since some of these compounds are used clinically in conjunction with ultraviolet A (UVA) radiation, information on the carcinogenic effects of UVR alone was provided in an appendix; however, no evaluation was made at that time.

Solar radiation is largely optical radiation (UV, visible and infrared), although both shorter wavelength (ionizing) and longer wavelength (microwaves and radiofrequency) radiation is present. UVR lies in the interval 100–400 nm and is further subdivided into UVA (315–400 nm), UVB (280–315 nm) and UVC (100–280 nm). The UV component of terrestrial radiation from the sun comprises about 95% UVA and 5% UVB; UVC is removed from extraterrestrial radiation by stratopheric ozone. Before the beginning of this century, the sun was essentially the only source of UVR; with the advent of artificial sources, the opportunity for additional exposure, not only to UVA and UVB but also to UVC, has increased. It should be stressed that the distinction of UVR into UVA, UVB and UVC ranges has no biological basis, and the potential of UVR for causing damage to biomolecules, cells, tissues and organisms varies enormously over the spectral region from 250 to 400 nm.

UVA radiation is one of the components of solar emissions and of emissions from medical lamps and lamps used for cosmetic purposes. UVB radiation is present in solar emissions, from lamps used in medicine and for cosmetic purposes and in certain lamps used for general illumination, such as unshielded fluorescent and tungsten–halogen lamps. It causes sunburn relatively easily and is immunosuppressive; it can cause ocular cataracts. The possibility that the UVB component of solar radiation will increase as a result of depletion of the ozone layer is a matter of concern. This question was not addressed in the present volume.

Human exposure to UVC radiation is uncommon and is related to the use of germicidal and tungsten–halogen lamps, phototherapy and welding arcs. Thus, very little is known about the effects of UVC on humans, although a great deal of information is available on the effects of radiation in this range on biomolecules, cells and viruses.

In the USA, skin melanoma has been second only to lung cancer in its rate of increase in incidence over the last 40 years: the incidence has been increasing by about 5% per year. The major sites have been male trunk and female leg. Mortality from melanoma may now be falling in younger generations (at least in the USA) due, possibly, to changes in sun exposure (Scotto *et al.*, 1991). There is also evidence that the incidence of nonmelanocytic skin cancer is increasing in some white-skinned populations (Gallagher *et al.*, 1990). Constitutional risk factors, e.g., skin type, hair and eye colour and specific subtypes of exposure (for example, occupational and recreational), have been assessed in individual studies or sections of the monographs but have not been included in the evaluations.

UVR is ubiquitous and cannot be totally avoided. An appendix to this volume presents a discussion on the use of topical sunscreens, taking into consideration both potentially beneficial, protective effects and possible adverse reactions. The biological effects of combinations of psoralens and UVR were not considered since these were the subjects of separate monographs (IARC, 1980, 1986, 1987) in the *IARC Monographs* series.

References

Gallagher, R.P., Ma, B., McLean, D.I., Yang, C.P., Ho, V., Carruthers, J.A. & Warshawski, L.M. (1990) Trends in basal cell carcinoma, squamous cell carcinoma, and melanoma of the skin from 1973 through 1987. *J. Am. Acad. Dermatol.*, **23**, 413–421

IARC (1980) *IARC Monographs on the Evaluation of the Carcinogenic Risk of Chemicals to Humans*, Vol. 24, *Some Pharmaceutical Drugs*, Lyon, pp. 101–124

IARC (1986) *IARC Monographs on the Evaluation of the Carcinogenic Risk of Chemicals to Humans*, Vol. 40, *Some Naturally Occurring Synthetic Food Components, Furocoumarins and Ultraviolet Radiation*, Lyon, pp. 317–371

IARC (1987) *IARC Monographs on the Evaluation of Carcinogenic Risks to Humans*, Suppl. 7, *Overall Evaluations of Carcinogenicity: An Updating of* IARC Monographs *Volumes 1 to 42*, Lyon, pp. 242–245

Scotto, J., Pitcher, H. & Lee, J.A.H. (1991) Indications of future decreasing trends in skin-melanoma mortality among whites in the United States. *Int. J. Cancer*, **49**, 490–497

SOLAR AND ULTRAVIOLET RADIATION

1. Exposure Data

1.1 Nomenclature

1.1.1 *Optical radiation*

Optical radiation is radiant energy within a broad region of the electromagnetic spectrum that includes ultraviolet (UV), visible (light) and infrared radiation. Ultraviolet radiation (UVR) is characterized by wavelengths between 10 and 400 nm—bordered on the one side by x rays and on the other by visible light (Fig. 1). Solar radiation is largely optical radiation, although ionizing radiation (i.e., cosmic rays, gamma rays and x rays, which have wavelengths less than approximately 10 nm) and radio-frequency radiation (i.e., wavelengths greater than 1 mm: microwaves and longer radio waves) are also present in the spectrum.

The optical radiation spectrum is generally considered to fall between 10 nm and 1 mm, and several different conventions have been developed to describe different bands within this spectrum. It is important to recognize that no single convention is uniquely 'correct' but that each may be useful for a particular branch of science and technology. For example, in optics, it is convenient to separate the spectrum into different bands on the basis of the transmission and absorption properties of optical materials (e.g., glass and quartz). In one optical convention, shown in Figure 1, UVR is divided into vacuum UV, extending from 10 to 180 nm; middle UV, from 180 nm to 300 nm; and near UV, from 300 nm to 380 or 400 nm. Meteorological scientists typically define optical spectral regions on the basis of atmospheric windows. Some spectral designations are based on uses, e.g., 'germicidal' and 'black-light' regions.

For the purposes of this monograph, the photobiological designations of the Commission Internationale de l'Eclairage (CIE, International Commission on Illumination) are the most relevant and are used throughout to define the approximate spectral regions in which certain biological absorption properties and biological interaction mechanisms may dominate (Commission Internationale de l'Eclairage, 1987). The CIE bands are: UVC (100–280 nm), UVB (280–315 nm) and UVA (315–400 nm). Visible light is the region between 400 nm and 780 nm.

It is important to recognize that these spectral band designations are merely short-hand notations and cannot be considered to designate fine dividing lines below which an effect is present and above which it does not occur. The reader should also be alerted to the fact that the CIE nomenclature is not always followed rigorously and that some authors introduce slight variations; for example, distinguishing between UVB and UVA at 320 rather than 315 nm (frequently used in the USA) and defining UVC as 200–280 nm (Moseley, 1988). The German Industrial Standard (DIN 5031) defines UVA as radiation between 315 and 380 nm (Mutzhas, 1986).

Figure 1. Electromagnetic spectrum with enlargement of ultraviolet (UV) region

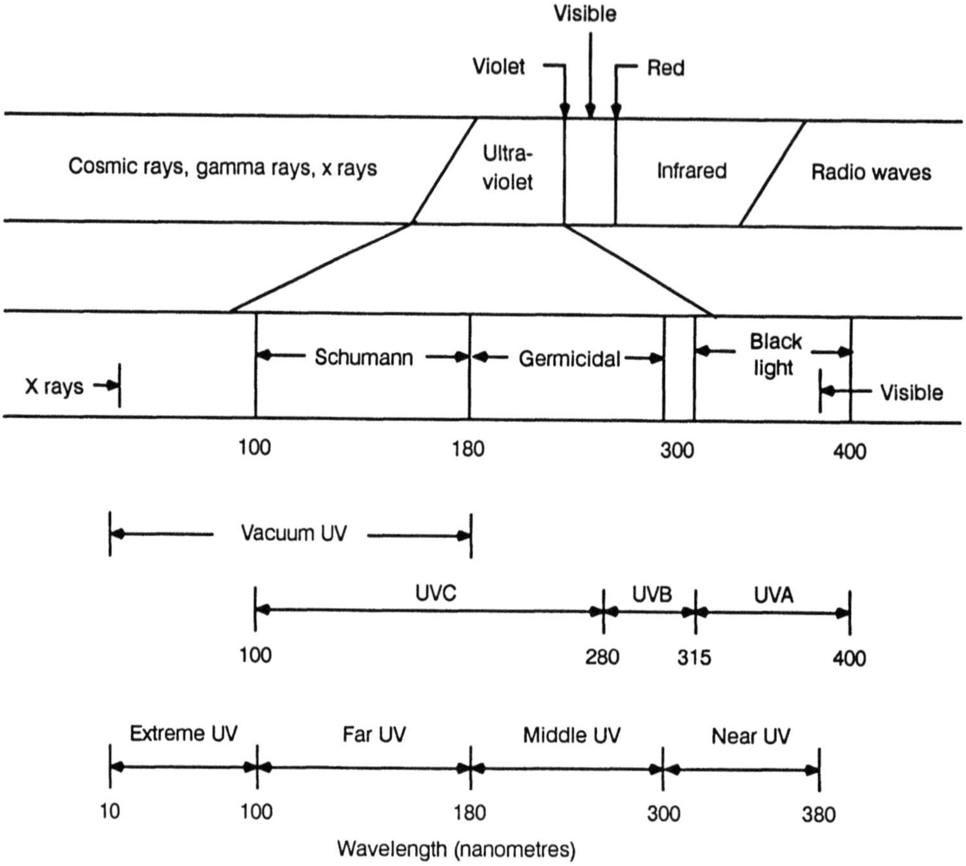

Adapted from WHO (1979), Morison (1983a), Sylvania (undated)

From the viewpoint of photochemistry and photobiology, interactions of optical radiation with matter are considered to occur when one photon interacts with one molecule to produce a photochemically altered molecule or two dissociated molecules (Phillips, 1983; Smith, 1989). In any photochemical interaction, the energy of the individual photon is important, since this must be sufficient to alter a molecular bond. The photon energy is generally expressed in terms of electron volts (eV). A wavelength of 10 nm corresponds to a photon energy of 124 eV, and 400 nm to an energy of 3.1 eV (WHO, 1979). The number of altered molecules produced relative to the number of absorbed photons is referred to as the 'quantum yield' (Phillips, 1983). The efficacy of photochemical interaction per incident quantum and the photobiological effects per unit radiant exposure typically vary widely with wavelength. A quantitative plot of such spectral variation, usually normalized to unity at the most effective wavelength, is referred to as an 'action spectrum' (Jagger, 1985).

1.1.2 *Quantities and units*

Two systems of quantities and units are used to describe the characteristics of light and light sources: the radiometric and the photometric systems. Radiometry can be applied to all optical sources and to all exposures to optical radiation (including solar radiation and UVR). Photometry can be used only to describe visible light sources, and photometric quantities are used in illumination engineering. The basic photometric unit is the lumen, which is defined in terms of the spectral response of the human eye (specifically, the spectral response of the CIE 'standard observer'), i.e., the action spectrum of vision, which is initially a photochemical process. It is important to recognize that radiometric quantities and units are absolute, while photometric quantities and units are related to standardized human perception; the relationship between the two sets of units varies significantly with the spectrum of radiation. The effects of optical radiation (including light), other than vision, must therefore be measured and quantified in terms of radiometric units and spectral characteristics rather than photometric units. This is particularly important in relation to the photobiological effects of UVR. Most lamps used for illumination are rated by manufacturers only in photometric terms (e.g., lumen output) and not in terms of UVR emission (Phillips, 1983).

The most important radiometric quantities and units commonly used to describe optical radiation are given in Table 1. Certain terms are used primarily to describe source characteristics, e.g., radiance, radiant intensity; whereas other terms are generally used to describe exposure (irradiance, radiant exposure). The term 'spectral' placed before any of the quantities implies restriction to a unit wavelength band, e.g., spectral irradiance (watts per square metre per nanometre) (Moseley, 1988). For a more detailed discussion of these parameters, see various standard textbooks on radiometry, such as Boyd (1983).

The quantities of radiometry are expressed in terms of absolute energy (Jagger, 1985). Radiant intensity is the power emitted per unit solid angle of a source. Radiance is the radiant intensity per unit area of source. Thus, a fluorescent lamp does not have very high radiance in comparison to the filament of a flashlight bulb, even though it has a high radiant power output. The radiometric term expressed in units of watts per square metre (dose rate) is irradiance, which is also the power striking a unit area of surface.

The energy of UVR falling on a unit surface area of an object was defined in 1954 by the First International Congress of Photobiology as the 'dose'; it has also been referred to as 'exposure dose'. The equivalent radiometric quantity is radiant exposure, expressed in joules per square centimetre or per square metre. Radiant exposure has been referred to as 'energy fluence' in some texts; however, fluence is a radiometric quantity, with the same units as radiant exposure, but referring to energy arriving at a plane of unit area from all directions, including backscatter. Thus, fluence is quite correctly of value in describing an exposure dose at a depth inside tissue; it has, however, seldom been calculated in photobiological studies of the effects of UVR, in which the radiant exposure incident upon the skin is normally measured. Radiant exposure is the amount of energy crossing a unit area of space normal to the direction of propagation of a beam of UVR. If the radiant energy arrives from many directions, as from the sky, then the fluence at one point is the sum of all the component fluences entering a unit sphere of space. The energy fluence rate is the power that crosses a unit area normal to the direction of propagation, or the energy per unit area per unit time

Table 1. Some basic terminology used to quantify optical radiation

Term	International symbol	Definition	SI unit	Synonyms and comments
Wavelength	λ		nm	Nanometre = 10^{-9} m (also called millimicron, mμ)
Radiant energy	Q_e	$\Sigma (P_e \times dt)$	J	Joule; 1 joule = 1 watt × second; total energy contained in a radiation field or total energy delivered to a given receiver by such a radiation field
Radiant flux	P_e	dQ_e/dt	W	Watt; rate of delivery of radiant energy ('radiant power'); also expressed as ϕ
Irradiance	E_e	dP_e/dA	W/m²	Radiant flux arriving over a given area ('fluence rate', 'dose rate', 'intensity', 'radiant incidence'). In photobiology, has also been expressed in W/cm², mW/cm² and µW/cm²
Radiant intensity	I_e	$dP_e/d\Omega$	W/sr	Watt/steradian; radiant flux emitted by source into a given solid angle (solid angle expressed in steradians)
Radiance	L_e	$dP_e/dA \times d\Omega$	W/m² × sr	Watt/m² × steradian; radiant flux per unit solid angle per unit area emitted by an extended source
Radiant exposure	H_e	$E_e \times t$	J/m²	Radiant energy delivered to a given area ('fluence', 'exposure dose', 'dose'); t = time in seconds. Has also been expressed as J/cm², mJ/cm² and µJ/cm²

Adapted from WHO (1979), Boyd (1983), Jagger (1985), Hoffman (1987) and Weast (1989)

(J/m²/s or W/m²). The terms dose (J/m²) and dose rate (W/m²) pertain to the energy and power, respectively, striking a unit surface area of an irradiated object (Jagger, 1985).

In terms of visible light perceived by humans, the photometric analogue of the radiance of a source is luminance (brightness), and irradiance is illuminance (measured in 'lux' or lumen per square metre). In photometry, the lumen is the unit of luminous power (Jagger, 1985).

1.1.3 *Units of biologically effective ultraviolet radiation*

In addition to general radiometric quantities, specialized quantities of effective irradiance relative to a specified photochemical action spectrum are used in photochemistry and photobiology. Effective radiant exposures to produce erythema (Jagger, 1985) or photokeratitis are examples. Effective irradiance or radiant exposure is not limited to photobiology, and a similar approach has been used to quantify the photocuring of inks, in photopolymerization (Phillips, 1983) and in assessing the hazards of UVR. In order to weight a source spectrally, the general formula involves an action spectrum and a spectral radiometric quantity. The effective irradiance of a given photobiological process is defined as:

$$\sum_{\lambda_1}^{\lambda_2} E_\lambda \times S_\lambda \times \Delta_\lambda$$

expressed in W/m², where E_λ is the spectral irradiance (W/m² × nm) at wavelength λ (nm) and Δ_λ is the wavelength interval ($\lambda_1 \rightarrow \lambda_2$) used in the summation (in nm). S_λ is a measure of the effectiveness of radiation of wavelength λ (nm), relative to some reference wavelength, in producing a particular biological end-point. As it is a ratio, S_λ has no units (American Conference of Governmental Industrial Hygienists, 1991).

Effective irradiance is equivalent to a hypothetical irradiance of monochromatic radiation with a wavelength at which S_λ is equal to unity. The time integral of effective irradiance is the effective radiant exposure (also called the 'effective dose').

A unit of effective dose commonly used in cutaneous photobiology is the 'minimal erythema dose' (MED). One MED has been defined as the lowest radiant exposure to UVR that is sufficient to produce erythema with sharp margins 24 h after exposure (Morison, 1983a). Another end-point often used in cutaneous photobiology is a just-perceptible reddening of exposed skin; the dose of UVR necessary to produce this 'minimal perceptible erythema' is sometimes also referred to as an MED. In unacclimatized, white-skinned populations, there is an approximately four-fold range in the MED of exposure to UVB radiation (Diffey & Farr, 1989). When the term MED is used as a unit of exposure dose, however, a representative value is chosen for sun-sensitive individuals. If, in the above expression for effective irradiance, S_λ is chosen as the reference action spectrum for erythema (McKinlay & Diffey, 1987) and a value of 200 J/m² at wavelengths for which S_λ is equal to unity is assumed for the MED, the dose (expressed in MED) received after an exposure period of t seconds is

$$t \times \Sigma\, E_\lambda \times S_\lambda \times \Delta_\lambda/200.$$

Notwithstanding the difficulties of interpreting accurately the magnitude of such an imprecise unit as the MED, it has the advantage over radiometric units of being related to the biological consequences of the exposure.

1.2 Methods for measuring ultraviolet radiation

UVR can be measured by chemical or physical detectors, often in conjunction with a monochromator or band-pass filter for wavelength selection. Physical detectors include radiometric devices, which depend for their response on the heating effect of the radiation, and photoelectric devices, in which incident photons are detected by a quantum effect such as the production of electrons. Chemical detectors include photographic emulsions, actinometric solutions and UV-sensitive plastic films.

1.2.1 *Spectroradiometry*

The fundamental way of characterizing a source of UVR is on the basis of its spectral power distribution in a graph (or table) which indicates the radiated power as a function of wavelength. The data are obtained by a technique known as spectroradiometry. Spectral measurements are often not required as ends in themselves but are used to calculate biologically weighted radiometric quantities. A spectroradiometer comprises three essential components (Gibson & Diffey, 1989):

(i) input optics, such as an integrating sphere or Teflon diffuser, which collects the incident radiation and conducts it to

(ii) the entrance slit of a monochromator, which disperses the radiation by means of one or two wavelength dispersive devices (either diffraction grating or prism). The monochromator also incorporates mirrors to guide the radiation from the entrance slit to the dispersion device and on to the exit slit, where it is incident on

(iii) a radiation detector, normally a photodiode or, for higher sensitivity, a photomultiplier tube.

Spectroradiometry is generally considered to be the best way of specifying UV sources, although the accuracy of spectroradiometry, particularly with respect to the UVB waveband of terrestrial radiation, is affected by a number of parameters including wavelength calibration, band width, stray radiation, polarization, angular dependence, linearity and calibration sources. It is therefore essential to employ a double monochromator for accurate characterization of terrestrial UVR and particularly UVB (Garrison *et al.*, 1978; Kostkowski *et al.*, 1982; Gardiner & Kirsch, 1991).

1.2.2 *Wavelength-independent (thermal) detectors*

General-purpose radiometers incorporate detectors that have a flat response over a wide range of wavelengths. Such thermal detectors operate on the principle that incident radiation is absorbed by a receiving element, and the temperature rise of the element is measured, usually by a thermopile or a pyroelectric detector. A thermopile, which comprises several thermocouples connected in series for improved sensitivity, must have a window made of fused silica for measuring UVR at wavelengths down to at least 250 nm. Pyroelectric detectors rely on a voltage generated by temperature changes in a lithium tantalate crystal. Thermal detectors are normally used to measure the total radiant power of a source rather than just the UV component (Moseley, 1988).

Instruments for measuring broad-band solar radiation fall into three categories: pyroheliometers, pyranometers and pyranometers with a shading device (Iqbal, 1983). These types of instrument find their applications in meteorology rather than in UV photobiology.

1.2.3 *Wavelength-dependent detectors*

Detectors of this type have a spectral response that varies widely depending on the types of detector and filters that may be incorporated. Detectors can be designed to have a spectral response that matches a particular action spectrum for a photobiological end-point. The success with which this is achieved is variable. The most widely used device, particularly for measuring solar UVR, has been the Robertson–Berger meter (Robertson, 1972; Berger, 1976), which incorporates optical filters, a phosphor and a vacuum phototube or photovoltaic cell. This device measures wavelengths of less than 330 nm in the global spectrum with a spectral response that rises sharply with decreasing wavelength. It has been used to monitor natural UVR continuously at several sites throughout the world (Berger & Urbach, 1982; Diffey, 1987a).

Detectors incorporating a photodiode or vacuum photocell in conjunction with optical filter(s) and suitable input optics (e.g., a quartz hemispherical detector) have been produced to match a number of different action spectra. One such detector is the International Light Model 730 UV Radiometer, which has a spectral response close to the action spectrum designated by the American Conference of Governmental Industrial Hygienists for evaluating the hazard to health of exposure to UVR, and has been used to measure irradiance over different terrains (Sliney, 1986).

Wavelength-dependent detectors with spectral responses largely in the UVA waveband are used, for example, in measuring the output of irradiation units for the treatment of psoriasis by psoralen photochemotherapy (Morison, 1983a).

A different yet complementary approach is the use of various photosensitive films as UV dosimeters. The principle is to relate the degree of deterioration of the films, usually in terms of changes in their optical properties, to the dose of incident UVR. The principal advantages of the film dosimeter are that it provides a simple means of integrating exposure continuously and allows simultaneous comparison of numerous sites that are inaccessible to bulky, expensive instruments (Diffey, 1987a). The most widely used photosensitive film is polymer polysulfone (Diffey, 1989a). Personal dosimeters of polysulfone film have been developed and used in a number of dosimetric studies (Challoner *et al.*, 1976, 1978; Leach *et al.*, 1978; Holman *et al.*, 1983a; Larkö & Diffey, 1983; Diffey, 1987a; Schothorst *et al.*, 1987a; Slaper, 1987; Rosenthal *et al.*, 1990).

It is difficult to achieve a prescribed UVR spectral response with wavelength-dependent detectors. Accurate results can be achieved only if the detectors are calibrated against the appropriate source spectrum using a spectroradiometer (Gibson & Diffey, 1989). Unless this is done, severe dosimetric errors can arise, particularly with measurements of solar UVR (Diffey, 1987a; Sayre & Kligman, 1992).

Accurate measurement of UVB radiation is far more difficult than would appear initially. The primary problem is that the UVB produced by most optical sources—the sun as well as incandescent and fluorescent lamps used for illumination—is only a very small fraction (i.e., less than 0.3%) of the total radiant energy emitted. Additionally, biological action spectra (e.g., for erythema and photokeratitis) typically decrease dramatically within the same waveband in which the source spectrum increases (Diffey & Farr, 1991a). This means that either a spectroradiometer or a direct-reading filtered 'erythemal' or 'hazard' meter must reject out-of-band radiant energy to better than one part in 10^4 or even 10^5. The spectral band-width of a monochromator can also greatly affect measurement error: too large a band-width can reduce the steepness of reported action spectra.

1.3 Sources and exposures

In the broadest sense, UVR may be produced when a body is heated (incandescence) or when electrons that have been raised to an excited state return to a lower energy level, as occurs in fluorescence, in an electric discharge in a gas and in electric arcs (optical plasma) (Sliney & Wolbarsht, 1980; Phillips, 1983; Moseley, 1988). The characteristics of exposures to both terrestrial solar radiation (an incandescent source) and artificial light sources are discussed in the following sections.

1.3.1 Solar ultraviolet radiation

Optical radiation from the sun is modified significantly as it passes through the Earth's atmosphere (Fig. 2), although about two-thirds of the energy from the sun that impinges on the atmosphere penetrates to ground level. The annual variation in extra-terrestrial radiation is less than 10%, but the variation in the modifying effect of the atmosphere is far greater (Moseley, 1988). Measurements corrected for atmospheric absorption show that the visible portion comprises approximately 40% of the total radiation received at the surface of the Earth. While UVR comprises only a small proportion of the total radiation (approximately 5%), this component is extremely important in various biological processes. The principal effect of infrared radiation is to warm the earth; approximately 55% of the solar radiation received at the surface of the earth is infrared (Foukal, 1990).

Fig. 2. Spectral irradiance from the sun outside the Earth's atmosphere (upper curve) and at sea level (lower curve)

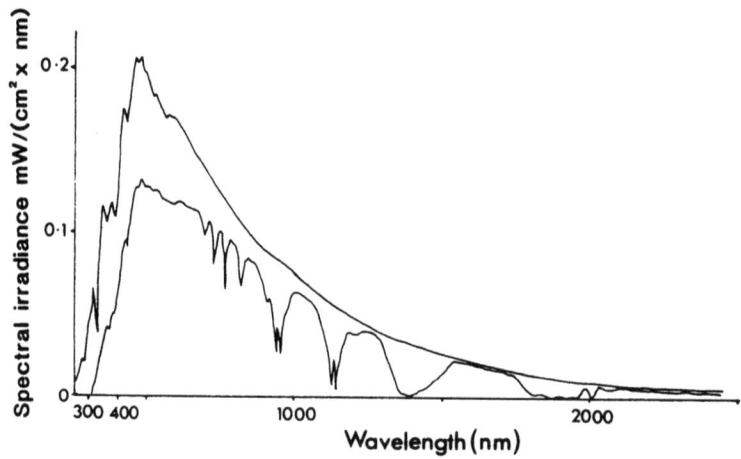

From Moseley (1988)

On its path through the atmosphere, solar radiation is absorbed and scattered by various constituents of the atmosphere. It is scattered by air molecules, particularly oxygen and nitrogen (Rayleigh scattering), which produce the blue colour of the sky. It is also scattered by aerosol and dust particles (Mie scattering) and is scattered and absorbed by atmospheric pollution. Total solar irradiance and the relative contributions of different wavelengths vary with altitude. Clouds attenuate solar radiation, although their effect on infrared radiation is greater than on UVR. Reflection of sunlight from certain ground surfaces may contribute significantly to the total amount of scattered UVR. An effective absorber of solar UVR is ozone in the stratosphere (Moseley, 1988). An equally important absorber in the longer wavelengths (infrared) is water vapour (Diffey, 1991); a secondary absorber in this range is carbon dioxide. These two filter out much of the solar energy with wavelengths longer than 1000 nm (Sliney & Wolbarsht, 1980).

The quality (spectral distribution) and quantity (total UV irradiance) of UVR reaching the Earth's surface depend on the radiated power from the sun and the transmitting properties of the atmosphere. Although UVC exists in the extra-terrestrial solar spectrum, it is filtered out completely by the ozone layer in the atmosphere. UVB radiation, which represents about 5% of the total solar UVR that reaches the Earth (Sliney & Wolbarsht, 1980), has been considered to be the most biologically significant part of the terrestrial UV spectrum. The levels of UVB radiation reaching the surface of the Earth, although heavily attenuated, are also largely controlled by the ozone layer.

Ozone (O_3) is a gas which comprises approximately one molecule out of every two million in the atmosphere. It is created by the reaction of molecular oxygen (O_2) with atomic oxygen (O), formed by the dissociation of O_2 by short-wavelength UVR (< 242 nm) in the stratosphere at altitudes between about 25 and 100 km. Absorption of UVR at wavelengths up to about 320 nm converts the ozone back to O_2 and O, and it is this dissociation of ozone that is responsible for preventing radiation at wavelengths less than about 290 nm from reaching the Earth's surface (Moseley, 1988; Diffey, 1991). Molina and Rowland (1974) first proposed that chlorofluorocarbons and other gases released by human activity could alter the natural balance of creative and destructive processes and lead to depletion of the stratospheric ozone layer. Substantial reductions, of up to 50%, in the ozone column observed in the austral spring over Antarctica were first reported in 1985 and may continue. There are, however, serious limitations in our current understanding of and ability to quantify ozone depletion at the present levels of contaminant release and in our ability to predict the effects on stratospheric ozone of any further increases (United Nations Environment Programme, 1989; United Kingdom Stratospheric Ozone Review Group, 1991).

A number of factors influence terrestrial UVR levels:
- *Variations in stratospheric ozone with latitude and season* (United Nations Environment Programme, 1989)
- *Time of day*: In summer, about 20–30% of the total daily amount of UVR is received between 11:00 and 13:00 h and 75% between 9:00 and 15:00 h (Diffey, 1991; Table 2 and Fig. 3). Although the amount of visible light falling on the ground in the summer may vary by only 30% between 12:00 and 15:00 h (local solar time), the short-wavelength component of the UVB spectrum undergoes a dramatic change during

Table 2. Percentage of daily UVB and UVA radiation received during different periods of a clear summer's day. Solar noon is assumed to be at 12:00 h, i.e., no allowance is made for daylight saving time

Latitude (°N)	UVB		UVA	
	11:00–13:00 h	9:00–15:00 h	11:00–13:00 h	9:00–15:00 h
20	30	78	27	73
40	28	75	25	68
60	26	69	21	60

From Diffey (1991)

Fig. 3. Daily variation in ultraviolet radiation: erythemal effective irradiance falling on a horizontal earth surface at Denver, CO, USA, on one summer's day

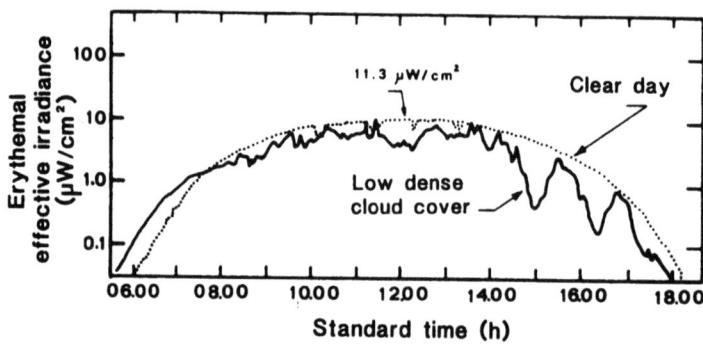

From Machta et al. (1975)

this period. At a wavelength of 300 nm, the spectral irradiance decreases by 10 fold, from approximately 1.0 to 0.1 µW/(cm² × nm) (Sliney, 1986).

- *Season*: Seasonal variation in terrestrial UV irradiance, especially UVB, at the Earth's surface is significant in temperate regions but much less nearer the equator (Table 3).

Table 3. Typical values for ambient daily and annual UVB radiation expressed in minimal erythema dose (MED)

Latitude (°N)	Diurnal UVB (MED)			
	Winter	Spring/Autumn	Summer	Annual
20 (Hawaii, USA)	14	20	25	6000
30 (Florida, USA)	5	12	15	4000
40 (Spain)	2	7	12	2500
50 (Belgium)	0.4	3	10	1500

From Diffey (1991)

- *Geographical latitude*: Annual UVR exposure dose decreases with increasing distance from the equator (Table 3).
- *Clouds*: Clouds reduce UV ground irradiance; changes in UVR are smaller than those of total irradiance because water in clouds attenuates solar infrared radiation much more than UVR. Even with heavy cloud cover, the scattered UVB component of sunlight (often called skylight) is seldom less than 10% of that under clear sky; however, very heavy cloud cover can virtually eliminate UVB even in summer. Light clouds scattered over a blue sky make little difference in sunburning effectiveness unless they directly cover the sun. Complete light cloud cover prevents about 50% of UVB energy, relative to that from a clear sky, from reaching the surface of the Earth (Diffey, 1991).

- *Surface reflection*: The contribution of reflected UVR to a person's total UVR exposure varies in importance with a number of factors (Table 4). A grass lawn scatters about 3% of incident UVB radiation. Sand reflects about 10–15%, so that sitting under an umbrella on the beach can lead to sunburn both from scattered UVB from the sky and reflected UVB from the sand. Fresh snow has been reported to reflect up to 85–90% of incident UVB radiation, although reflectance of about 30–50% is probably more typical. Ground reflectance is important, because parts of the body that are normally shaded are exposed to reflected radiation (Diffey, 1990a).

Table 4. Representative terrain reflectance factors for horizontal surfaces measured with a UVB radiometer at 12:00 h (290–315 nm) in the USA

Material	Reflectance (%)
Lawn grass, summer, Maryland, California and Utah	2.0–3.7
Lawn grass, winter, Maryland	3.0–5.0
Wild grasslands, Vail Mountain, Colorado	0.8–1.6
Lawn grass, Vail, Colorado	1.0–1.6
Flower garden, pansies	1.6
Soil, clay/humus	4.0–6.0
Sidewalk, light concrete	10–12
Sidewalk, aged concrete	7.0–8.2
Asphalt roadway, freshly laid (black)	4.1–5.0
Asphalt roadway, two years old (grey)	5.0–8.9
House paint, white, metal oxide	22
Boat dock, weathered wood	6.4
Aluminium, dull, weathered	13
Boat deck, wood, urethane coating	6.6
Boat deck, white fibreglass	9.1
Boat canvas, weathered, plasticized	6.1
Chesapeake Bay, Maryland, open water	3.3
Chesapeake Bay, Maryland, specular component of reflection at Z = 45 °N	13
Atlantic Ocean, New Jersey coastline	8.0
Sea surf, white foam	25–30
Atlantic beach sand, wet, barely submerged	7.1
Atlantic beach sand, dry, light	15–18
Snow, fresh	88
Snow, two days old	50

From Sliney (1986)

- *Altitude*: In general, each 300-m increase in altitude increases the sunburning effectiveness of sunlight by about 4%. Conversely, places on the Earth's surface below sea level have lower UVB exposures than nearby sites at sea level (Diffey, 1990a).

- *Air pollution*: Tropospheric ozone and other pollutants can decrease UVR, particularly in urban areas (Frederick, 1990).

(a) *Measurements of terrestrial solar radiation*

Since UVR wavelengths between about 295 and 320 nm (UVB radiation) in the terrestrial solar spectrum are thought to be those mainly responsible for adverse health effects, a number of studies have concentrated on measuring this spectral region (Sliney, 1986). Accurate measurements of UVR in this spectral band are difficult to obtain, however, because the spectral curve of terrestrial solar irradiance increases by a factor of more than five between 290 and 320 nm (Fig. 4). Nevertheless, extensive measurements of ambient

Fig. 4. Action spectrum designated by the American Conference of Governmental Industrial Hygienists (ACGIH) for assessing the hazard of ultraviolet radiation (very similar to erythemal action spectrum from 300–230 nm) and the solar spectrum. The ACGIH action spectrum, which is unitless, is closely fit by some radiometers; however, because of the small overlap of the terrestrial solar spectrum with the action spectrum, problems of stray light must be dealt with by constant checks with a filter that blocks wavelengths of less than 320 nm

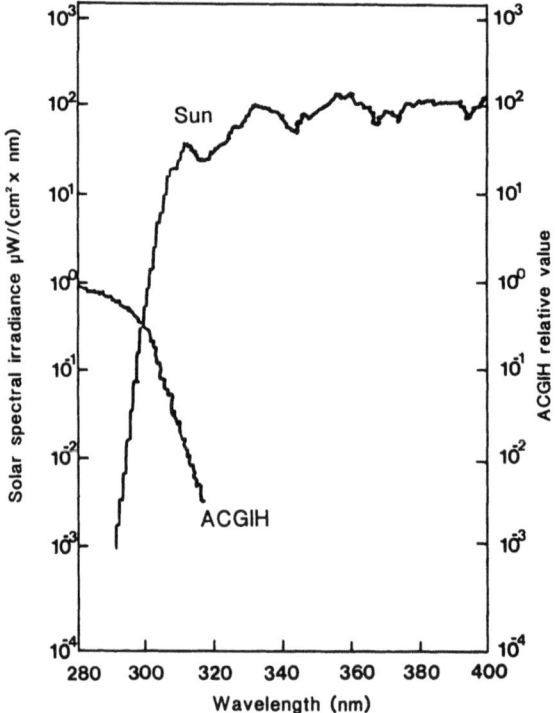

Adapted from Sliney *et al.* (1990)

UVR in this spectral band have been performed worldwide (Schulze, 1962; Schulze & Gräfe, 1969; Henderson, 1970; Sundararaman et al., 1975; Garrison et al., 1978; Doda & Green, 1980; Mecherikunnel & Richmond, 1980; Kostkowski et al., 1982; Ambach & Rehwald, 1983; Blumthaler et al., 1983; Livingston, 1983; Blumthaler et al., 1985a,b; Kolari et al., 1986; Hietanen, 1990; Sliney et al., 1990). Longer-wavelength UVR (UVA) was measured at the same time in many of these studies. Measurements of terrestrial solar UVA radiation are less subject to error than measurements of UVB, since the spectrum does not vary widely with zenith angle and the spectral irradiance curve is relatively flat.

Maps of annual UVR exposure, such as that shown in Figure 5, have been compiled for epidemiological studies of skin cancer and other diseases (Schulze, 1962, 1970; Scotto et al., 1976). Despite the large numbers of measurements, their interpretation in relation to human exposure has been complicated by three factors: (i) the considerable variation in UVB spectral irradiance with solar position throughout the day and with season; (ii) the effect of the geometry of exposure of individuals; and (iii) variation between humans in outdoor exposure and the parts of their bodies that are exposed.

Fig. 5. Global distribution of ultraviolet radiation

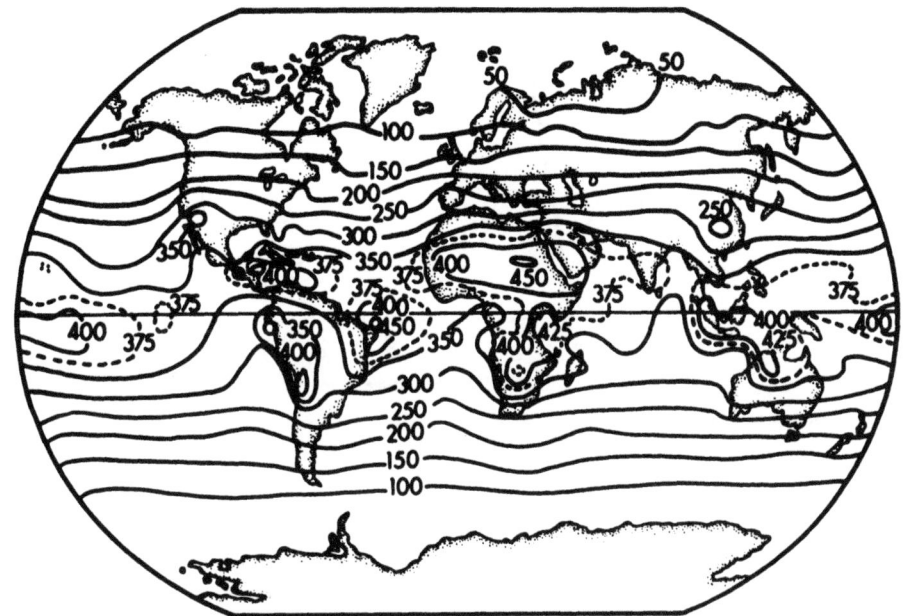

From Schulze (1970); WHO (1979)

The total solar radiation that arrives at the Earth's surface is termed 'global radiation', and measurements of terrestrial UVR most frequently pertain to this quantity, i.e., the radiant energy falling upon a horizontal surface from all directions (both direct and scattered radiation). Global radiation comprises two components, referred to as 'direct' and 'diffuse'.

Approximately 70% of the UVR at 300 nm is in the diffuse component rather than in the direct rays of the sun (Fig. 6). The ratio of diffuse to direct radiation increases steadily from less than 1.0 at 340 nm to at least 2.0 at 300 nm (Garrison *et al.*, 1978).

Fig. 6. Diffuse and direct solar spectral irradiance (solar zenith angle, 45°)

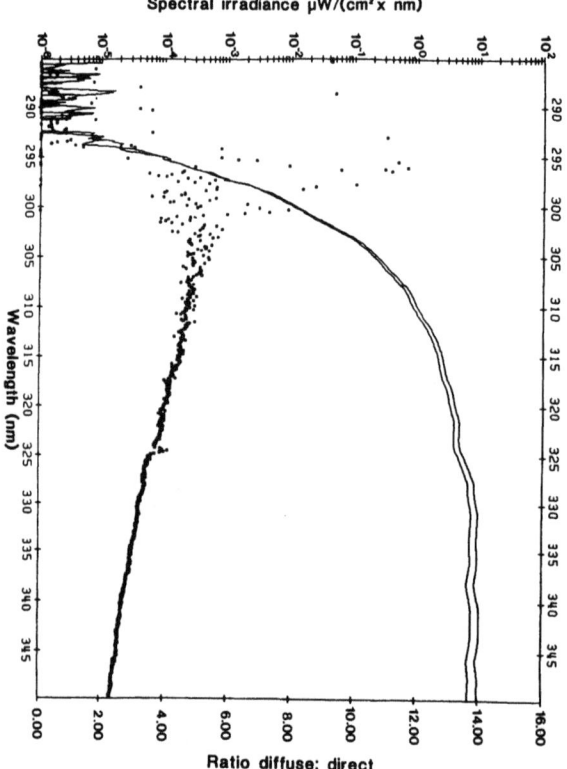

From Garrison *et al.* (1978)

UVR reflected from the terrain (the albedo) may also be important; however, essentially all measurement programmes have been limited to the direct and total diffuse components of sunlight. While such measurements are of interest in calculating the exposure dose of UVR of a prone individual, they are of very limited value in estimating exposure of the eye and shaded skin surfaces (e.g., under the chin), where the UVB radiation incident upon the body from terrain reflectance and horizon sky is of far greater importance. Sliney (1986) and Rosenthal *et al.* (1988) reported measurements of outdoor ambient UVR that included the reflected component to the eye. Exposure data for different anatomical sites is of value in developing biological dose–response relationships (Diffey *et al.*, 1979). The fact that ocular exposure differs significantly from cutaneous exposure is emphasized by the finding that photokeratitis is seldom experienced during sunbathing yet the threshold for UV photokeratitis is less than that for erythema of the skin (Sliney, 1986).

Measurements of the angular distribution of UVR relative to solar position and cloud distribution have been reported (Sliney, 1986; Fig. 7). A cloud obscuring the sun had no effect upon the UV radiance of open blue sky or the horizon sky; however, when the sun was 'out' (i.e., in an open sky), clouds near the horizon opposite the sun apparently reflected more UVR than would otherwise be present from the blue sky. This confirms the findings of studies of photographs of the sky taken through a narrow-band filter at 320 nm (Livingston, 1983), which revealed that the sky looks almost uniformly bright even when clouds are present and the clouds disappear into a uniformly hazy sky. Only the sun stands out, as would be expected from the plots on Figure 7. When the sun is near the horizon and can be looked at without great discomfort (i.e., at $Z = 75$–$90°$), the effective UV irradiance is again of the order of 0.3 μW/cm2, e.g., about 0.08–1.1 μW/cm^2 at an elevation angle of 12–15 ° (Sliney, 1986).

Fig. 7. Semilogarithmic plots of the angular dependence of skylight for 290–315 nm ultraviolet radiation (UVR) with the sun at zenith angle of about 45 °. A narrow field-of-view detector was scanned from zenith to the horizon. Uppermost curves show that direct UVR from the sun is more than 10 times greater than scattered UVR normally incident upon the eye at near-horizon angles where the zenith angle $Z = 70$–$90°$. Most surprising is the similarity of blue sky and cloudy sky UV irradiances at zenith or near the horizon.

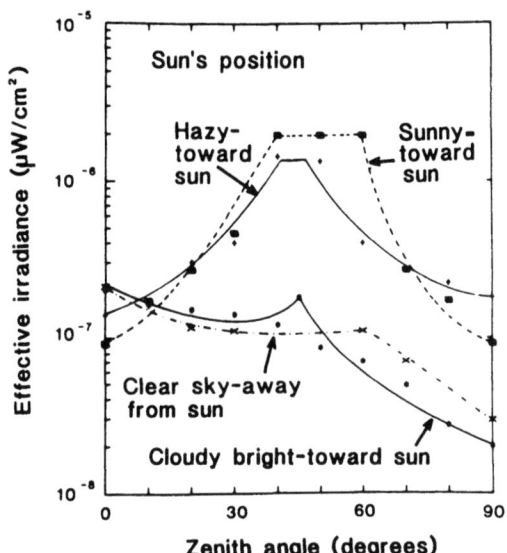

Adapted from Sliney (1986)

(b) Personal exposures

The exposure of different anatomical sites to solar UVR depends not only on ambient UVR and orientation of sites with respect to the sun but also on cultural and social behaviour, type of clothing and whether spectacles are worn.

Measurements of ambient UVR are useful in that they provide upper limits on human exposure (Scotto et al., 1976). They are of lesser value for assessing exposure doses received by groups of individuals. Polysulfone film has been used to monitor personal exposure to solar UVR (see p. 49). The wide variations in recorded exposure doses reflect diversity of behaviour and, in most cases, the small numbers (< 30) of subjects monitored. Nevertheless, it can be estimated that recreational (excluding vacations) exposure to the sun of people in northern Europe (where most of these studies were carried out) results in an annual solar exposure dose to the face of 20-100 MED, depending on the propensity for outdoor pursuits. The annual weekday UV exposure dose of indoor workers is around 30 MED; as a two-week outdoor vacation can result in a further 30-60 MED, the total annual exposure dose to the face of most indoor workers is probably in the range 40-160 MED. Outdoor workers at the same latitudes receive about two to three times these exposure doses, typically around 250 MED (Diffey, 1987b; Slaper, 1987).

An alternative approach to estimating personal exposure is to combine measured data on ambient UVR with a behavioural model of exposure. This approach was applied to a group of more than 800 outdoor workers in the USA (40 °N) by Rosenthal et al. (1991). These investigators estimated annual facial exposure doses of 30-200 MED, which are considerably lower than those estimated for outdoor workers in northern Europe, perhaps because Rosenthal et al. assumed facial exposure to be about 5-10% of ambient. A number of researchers have used polysulfone film badges on both human subjects (Holman et al., 1983a; Rosenthal et al., 1990) and mannequins (Diffey et al., 1977, 1979; Gies et al., 1988) to measure solar UVR exposure on the face relative to ambient exposure. The results vary considerably, reflecting factors such as positioning of film badges, behaviour of individuals, solar altitude and the influence of shade. Examination of the data suggests, however, that the exposure of an unprotected face is probably close to 20% of the ambient. Using this estimate, the annual facial exposure doses in the outdoor worker group studied by Rosenthal et al. (1991) would be about 80-500 MED. These data demonstrate clearly the current uncertainties associated with estimates of population exposure doses.

1.3.2 *Exposure to artificial sources of ultraviolet radiation*

(a) *Sources*

Six artificial sources that often produce UVR incidental to the production of visible light (Sliney & Wolbarsht, 1980; Phillips, 1983; Moseley, 1988) are described below.

(i) *Incandescent sources*

Optical radiation from an incandescent source appears as a continuous spectrum. Incandescent sources are usually ascribed a certain 'colour temperature', defined as the temperature of a black body that emits the same relative spectral distribution as the source. UVR is emitted in significant quantity when the colour temperature exceeds 2500 °K (2227 °C). Tungsten-halogen lamps in a quartz envelope (colour temperature, 3000 °K [2727 °C]) may emit significant UVR, whereas the UVR emission of an ordinary tungsten light bulb is negligible.

(ii) *Gas discharge lamps*

Another method of producing optical radiation is to pass an electric current through a gas. The emission wavelengths are determined by the type of gas present in the lamp and appear as spectral lines. The width of the lines and the amount of radiation in the interval between them (the continuum) depend on the pressure in the lamp. At low pressures, fine lines with little or no continuum are produced; as pressure is increased, the lines broaden and their relative amounts alter. Low-pressure discharge lamps, commonly containing mercury, argon, xenon, krypton or neon, are useful for spectral calibration. Medium-pressure mercury lamps operate at an envelope temperature in the region of 600–800 °C.

(iii) *Arc lamps*

Arc lamps operate at high pressures (20–100 atm [2020–10133 kPa]) and are very intense sources of UVR. Commonly available lamps contain xenon, mercury or a mixture of the two elements, which are effective sources of UVR. Xenon arc lamps operate at a colour temperature of 6000 °K (5727 °C); they are often used as the light source in solar simulation or are combined with a monochromator in spectral illumination systems. Deuterium arc lamps provide a useful source of UVC radiation and find their main use in spectrophotometers and as a calibration source for spectroradiometers.

(iv) *Fluorescent lamps*

The primary source of radiation in a fluorescent lamp arises from a low-pressure mercury discharge, which produces a strong emission at 254 nm, which in turn excites a phosphor-coated lamp to produce fluorescence. By altering the composition and thickness of the phosphor and the glass envelope, a wide variety of emission spectral characteristics can be obtained. The output is thus chiefly the fluorescent emission spectrum from the coating, with a certain amount of breakthrough of UVB mercury lines at 297, 303 and 313 nm, as well as those in the UVA and visible regions (WHO, 1979).

(v) *Metal halide lamps*

The addition of other metals (as halide salts) to a mercury discharge lamp allows for the addition of extra lines to the mercury emission spectrum. Most such tubes are basically medium-pressure discharge lamps with one or more metal halide additives, usually iodide. Advantage has been taken of the strong lead emission lines at 364, 368 and 406 nm in the lead iodide lamp, in which there is a 50% increase in output in the region between 355 and 380 nm compared to a conventional mercury lamp. Antimony and magnesium halide lamps provide spectral lines in the UVB and UVC regions.

(vi) *Electrodeless lamps*

A type of lamp recently introduced on a large scale is the electrodeless lamp. In this design, the discharge tube absorbs microwave energy fed, via waveguides, into a microwave chamber containing the tube. Two 1500-W magnetrons generate microwave energy at 2450 MHz. The life of such lamps is longer than that of electrode lamps, and a greater range of metal halides is available. Electrodeless lamps are used extensively for UV curing of inks and coatings, particularly when a short lamp length is adequate for the area to be irradiated. They have often been the first choice for curing prints on containers such as two-piece cans, plastic pots and bottles, and tubes.

(b) Human exposure

Although the sun remains the main source of UVR exposure for humans, the advent of artificial UVR sources has increased the opportunity for both intentional and unintentional exposure.

Intentional exposure is most often to acquire a tanned skin, frequently using sunbeds and solaria emitting principally UVA (315–400 nm) radiation (Diffey, 1987c). Another reason for intentional exposure to artificial UVR is the treatment of skin diseases, notably psoriasis.

Unintentional exposure is most often the result of occupation, and workers in many industries (see p. 66) may be exposed to UVR from artificial sources. The general public is exposed to low levels of UVR from sources such as fluorescent lamps used for indoor lighting and may be exposed in shops and restaurants where UVA lamps are employed in traps to attract flying insects.

(i) Cosmetic use

To some individuals, a tanned skin is socially desirable. A 'suntanning industry' has grown up, particularly in northern Europe and North America, in which artificial sources of UVR supplement exposure to sunlight.

Description of UVR sources used for tanning: Prior to the mid-1970s, the source of UVR was usually an unfiltered, medium- or high-pressure mercury arc lamp which emitted a broad spectrum of radiation, from UVC through to visible and infrared radiation (Diffey & Farr, 1991b). The units often incorporated one or more infrared heaters and were commonly called 'sunlamps' or 'health lamps' (Anon., 1979). One disadvantage of this type of unit was that the area of irradiation was limited to a region such as the face and so whole-body tanning was tedious. By incorporating several mercury arc lamps into a 'solarium', whole body exposure was achieved. Tanning devices based on mercury arc lamps emit relatively large quantities of UVB and UVC radiation, resulting in a significant risk of burning and acute eye damage. Solaria that incorporate unfiltered mercury arc lamps are therefore now less popular (Diffey, 1990a).

So-called UVB fluorescent lamps (e.g., Westinghouse FS Sunlamp, Philips TL12) emit approximately 55% of their UV energy in the UVB and approximately 45% in the UVA regions (Diffey & Langley, 1986). They were often used in tanning booths, more commonly in the USA than in Europe.

Sunbeds, incorporating high-intensity UVA fluorescent lamps, were developed in the 1970s. These devices consist of a bed and/or canopy incorporating 6–30 fluorescent lamps 150–180 cm in length. The earliest type of UVA lamp used in sunbeds is typified by the Philips TL09, Wotan L100/79 and Wolff Solarium lamps (Diffey, 1987c). The spectral power distribution from this type of lamp is shown in Figure 8a. The emission spectrum comprises the fluorescence continuum, extending from about 315 to 400 nm and peaking at 350–355 nm, together with the characteristic lines from the mercury spectrum down to 297 nm (UVB) (Diffey & McKinlay, 1983). The UVA irradiance at the skin surface from a typical sunbed or suncanopy containing these lamps is between 50 and 150 W/m^2 (Bowker & Longford, 1987; Bruyneel-Rapp *et al.*, 1988).

Fig. 8. Spectral emissions of different lamps used for cosmetic tanning: (a) Philips TL09 (Diffey, 1987c); (b) Philips TL10R (Diffey, 1987c); (c) Wolff Bellarium S (B.L. Diffey, unpublished data); (d) optically filtered high-pressure metal halide lamp (Diffey, 1987c)

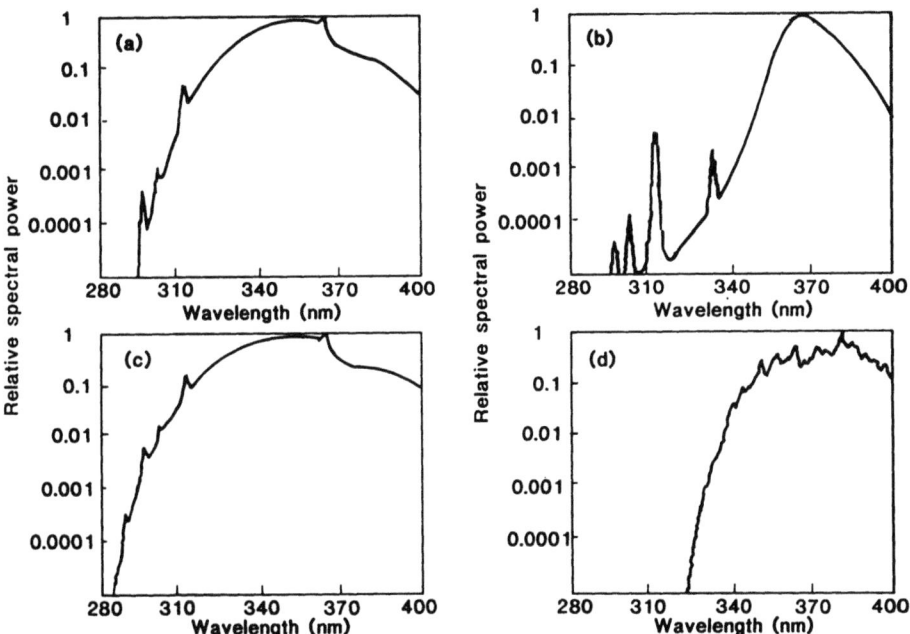

In the mid-1980s, another type of UVA fluorescent lamp (Philips TL10R) was introduced especially for cosmetic tanning. The principal features of this type of lamp were a reflector intrinsic to the lamp envelope and a fluorescence spectrum extending from about 340 to 400 nm, peaking at 370 nm (Fig. 8b); note also the presence of characteristic mercury lines in the UVB region. The skin surface irradiance from a sunbed or suncanopy incorporating Philips TL10R lamps is typically around 250 W/m^2 (Diffey, 1987c).

Another type of UV fluorescent lamp that has been used in sunbeds is the so-called 'fast tan' tube. This type of lamp is typified by the Wolff Bellarium S, the spectral power distribution of which is shown in Figure 8(c). The spectrum extends from about 290 to 400 nm and peaks at around 350 nm (Diffey & Farr, 1987).

Optically filtered, high-pressure mercury lamps doped with metal halide additives are also used in cosmetic tanning. The spectral emission lies entirely within the UVA waveband (Fig. 8d), and irradiances at the skin surface of more than 1000 W/m^2 can be achieved. The best known of this type of unit is probably the UVASUN (Mutzhas, 1986).

A summary of the physical and photobiological emissions from these different types of lamps is given in Table 5 (Diffey & Farr, 1991a).

Table 5. Characteristics of different ultraviolet (UV) lamps used for tanning

Lamp	Radiation emission (%)			Contribution to tanning (%)		
	UVA	UVB	UVC	UVA	UVB	UVC
Mercury arc sunlamp	40	40	20	0	35	65
Simulated sunlight lamp	95	5	0	20	80	0
Type I UVA lamp	99	1	0	60	40	0
Type II UVA lamp	> 99.9	< 0.1	0	> 90	< 10	0
Optically filtered high-pressure lamp[a]	100	0	0	100	0	0
Summer UV sunlight[b]	95	5	0	20	80	0

From Diffey & Farr (1991b) unless otherwise specified
[a]From Mutzhas (1986)
[b]From Sliney & Wolbarsht (1980)

Exposure to UVR sources used for tanning: Telephone surveys carried out in the Netherlands (Bruggers *et al.*, 1987) and in the United Kingdom (Anon., 1987) in the mid-1980s showed that 7–9% of the adult population in each country had used sunbeds in the previous one to two years. A more recent market survey in the United Kingdom (R. McLauchlan, personal communication), with a sample size of 5800, gave a slightly higher figure, with 10% of the population having used a sunbed during the previous year (1988) and 19% of the sample admitting to having used a sunbed at some time in the past. In these and other surveys in the United Kingdom (Diffey, 1986) and the USA (Dougherty *et al.*, 1988), women accounted for 60–85% of users, about half of the subjects being young women aged between 16 and 30. The commonest reason given for using tanning equipment was to acquire a pre-holiday tan (Anon., 1987; R. McLauchlan, personal communication); other reasons included perceived health benefits, reduction of stress and improved relaxation, protection of the skin before going on holiday, sustaining a holiday tan and treatment of skin diseases such as psoriasis and acne (Diffey, 1986; Dougherty *et al.*, 1988).

In the Dutch survey (Bruggers *et al.*, 1987), about half of the users interviewed used tanning equipment at home and the other half used facilities at commercial premises, such as tanning salons, hairdressers, sports clubs and swimming pools. Most people had used UVA equipment; 24% had used either UVB mercury arc sunlamps or solaria incorporating these lamps. A more recent survey in the United Kingdom (McLauchlan, 1989) confirmed the Dutch finding that the amount of use at home and at commercial premises was approximately the same. A survey carried out at commercial establishments in the United Kingdom indicated that all the equipment used emitted primarily UVA radiation, mostly from fluorescent UVA lamps and 10% from optically filtered high-pressure metal halide lamps (Diffey, 1986). Sales of tanning appliances in the United Kingdom increased rapidly during the 1980s, but by the end of the decade there appeared to be a steady, or possibly reduced, level of sales (Diffey, 1990a).

The mean number of tanning sessions per year in the Dutch study was 23 (Bruggers *et al.*, 1987). In the United Kingdom, half-hour sessions were the most popular (Diffey, 1986). Each tanning session with UVA equipment normally results in an erythemally-weighted exposure of about 0.8 MED (150 J/m^2), whereas exposure to mercury arc lamps results in about 2 MED per session (400 J/m^2). In the Dutch survey, it was estimated that the median annual exposure was 24 MED (4.8 kJ/m^2) (Bruggers et al., 1987).

(ii) *Medical and dental applications*

UVR has both diagnostic and therapeutic applications in medicine and dentistry. The diagnostic uses are confined largely to fluorescing of skin and teeth, and the UVR source is normally an optically filtered medium-pressure mercury arc lamp producing radiation mainly at 365 nm (so-called 'Wood's lamps') (Caplan, 1967). Radiation exposure is limited to small areas (< 15 cm in diameter), and the UVA radiation dose per examination is probably no more than 5 J/cm^2. The therapeutic uses of UVR, which result in considerably higher doses, are mainly in the treatment of skin diseases and occasionally the symptomatic relief of pruritus.

Phototherapy: The skin diseases that are most frequently treated with UVR are psoriasis and eczema. Phototherapy of psoriasis at hospital may include the use of tar and related derivatives and other substances, such as anthralin, on the skin (Morison, 1983a; see also IARC, 1987a).

The first treatment of psoriasis with an artificial source of UVR is credited to Sardemann, who used a carbon arc lamp of the type developed by Finsen at around the turn of the century. These lamps were unpopular in clinical practice because they emitted noise, odour and sparks, and they were superseded by the development of the medium-pressure mercury arc lamp. In the 1960s, a variety of metal halides were added to mercury lamps to improve emissions in certain regions of the UV and visible spectra. Fluorescent lamps were developed in the late 1940s; since then, a variety of phosphor and envelope materials have been used to produce lamps with emissions in different regions of the UV spectrum, such that, today, there exists a wide range of lamps for the phototherapy of skin diseases (Diffey & Farr, 1987).

Lamp systems can be classified into one of five categories in terms of suitability for phototherapy (Diffey, 1990b):

Type A: a single, medium-pressure mercury arc or metal halide lamp;

Type B: one or more vertical columns containing five or six optically filtered high-pressure metal halide lamps;

Type C: a canopy or cubicle containing fluorescent sunlamps which emit predominantly UVB but also significant amounts of radiation at wavelengths below 290 nm (e.g., Westinghouse FS sunlamp, Philips TL12 and Sylvania UV21 lamps);

Type D: a canopy, sunbed or cubicle incorporating fluorescent lamps which emit predominantly UVB radiation and negligible amounts of radiation at wavelengths below 290 nm (e.g., the Wolff Helarium);

Type E: a newly developed fluorescent lamp that emits a narrow band of radiation around 311–312 nm (Philips TL01).

The spectral power distributions characteristic of each of these five types of lamp are shown in Figure 9. The therapeutic radiation for psoriasis lies principally within the UVB waveband (Parrish & Jaenicke, 1981), and the cumulative UVB dose required for clearing

Fig. 9. Spectral power distributions of different types of phototherapy lamp (Diffey, 1990b). Type A: unfiltered medium-pressure mercury arc lamp; type B: optically filtered iron iodide lamp; type C: fluorescent sunlamp (Philips TL12); type D: Wolff Helarium lamp; type E: narrow-band UVB fluorescent lamp (Philips TL01)

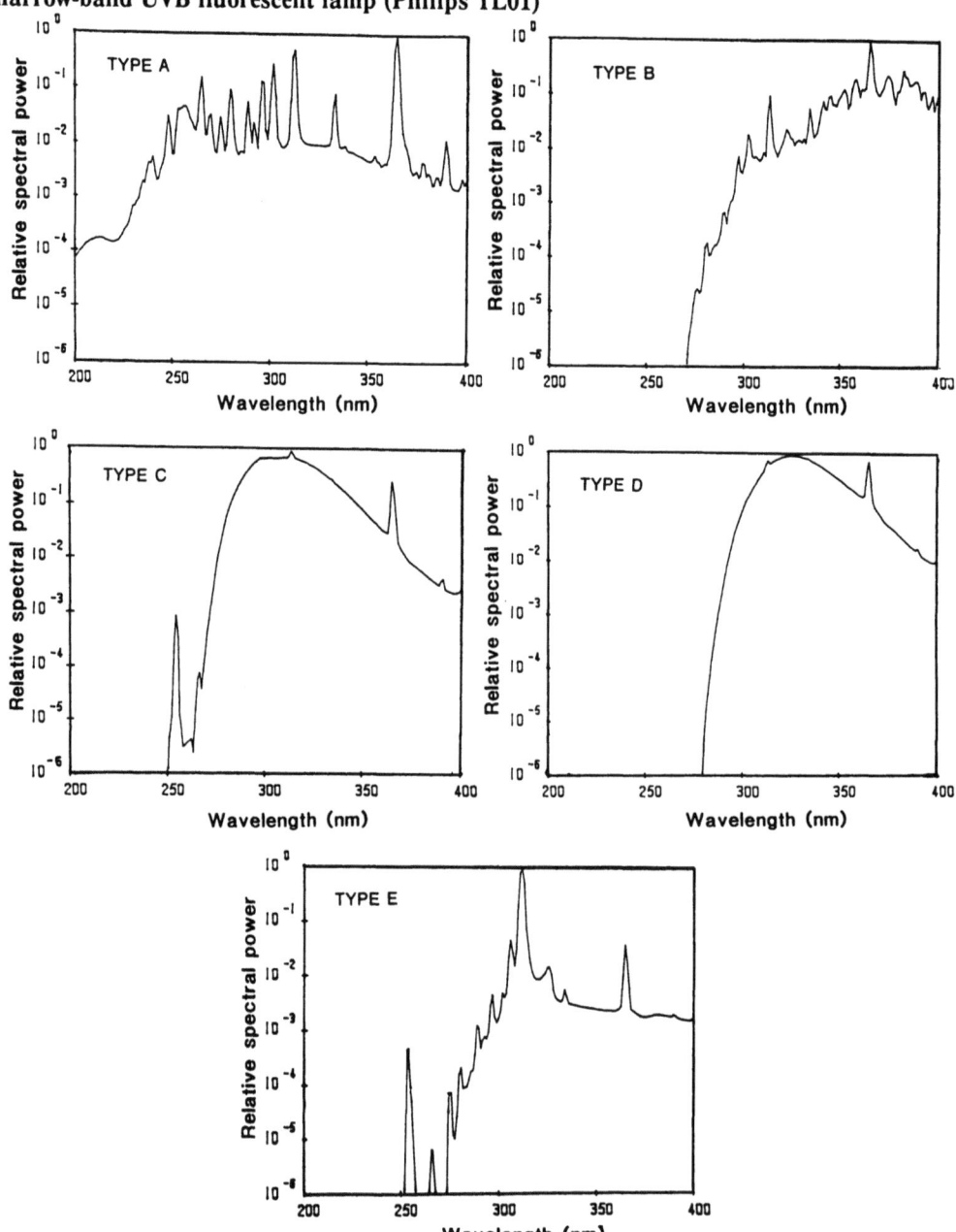

psoriasis is typically 100–200 MED (Diffey, 1990a), usually delivered over a course consisting of 10–30 exposures over 3–10 weeks (van der Leun & van Weelden, 1986).

Annual doses received by 90% of patients given UVB phototherapy for psoriasis range from about 60 to 670 MED, with a typical dose in a single course being between 200 and 300 MED (Slaper, 1987).

Psoralen photochemotherapy (see also IARC, 1980, 1986a, 1987b): This form of treatment, known colloquially as PUVA, involves the combination of photoactive drugs, psoralens (P), with long-wave UVR (UVA) to produce a beneficial effect. Psoralen photochemotherapy has been used to treat many skin disease in the past decade, although its principal success has been in the management of psoriasis (Parrish *et al.*, 1974), a disorder characterized by an accelerated cell cycle and rate of DNA synthesis. Psoralens may be applied to the skin either topically or systemically; the latter route is generally preferred, and the psoralen most commonly administered is 8-methoxypsoralen. The patient is usually exposed to UVA radiation from banks of fluorescent lamps with the spectral power distribution shown in Figure 8a. Values for UVA irradiance in clinical treatment cubicles have been found to range from 16 to 140 W/m^2 (Diffey *et al.*, 1980; Diffey, 1990b), although an irradiance of 80 W/m^2 is probably typical. The UVA dose per treatment session is usually in the range 1–10 J/cm^2 (Diffey *et al.*, 1980).

Generally, approximately 25 treatments over a period of 6–12 weeks, with a cumulative UVA dose of 100–250 J/cm^2, are required to clear psoriatic lesions (Melski *et al.*, 1977; Henseler *et al.*, 1981). PUVA therapy is not a cure for psoriasis, and maintenance therapy is often needed at intervals of between once a week to once a month to prevent relapse (Gupta & Anderson, 1987).

Neonatal phototherapy for hyperbilirubinaemia: Phototherapy is sometimes used in the treatment of neonatal jaundice or hyperbilirubinaemia. The preferred method of treatment is to irradiate the baby for several hours a day for up to one week with visible light, particularly blue light (Sisson & Vogl, 1982). The lamps used for phototherapy, although intended to emit only visible light, may also have a UV component: One commercial neonatal phototherapy unit was found to emit not only visible light and UVA but also radiation at wavelengths down to 265 nm (Diffey & Langley, 1986).

Fluorescence in cutaneous and oral diagnosis: Wood's light—a source of UVA obtained by filtering optically a mercury arc lamp with 'blackglass'—is used by dermatologists as a diagnostic aid in skin conditions that produce fluorescence (Caplan, 1967; Diffey, 1990a). As irradiation of the oral cavity with a Wood's lamp can produce fluorescence under certain conditions, this has been used in the diagnosis of various dental disorders, such as early dental caries, the incorporation of tetracycline into bone and teeth, dental plaque and calculus (Hefferren *et al.*, 1971).

Polymerization of dental resins: Pits and fissures in teeth have been treated using an adhesive resin polymerized with UVA. The resin is applied with a fine brush to the surfaces to be treated and is hardened by exposure to UVA radiation at a minimal irradiance of 100 W/m^2 for 30 s or so (Eriksen *et al.*, 1987; Diffey, 1990a).

(iii) *Occupational exposures*

Artificial sources of UVR are used in many different ways in the working environment. In some cases, the UV source is well contained within an enclosure and, under normal circumstances, presents no risk of exposure to personnel. In other applications of UVR, it is inevitable that workers are exposed to some radiation, normally by reflection or scattering from adjacent surfaces. Occupational exposure to UVR is also a consequence of exposure to general lighting in the workplace.

Industrial photoprocesses: Many industrial processes involve a photochemical component. The large-scale nature of these processes often necessitates the use of high-power (several kilowatts) lamps such as high-pressure metal halide lamps (Diffey, 1990a).

The principal industrial applications of photopolymerization include the curing of protective coatings and inks and photoresists for printed circuit boards. The curing of printing inks by exposure to UVR is now widespread; as the cure takes only a fraction of a second, UV drying units can be installed between printing stations on a multicolour line, so that each colour is dried before the next is applied. Another major use of UV curing has been for metal decorating in the packaging industry (Phillips, 1983). UVA is also used to inspect printed circuit boards and integrated circuits in the electronics industry (Pauw & Meulemans, 1987).

Artificial sources of UVR are used to test the weathering capability of materials such as polymers. Xenon-arc lamps are often the light source because their emission spectra is similar to the spectrum of terrestrial sunlight, although some commercial weathering chambers incorporate carbon-arc lamps, high-pressure metal halide lamps or fluorescent sunlamps (Davis & Sims, 1983).

Sterilization and disinfection: Radiation with wavelengths in the range 260–265 nm is the most effective for this use, since it corresponds to a maximum in the DNA absorption spectrum. Low-pressure mercury discharge tubes are thus often used as the radiation source, as more than 90% of the radiated energy lies in the 254 nm line. These lamps are often referred to as 'germicidal lamps', 'bactericidal lamps' or simply 'UVC lamps' (Diffey, 1990a).

UVC radiation has been used to disinfect sewage effluents, drinking-water, water for the cosmetics industry and swimming pools. Germicidal lamps are sometimes used inside microbiological safety cabinets to inactivate airborne and surface microorganisms (Diffey, 1990a). The combination of UVR and ozone has a very powerful oxidizing action and can reduce the organic content of water to extremely low levels (Phillips, 1983).

Welding (see also IARC, 1990): Welding equipment falls into two broad categories: gas welding and electric arc welding. Only the latter process produces significant levels of UVR, the quality and quantity of which depend primarily on the arc current, shielding gas and metals being welded (Sliney & Wolbarsht, 1980).

Welders are almost certainly the largest occupational group with exposure to artificial sources of UVR. It has been estimated (Emmett & Horstman, 1976) that there may be as many as half a million welders in the USA alone. The levels of UV irradiance around electric arc welding equipment are high; effective irradiance (relative to the action spectrum of the American Conference of Governmental Industrial Hygienists) at 1 m at an arc current of 400 A ranged from 1 to 50 W/m^2 (Table 6), and the unweighted UVA irradiance ranged from 3 to

70 W/m², depending on the type of welding and the metal being welded (Cox, 1987; Mariutti & Matzeu, 1987). It is not surprising therefore that most welders at some time or another experience 'arc eye' or 'welder's flash' (photokeratitis) and skin erythema. The effective irradiance at 0.3 m from many types of electric welding arcs operating at 150 A is such that the maximum permissible exposure time for an 8-h working period on unprotected eyes and skin varies from a few tenths of a second to about 10 s, depending on the type of welding process and the material used (Cox, 1987).

Table 6. Limits of exposure to ultraviolet radiation and radiation effectiveness

Wavelength (nm)	Exposure limit (J/m²)	Relative spectral effectiveness (S_λ)[a]
180	2500	0.012
190	1600	0.019
200	1000	0.030
205	590	0.051
210	400	0.075
215	320	0.095
220	250	0.120
225	200	0.150
230	160	0.190
235	130	0.240
240	100	0.300
245	83	0.360
250	70	0.430
254[b]	60	0.500
255	58	0.520
260	46	0.650
265	37	0.810
270	30	1.000
275	31	0.960
280[b]	34	0.880
285	39	0.770
290	47	0.640
295	56	0.540
297[b]	65	0.460
300	100	0.300
303[b]	250	0.120
305	500	0.060
308	1200	0.026
310	2000	0.015
313[b]	5000	0.006
315	1.0×10^4	0.003
316	1.3×10^4	0.0024
317	1.5×10^4	0.0020
318	1.9×10^4	0.0016
319	2.5×10^4	0.0012

Table 6 (contd)

Wavelength (nm)	Exposure limit (J/m^2)	Relative spectral effectiveness $(S_\lambda)^a$
320	2.9×10^4	0.0010
322	4.5×10^4	0.00067
323	5.6×10^4	0.00054
325	6.0×10^4	0.00050
328	6.8×10^4	0.00044
330	7.3×10^4	0.00041
333	8.1×10^4	0.00037
335	8.8×10^4	0.00034
340	1.1×10^5	0.00028
345	1.3×10^5	0.00024
350	1.5×10^5	0.00020
355	1.9×10^5	0.00016
360	2.3×10^5	0.00013
365[b]	2.7×10^5	0.00011
370	3.2×10^5	0.000093
375	3.9×10^5	0.000077
380	4.7×10^5	0.000064
385	5.7×10^5	0.000053
390	6.8×10^5	0.000044
395	8.3×10^5	0.000036
400	1.5×10^6	0.000030

From American Conference of Governmental Industrial Hygienists (1991); wavelengths chosen are representative, and other values should be interpolated at intermediate wavelengths.
[a]For explanation, see pp. 46–47
[b]Emission lines of a mercury discharge spectrum

In a survey of electric arc welders in Denmark, 65% of those questioned had experienced erythema; however, as no indication of the frequency of skin reactions was reported, it is not possible to estimate annual exposure (Eriksen, 1987). Monitoring of the exposure to UVR of non-welders working in the vicinity of electric arc welding apparatuses showed that their daily exposure dose exceeded the maximum permissible exposure limits by almost an order of magnitude (Barth et al., 1990).

Phototherapy: Although there is a trend to the use of enclosed treatment cubicles, some of the lamps used to treat skin disease (see the section on medical and dental applications) are unenclosed, emit high levels of UVR and can present a marked hazard to staff; at 1 m from these lamps, the recommended 8-h occupational exposure limits can be exceeded in less than 2 min (Diffey & Langley, 1986).

In a study of the exposure of staff in hospital phototherapy departments (Larkö & Diffey, 1986), annual exposure to UVR could be estimated from the number of occasions per year on which staff had experienced at least minimal erythema (Diffey, 1989b). Estimated annual occupational exposures to UVR were 15, 92 and 200 MED, corresponding to a frequency of erythema of once per year, once per month and once per week, respectively.

Operating theatres: UVC lamps have been used since the 1930s to decrease the levels of airborne bacteria in operating theatres (Berg, 1987). The technique requires complete protection of the eyes and skin of staff and patients; for this and other reasons, filtered air units are often preferred.

Research laboratories: Sources of UVR are used by most experimental scientists engaged in aspects of photobiology and photochemistry and in molecular biology. These applications, in which the effect of UV irradiation on biological and chemical species is of primary interest to the researcher, can be differentiated from UV fluorescence by absorption techniques where the effect is of secondary importance (Diffey, 1990a).

UV photography: There are two distinct forms of UV photography: reflected or transmitted UV photography and UV fluorescence photography. In both applications, the effective radiation lies within the UVA waveband (Lunnon, 1984).

UV lasers: High-power lasers which emit in the UV region, used in nuclear and other research laboratories, are far less common than those that emit in the visible or infrared regions of the electromagnetic spectrum.

Nitrogen lasers emit at a wavelength of 337 nm (Phillips, 1983), and instruments with a peak power output of up to 2.3 MW per pulse are available. Nitrogen lasers can be used in conjunction with fluorescent dyes to produce spectral emissions of 360–900 nm, with a power pulse of 200–480 kW. If frequency doubling crystals are used in conjunction with a nitrogen laser, UV emissions down to 260 nm are possible.

An alternative laser source of UVR is the excimer laser. (The term 'excimer' denotes a homonuclear molecule which is bound in an electronically excited state but is dissociative in the ground state [Phillips, 1983].) The wavelength of the pulsed UVR from this type of laser depends on the excimer molecules, such as ArF, F_2, XeCl and KrF, which emit at 193, 157, 308 and 248 nm, respectively (Phillips, 1983; Bos & de Haas, 1987). On the basis of worst-case assumptions, the estimated annual risk for skin cancer for workers exposed to UV lasers in medical applications is equivalent to about one additional day of sunbathing, and that for workers exposed to UV lasers in laboratories is comparable to the risk for outdoor workers (Sterenborg *et al.*, 1991).

Quality assurance in the food industry: Many contaminants of food products can be detected by UV fluorescence techniques. For example, the bacterium *Pseudomonas aeruginosa*, which causes rot in eggs, meat and fish, can be detected by its yellow-green fluorescence under UVA irradiation. One of the longest established uses of UVA fluorescence in public health is to demonstrate contamination with rodent urine, which is highly fluorescent (Ultra-Violet Products, Inc., 1977).

Insect traps: Many flying insects are attracted by UVA radiation, particularly in the region around 350 nm. This phenomenon is the principle of electronic insect traps, in which a UVA fluorescent lamp is mounted in a unit containing a high-voltage grid. The insect, attracted by the UVA lamp, flies into the unit and is electrocuted in the air gap between the high-voltage grid and a grounded metal screen. Such units are commonly found in areas where food is prepared and sold to the public (Diffey, 1990a).

Sunbed salons and shops: The continuing popularity of UVA sunbeds and suncanopies for cosmetic tanning has resulted in the establishment of a large number of salons and shops selling sunbeds for use at home. Some shops may have 20 or more UVA tanning appliances, all switched on, thus exposing members of the public and staff to high levels (> 20 W/m^2) of UVA radiation (Diffey, 1990a).

Discotheques: UVA 'blacklight' lamps are sometimes used in discotheques to induce fluorescence in the skin and clothing of dancers. The levels of UVA emitted are usually low (< 10 W/m^2) (Diffey, 1990a).

Offices: Signatures can be verified by exposing a signature obtained with colourless ink to UVA radiation, under which it fluoresces. UVA exposure of office staff is normally to hands, and irradiance is low (< 10 W/m^2) (Diffey, 1990a).

(iv) *General lighting*

Fluorescent lamps used for general lighting in offices and factories emit small quantities of both UVA and UVB. A UVA irradiance of 30 mW/m^2 (Diffey, 1990a) and a UVB irradiance of 3 mW/m^2 (McKinley & Whillock, 1987) were found for bare fluorescent lamps with a typical illuminance of 500 lux. These UV levels give rise to an annual exposure of indoor workers to no more than 5 MED, and this dose can be reduced appreciably by the use of plastic diffusers (McKinlay & Whillock, 1987). A study of the personal doses of UVR received by workers in the car manufacturing industry who were engaged in inspecting paintwork of new cars under bright fluorescent lamps indicated a similar annual exposure (Diffey *et al.*, 1986). Most plastic diffusers reduce erythemally effective irradiance to 0.2% or less of that of the bare lamp. An exception is clear acrylic diffusers, which absorb only about 20% of the erythemally effective radiation. The absorption of UVA radiation by diffusers is less effective, transmission ranging from 1% for opal polycarbonate to 74% for clear acrylic (McKinlay & Whillock, 1987). Spectroradiometric measurements of the UV levels from indoor fluorescent lamps carried out in the USA, however, indicated much higher annual doses for people exposed occupationally for 2000 h per year: The annual estimated exposure dose ranged from 8 to 30 MED for an illuminance level of 500 lux from bare lamps (Cole *et al.*, 1985).

Desk-top lights which incorporate tungsten–halogen (quartz) lamps may result in exposure to UVR of the hands and arms, if the lamps are used in excess of recommended occupational exposure levels (McKinlay *et al.*, 1989). Experimental studies have shown that erythema can be induced in susceptible individuals after a 15-min exposure at 10 cm from a 100-W tungsten–halogen source, principally by the UVB component of the emission (Cesarini & Muel, 1989). Tungsten–halogen lamps are also used for general lighting (e.g., spotlights, indirect lighting, floor lamps) in some countries.

(c) *Regulations and guidelines*

(i) *Cosmetic use*

The most comprehensive guidelines for the use of sunlamps and sunbeds in cosmetic tanning are those published by the International Electrotechnical Commission (1987, 1989). The guidelines classify tanning appliances into one of four types according to the effective irradiance at short ($\lambda \leq 320$ nm) and long ($320 < \lambda \leq 400$ nm) UV wavelengths (Table 7).

Table 7. Classification of tanning appliances

Type	Effective irradiance (W/m^2)	
	$\lambda \leq 320$ nm	320 nm $< \lambda \leq 400$ nm
1	< 0.0005	≥ 0.15
2	0.0005–0.15	≥ 0.15
3	< 0.15	< 0.15
4	≥ 0.1	< 0.15

From International Electrotechnical Commission (1989)

Effective radiance is defined as:

$$\sum_{250}^{400} E_\lambda \times S_\lambda \times \Delta_\lambda,$$

where E_λ is the spectral irradiance (W/m^2 × nm) at wavelength λ (nm) at the shortest recommended exposure distance; Δ_λ is the wavelength interval used in the summation; and S_λ is the relative erythemal effectiveness recently adopted by the Commission Internationale de l'Eclairage (McKinlay & Diffey, 1987), specified as shown in Table 8. The guidelines recommend that the exposure time for the first session on untanned skin should correspond to an effective dose not exceeding 100 J/m^2; this is approximately equivalent to 1 MED for subjects with sun-reactive skin type I. The annual exposure should not exceed an effective dose of 25 kJ/m^2 (International Electrotechnical Commission, 1989).

Table 8. Specifications of relative erythemal effectiveness

Wavelength (λ; nm)	Relative erythemal effectiveness (S_λ) (weighting factor)
$\lambda < 298$	1
$298 < \lambda < 328$	$10^{0.094(298-\lambda)}$
$328 < \lambda \leq 400$	$10^{0.015(139-\lambda)}$

From McKinlay & Diffey (1987); International Electrotechnical Commission (1989)

Although these guidelines form the basis of several national standards on sunlamp and sunbed use, it should be noted that variations exist; for example, in the Netherlands, Norway and Sweden, certain UV appliances are not permitted. Regulations concerning the use of tanning appliances are in force in only a few countries, but many others have published advice on sunbed use, including information on adverse effects, as well as guidelines on manufacturing standards.

(ii) *Occupational exposure*

Guidance on the maximal limits of exposure to UVR as a consequence of occupation is given by the International Non-ionizing Radiation Committee of the International Radiation

Protection Association. These exposure limits, which apply only to incoherent (i.e., non-laser) sources, represent conditions under which it is expected that nearly all individuals may be repeatedly exposed without adverse effects and are below levels which would be used for medical or cosmetic exposure to UVR. The limits for occupational exposure to UVR incident upon the skin or eye were considered separately for the UVA spectral region (315–400 nm) and the actinic UV spectral region (UVC and UVB, 180–315 nm). In 1984, the limit provided an equal spectral weighting between 315 and 400 nm, a maximal 1000-s radiant exposure of 10 KJ/m^2 and a maximal irradiance of 10 W/m^2 for longer periods (International Non-ionizing Radiation Committee of the International Radiation Protection Association, 1985). Studies of skin and ocular injury resulting from exposure to UVA led the Committee to issue revised exposure limits in 1988: For the UVA spectral region (315–400 nm), the total radiant exposure incident upon the unprotected eye should not exceed 1.0 J/cm^2 (10 kJ/m^2) within an 8-h period, and the total 8-h radiant exposure incident upon the unprotected skin should not exceed the values given in Table 6. Values for the relative spectral effectiveness S_λ are given up to 400 nm to expand the action spectrum into the UVA region for determining the exposure limit for skin exposure. For the actinic UV spectral region (UVC and UVB, 180–315 nm), the radiant exposure incident upon the unprotected skin or eye within an 8-h period should not exceed the values given in Table 6 (International Non-ionizing Radiation Committee of the International Radiation Protection Association, 1989).

The effective irradiance (E_{eff}) in W/m^2 of a broad-band source weighted against the peak of the spectral effectiveness curve (270 nm) is determined according to the formula:

$$E_{\text{eff}} = \Sigma\ E_\lambda \times S_\lambda \times \Delta_\lambda,$$

where E_λ is the spectral irradiance (W/m^2 × nm) from measurements, S_λ is the relative spectral effectiveness (Table 6) and Δ_λ is the band-width (nm) of the calculation or measurement interval (International Non-ionizing Radiation Committee of the International Radiation Protection Association, 1985).

The maximal permissible exposure time in seconds for exposure to UVR incident on the unprotected skin or eye within an 8-h period is computed by dividing 30 J/m^2 by the value of E_{eff} in W/m^2 (American Conference of Governmental Industrial Hygienists, 1991). A worker receiving the maximal permissible exposure of 30 J/m^2 per 8-h day will, in the course of a working year, have a cumulative dose of 60–70 MED (Diffey, 1988), a value comparable with the natural exposure of non-occupationally exposed indoor workers (Diffey, 1990a).

Occupational exposure limits to lasers were also defined by the International Non-Ionizing Radiation Committee of the International Radiation Protection Association in 1989, at 3 mJ/cm^2 and 40 mJ/cm^2 over 8 h for argon–fluoride and xenon–chloride lasers, respectively (Sliney, 1990).

2. Studies of Cancer in Humans

2.1 Solar radiation

2.1.1 *Nonmelanocytic skin cancer*

Nonmelanocytic skin cancer is classified into two major histological types: basal-cell carcinoma and squamous-cell carcinoma. Basal-cell carcinoma is the commoner type in white populations. No information was available to the Working Group on other types of nonmelanocytic skin cancer.

(a) Case reports

In general, case reports were not considered, owing to the availability of more informative data.

(i) *Studies of xeroderma pigmentosum patients*

Xeroderma pigmentosum is a rare autosomal-recessive genetic disease in which there is an excision repair defect, as observed in cultured skin fibroblasts damaged by UVR (Cleaver, 1968). Patients display cellular and clinical hypersensitivity to UVR (Kraemer, 1980). The disease is present in about one in 250 000 people in the USA and Europe (Cleaver & Kraemer, 1989), and as many as 1 in 100 000 (Takebe *et al.*, 1987) or even 1 in 40 000 (Cleaver & Kraemer, 1989) people may be affected in Japan.

In a survey of 830 cases located through published case reports (Kraemer *et al.*, 1987), 45% had malignant skin neoplasms. Most of the patients were young, and the median age of development of the first skin cancer in the 186 patients for whom information was available was eight years; this observation presumably represents a substantial excess over the expected number. Only 259 neoplasms were specifically categorized as basal- or squamous-cell carcinoma in the published reports. Of these, 97% were on constantly exposed sites (face, head and neck) by comparison with 80% of similar tumours in the US general population. [The Working Group recognized that data collected from previously published case reports is not uniform and may not be typical of a true incidence or prevalence series.]

(ii) *Studies of transplant recipients*

Australian renal transplant recipients were reported to have an increased risk for nonmelanocytic skin cancer (Hardie *et al.*, 1980). Among 875 male and 669 female Australasian recipients, aged 35–64, 47 squamous-cell carcinomas and 27 basal-cell carcinomas were observed among males and 27 squamous-cell and 15 basal-cell carcinomas were observed among females (Kinlen *et al.*, 1979). The rates/10^5 person-years for squamous-cell carcinoma were 2680 in males and 1710 in females, or 3.0 and 5.9 times the rates observed among residents of the same age distribution surveyed in Geraldton, Western Australia (Kricker *et al.*, 1990). For basal-cell carcinoma, the rates for 1540 (males) and 940 (females) were 1.154 and 1.150 times the Geraldton rates, respectively.

By February 1980, a registry in Denver, Colorado (USA), had received data on 906 organ transplant recipients who had developed 959 types of cancer: 42% arose in the skin, of which 47% were squamous-cell carcinomas (Penn, 1980). While several studies from areas with lower solar radiation are available (Boyle et al., 1984), neither singly nor collectively do they contain enough observations to permit a comparable calculation.

(b) Descriptive studies

Nonmelanocytic skin cancer is often not recorded in cancer registries (e.g., in the USA and in most parts of Australia), and when it is registered case ascertainment is likely to be incomplete since many patients are treated in consulting rooms, frequently without histological verification (Doll et al., 1970). Thus, descriptive studies of the incidence of nonmelanocytic skin cancer can be difficult to perform because of the absence of routinely collected data or difficult to interpret because of incomplete registration. Studies in Australia and the USA have relied upon special surveys, while in the United Kingdom and the Nordic countries data from cancer registries have been used. Studies of mortality rates are also difficult to interpret because nonmelanocytic skin cancer is rarely fatal, and many deaths are incorrectly attributed to skin cancer (Muir et al., 1987).

A number of features of the occurrence of nonmelanocytic skin cancer as revealed by descriptive studies have been taken as evidence that exposure to the sun is a major cause of the disease. These include features presumed to be related to sun exposure such as sex, anatomical site, latitude of residence (or annual dose of UVB radiation), migration from places of low insolation to places of high insolation, occupation and features related to sensitivity to the sun such as race (i.e., degree of skin pigmentation).

(i) *Host factors*

The occurrence of nonmelanocytic skin cancer according to host factors such as race provides indirect evidence that sunlight is a cause. In most white populations, nonmelanocytic skin cancer occurs more commonly in men than in women (Muir et al., 1987). The highest incidence rates have been recorded among Australians, who are largely of British (Celtic) descent (Giles et al., 1988). Populations with greater skin pigmentation have low rates of nonmelanocytic skin cancer, for instance, in South Africa (Oettlè, 1963) and Singapore (Shanmugaratnam et al., 1983).

Albinism is an inherited disorder of melanin metabolism, with a decrease or complete absence of melanin. Large numbers of skin cancers (mostly squamous-cell carcinomas) have been reported in albinos (Luande et al., 1985; Kromberg et al., 1989).

(ii) *Anatomical distribution*

The majority of cases of skin cancer recorded in cancer registries (Haenszel, 1963 [USA]; Whitaker et al., 1979 [United Kingdom]; Swerdlow, 1985 [United Kingdom]; Levi et al., 1988 [Switzerland]; Østerlind et al., 1988a [Denmark]; Moan et al., 1989 [Norway]) and in special surveys in the USA (Haenszel, 1963; Scotto et al., 1983) occurred on the head and neck. In contrast, in two studies in Australia—one of incidence (Giles et al., 1988) and the other of prevalence (Kricker et al., 1990)—the proportions of cancers on the head and neck were lower. [The Working Group noted that the contrasting results may be due to time differences.] In the incidence survey, 43% of squamous-cell carcinomas and 66% of basal-cell carcinomas were on the head and neck. In the prevalence survey, about one-third of all basal-cell carcinomas were on the head and neck, whereas the trunk accounted for about half of these lesions. The density of tumours was five times greater in men and eight times greater in women on usually exposed sites than on sites which were sometimes exposed. Squamous-cell carcinomas occurred almost exclusively on exposed sites. The site distributions of both types of nonmelanocytic skin type are generally similar in the two sexes (Østerlind *et al.*, 1988a; Moan *et al.*, 1989; Kricker *et al.*, 1990).

A distinctive feature of the site distribution of basal-cell carcinoma is a virtual absence on the dorsa of the hands and infrequent occurrence on the forearms, compared with the distribution of squamous-cell carcinoma (Haenszel, 1963; Silverstone & Gordon, 1966; Levi *et al.*, 1988; Magnus, 1991). Basal-cell carcinoma also occurs frequently on parts of the face that receive comparatively little sun exposure (Urbach *et al.*, 1966).

[The Working Group noted that cancers on the head and neck may be more likely to be diagnosed than cancers at other sites.]

(iii) *Geographical variation*

Nonmelanocytic skin cancer incidence and mortality have long been known to increase with increasing proximity to the equator. Gordon and Silverstone (1976) demonstrated a negative correlation between incidence of nonmelanocytic skin cancer in various countries and latitudes by tabulating the incidence according to latitudinal zones. Much of the early evidence came from surveys conducted in the USA. In the first of these, Dorn (1944a,b,c) reported the results of the US First National Cancer Survey conducted in 10 urban areas in 1937–38. [Nonmelanocytic] skin cancer incidence was greater among whites living in the south than in the north of the country. Blum (1948) subsequently reanalysed these data, substituting latitude for place of residence, and showed a strong inverse relationship between incidence of mostly nonmelanocytic skin cancer and latitude. No other cancer, with the exception of the buccal cavity (including the lip), showed a similar latitude gradient.

Auerbach (1961), using data from the US Second National Cancer Survey conducted in 1947–48 in the same areas as the previous survey, calculated that the age-adjusted rates for skin cancer doubled for each $3°48'$ (approximately 265 miles) of latitude towards the equator; similar gradients were seen for men and women and in all age groups. Haenszel (1963) reanalysed data from this survey for four southern and four northern cities. The inverse gradient with latitude was present for both basal-cell and squamous-cell carcinoma. In addition, there was some evidence that the gradient was strongest for head, neck and upper limbs (sites which are usually exposed).

A similar latitude gradient was seen in the US Third National Cancer Survey (Scotto *et al.*, 1974). Inverse latitude gradients have also been reported in Australia (Silverstone & Gordon, 1966; Giles *et al.*, 1988) and in the Nordic countries (Teppo *et al.*, 1980; Moan *et al.*, 1989; Magnus, 1991).

Several authors have correlated nonmelanocytic skin cancer incidence (or mortality) with estimates of UVR. Green *et al.* (1976) reported a positive correlation between estimates of annual UV dose and of incidence rates in the USA, the United Kingdom, Canada and Australia. Estimates of UV dose were derived from models relating latitudinal and seasonal ozone distributions, adjusted for cloud cover. [The Working Group noted that no allowance was made in the analysis for different methods of case ascertainment. It is not clear how well the predicted values were correlated with actual levels of UVR.]

A positive correlation, stated to be stronger than that for latitude, was seen between UVR, as measured by Robertson-Berger meters, and the incidence of nonmelanocytic skin cancer in four cities in the US Third National Cancer Survey (Scotto et al., 1982). Scotto et al. (1983) examined incidence data collected in eight cities in 1977-78 and again showed an inverse relationship with latitude and a positive correlation with measurements of UVR. The gradient was steeper for squamous-cell than for basal-cell carcinoma.

Moan et al. (1989) examined nonmelanocytic skin cancer incidence in six regions of Norway from 1976 to 1985, excluding the area around Oslo to reduce bias due to possible differences in reporting and diagnosis. Two measures of UVR, one weighted according to the action spectrum for erythema and the other according to the action spectrum for mutagenesis in cells in the basal layer of the skin, were derived from atmospheric models. Similar, positive relationships between UVR and nonmelanocytic skin cancer incidence were obtained with each method.

Elwood et al. (1974) conducted a study of mortality from nonmelanocytic skin cancer in the contiguous states of the USA and in all of the provinces of Canada in 1950-67. The correlation between latitude and mortality was as strong as that between mortality and an index of UVR derived from a model relating erythemal dose according to latitude with adjustments for cloud cover.

(iv) *Migration*

Studies of migrants to Australia (and other countries with high exposure to the sun) offer the opportunity to examine, indirectly, the effect of exposure to the sun. Most migrants to Australia come from higher latitudes which have lower levels of exposure to the sun than Australia. The effect of exposure to the sun is most readily examined in migrants from the British Isles to Australia, from whom most Australians are descended.

Armstrong et al. (1983) found that the age-adjusted mortality rate among men born in England or Wales was 0.55 (95% confidence interval (CI), 0.43-0.71) times that in Australian-born men. There was little evidence that rates in migrants increased with duration of residence in Australia, although the numbers of deaths were small and the rates unstable.

Giles et al. (1988) found age-adjusted incidence rates of 402 per 100 000 person-years among immigrants from the British Isles and 936 in the Australian-born population.

(v) *Occupation*

Death certificates for 1911-44 in England and Wales were used in an analysis of cancer of the skin, excluding melanomas, in male agricultural workers, miners and quarriers and professionals (Atkin et al., 1949). During part of the period (1911-16), cancers of the penis, scrotum and skin were classified together, and the numbers of cancers of the skin alone were estimated from the proportions occurring in the later period. The standardized mortality ratios (SMRs) were greater for those engaged in agriculture (142.4 [137.4-147.6]) than for those in mining (94.4 [88.8-100.3]), and lowest of all for professionals (47.5 [42.6-52.9]).

Whitaker et al. (1979) examined occupations among cases of squamous-cell carcinoma reported to the Manchester Regional Cancer Registry, United Kingdom, in 1967-69. The occupations of 23% of cases were not ascertained. In men, standardized registration ratios

(SRRs) were elevated for textile workers (238; $p < 0.001$) and farmers (243; $p < 0.001$). The SRR was also high for female farmers (690; $p < 0.001$). Male fishermen, chemical workers and paper/printing workers had high SRRs for squamous-cell carcinoma of the arm, and building workers for squamous-cell carcinoma of the ear.

The association between occupation and nonmelanocytic skin cancer was examined in England and Wales in 1970–75 in a 10% sample of all male incident cases for which occupation was recorded (Beral & Robinson, 1981). Individuals were assigned, on the basis of stated occupation, to one of three groups: outdoor workers, indoor office workers and other indoor workers, according to the classification of occupations of the Office of Population Censuses and Surveys. The SRRs for men aged 15–64 were 110 [95% CI, 109–116] for outdoor work, 97 [92–103] for office work and 92 [86–89] for other indoor work. Since place of work may be confounded with social class, the analyses were repeated for men aged 15–64 years in social class III; the SRRs were 112 [102–122] for outdoor work, 111 [100–123] for office work and 85 [78–92] for other indoor work.

Vågerö et al. (1986) linked cancer incidence data in Sweden from 1961 to 1979 with census data from 1960 to determine the occupations of cases of nonmelanocytic skin cancer. Occupations were classified into three main groups: office workers, other indoor workers and outdoor workers. SRRs standardized for age, county of residence and social class, were slightly higher for outdoor workers (106; 95% CI, 101–112) than for office workers (103; 96–110) and other indoor workers (95; 91–100). The authors noted that registration may have been more complete among high socioeconomic groups.

(c) Cross-sectional studies

Design features of cross-sectional studies of exposure to the sun are summarized in Table 9, and the results are shown in Table 10.

A population-based survey of the prevalence of nonmelanocytic skin cancer [types not separated] was conducted in County Galway, Ireland (O'Beirn et al., 1970). Exposed areas of skin were examined for the presence of cancers. In the 26 cases found, there was no significant association with frequent severe sunburn for basal-cell or squamous-cell skin cancer; among males, there was a positive relationship between cumulative hours of exposure to sunlight and the prevalence of nonmelanocytic skin cancer.

Silverstone and Gordon (1966) and Silverstone and Searle (1970) reported the results of three surveys in Queensland, Australia. Exposed areas of the skin were examined, and subjects were asked to report previously treated nonmelanocytic skin cancer [types not separated]. Women performing home duties were classified as indoor workers. Outdoor occupation showed a weakly positive association with past and present incidence in men and a negative association in women.

Holman et al. (1984a) conducted a population-based survey of 1216 subjects in western Australia. After controlling for age, cutaneous sun damage (as assessed by microtopography) was strongly related to a past history of nonmelanocytic skin cancer.

Engel et al. (1988) analysed data on basal-cell epithelioma (carcinoma) from the First National Health and Nutrition Examination Survey in the USA (1971–74). Dermatologists diagnosed skin cancers and assessed actinic skin (solar) damage, but histological confirmation of the diagnosis was not obtained routinely. Strong associations between the prevalence of basal-cell epithelioma and solar skin damage were seen in both men and women.

Green *et al.* (1988a) conducted a survey of the prevalence of nonmelanocytic skin cancer [types not separated for calculation of RR] in Queensland, Australia. Information about exposure to the sun was obtained from questionnaires; dermatologists diagnosed skin cancers and assessed signs of actinic damage (solar lentigines, telangiectasia of the face, solar elastosis of the neck and solar keratoses). After adjustment for age, sex, skin colour and ability to tan, outdoor occupation and number of sunburns were both weakly associated with increased prevalence. Stronger associations were seen for cutaneous indicators of sun exposure, particularly for solar lentigines on the hands and telangiectasia on the face. Recreational exposure was not associated independently with nonmelanocytic skin cancer.

In a later report (Green, 1991), the occurrence of nonmelanocytic skin cancer was positively correlated with grade of cutaneous microtopography.

In a subsequent study (Green & Battistutta, 1990), subjects were asked to report nonmelanocytic skin cancer treated between 1 December 1985 and 30 November 1987, around the survey in 1986. Medical records were searched to confirm the diagnoses. Subjects who had had a skin cancer diagnosed at the prevalence survey were excluded. Outdoor occupation, outdoor leisure activities and number of sunburns showed little association with basal-cell carcinoma in an analysis including past history of skin cancer. All three variables were related to incidence of squamous-cell carcinoma. [The Working Group noted that the exclusion of subjects found to have skin cancer during the prevalence survey makes interpretation of these results difficult. The inclusion of past history of skin cancer in the analysis would have weakened any association with exposure to the sun.]

Vitasa *et al.* (1990) conducted a survey of the occurrence of nonmelanocytic skin cancer among men engaged in traditional fishing practices ('watermen') in Maryland, USA. Subjects were examined by dermatologists and interviewed about their history of exposure to the sun. Estimates of individual annual and lifetime doses of UVB radiation were made by weighting the ambient UVR by a history of occupation and outdoor activities and by taking into account relative doses recorded by film dosimeters on the face. Patients with squamous-cell carcinoma aged 15–60 had had an 11% higher annual dose of UVB radiation and those with basal-cell carcinoma had had an 8% lower annual dose than that of age-matched watermen without cancers. The effect of cumulative UVB radiation was examined after adjustment for age, eye colour, childhood freckling and skin reaction to sunlight, all of which were positively associated with occurrence of both types of nonmelanocytic skin cancer. Cumulative UVB radiation dose was not associated with basal-cell carcinoma but was positively associated with squamous-cell carcinoma. The latter association was significant in a comparison of the top quarter of cumulative UVB *versus* the bottom three-quarters but not in a comparison of exposures above and below the median. [The Working Group noted that the results for the two types of cancer are not necessarily incompatible, both because of the small number of cases and the fact that the diagnosis was confirmed histopathologically in only 62%.]

Table 9. Design features of cross-sectional studies of sun exposure and nonmelanocytic skin cancer

Reference	Place	Period of diagnosis	Population	Sample size	Response rate	Cases	Histological confirmation
O'Beirn et al. (1970)	County Galway, Ireland	1960s	Population-based	1338	Approx. 81%	13 BCC; 13 SCC on exposed sites only	Incomplete; 57% had biopsies
Silverstone & Gordon (1966); Silverstone & Searle (1970)	Queensland, Australia	1961–63	Population-based	About 2200	87%	221 BCC or SCC on exposed surfaces	Incomplete
Holman et al. (1984a)	Busselton, Western Australia	1981	Population-based	1216		102, type not stated	No
Engel et al. (1988)	USA	1971–74	Population-based	20 637	74%	BCC, number not stated	Incomplete [small proportion]
Green et al. (1988a)	Nambour, Australia	1986	Population-based	2095	70–78%	42 BCC or SCC [90% of subjects examined on head/neck/hands/forearms only]	Yes
Green & Battistutta (1990)	Nambour, Australia	1985–87	Population-based	1770	84%	66 BCC; 21 SCC self-reported (confirmed from medical records)	Incomplete
Vitasa et al. (1990)	Maryland, USA	1985–86	Male fishermen > 30 years old	838	70%	33 BCC; 35 SCC	Incomplete

BCC, basal-cell carcinoma; SCC, squamous-cell carcinoma

Table 10. Summary of results of cross-sectional studies of nonmelanocytic skin cancer

Reference	Index of exposure	Categories	Odds ratio (95% CI)	Comments
O'Beirn et al. (1970)	Sunlight hours (lifetime)	< 30 000 h > 50 000 h	1.00 [8.10 (1.2–348.2)]	Mean aged > 60 years; calculated from raw data [p = 0.02]
Silverstone & Searle (1970)	Occupation	Indoors Outdoors	1.0 [1.29]	Men, chi-square = 1.4 [p > 0.1]; calculated from raw data, no adjustment
	Occupation	Indoors Outdoors	1.0 [0.6]	Women, chi-square = 0.3 [p > 0.1]; calculated from raw data, no adjustment
Holman et al. (1984a)	Cutaneous microtopography	Grades 1–3 Grade 4 Grade 5 Grade 6	1.0 3.9 3.6 9.2	p = 0.004, trend adjusted for age
Engel et al. (1988)	Solar skin damage	None Any None Any	1.0 [8.0] 1.0 [6.0]	BCC, men, age-adjusted prevalence ratio, p < 0.01 BCC, women, age-adjusted prevalence ratio, p < 0.01
Green et al. (1988a)	Occupational exposure	Indoors Indoors and outdoors Outdoors	1.00 1.01 (0.44–2.31) 1.76 (0.77–4.05)	Adjusted for age, sex, skin colour and propensity to sunburn
	Painful sunburns	None 1 2–5 ≥ 6	1.00 0.77 (0.22–2.61) 1.09 (0.41–2.95) 1.66 (0.59–4.64)	Adjusted for age, sex, skin colour and propensity to sunburn
	Solar lentigines on hands	None 1–10 11–20 ≥ 21	1.00 1.61 (0.78–3.35) 1.43 (0.43–4.77) 3.78 (1.06–13.41)	Adjusted for age, sex and other signs of actinic damage
	Telangiectasia on face	None Mild Moderate Severe	1.00 1.63 (0.58–4.57) 2.74 (0.89–8.40) 3.67 (0.79–17.11)	Adjusted for age, sex and other signs of actinic damage

Table 10 (contd)

Reference	Index of exposure	Categories	Odds ratio (95% CI)	Comments
Green et al. (1988a) (contd)	Actinic elastosis on neck	None Mild to moderate Severe	1.00 1.42 (0.53–3.80) 1.75 (0.56–5.45)	Adjusted for age, sex and other signs of actinic damage
	Solar keratoses on face	None 1–5 6–20 21–50 ≥ 51	1.00 1.55 (0.67–3.59) 1.86 (0.69–5.04) 3.00 (0.54–16.69) 2.72 (0.73–10.15)	Adjusted for age, sex and other signs of actinic damage
Green & Battistutta (1990)	BCC Occupational exposure	Mainly indoors Indoors and outdoors Mainly outdoors	1.0 1.5 (0.8–2.9) 1.3 (0.6–2.8)	Adjusted for age, sex, skin colour and past history of skin cancer
	Leisure exposure	Mainly indoors Indoors and outdoors Mainly outdoors	1.0 1.0 (0.4–2.2) 0.6 (0.3–1.3)	Adjusted for age, sex, skin colour and past history of skin cancer
	No. of painful sunburns	None 1 2–5 ≥ 6	1.0 0.5 (0.2–1.4) 0.6 (0.3–1.5) 1.0 (0.4–2.5)	Adjusted for age, sex, skin colour and past history of skin cancer
	SCC Occupational exposure	Mainly indoors Indoors and outdoors Mainly outdoors	1.0 4.4 (0.9–20.9) 5.5 (1.1–28.2)	Adjusted for age, sex, skin colour and past history of skin cancer
	Leisure exposure	Mainly indoors Indoors and outdoors Mainly outdoors	1.0 2.0 (0.2–19.9) 3.9 (0.5–30.9)	Adjusted for age, sex, skin colour and past history of skin cancer
	No. of painful sunburns	0–1 2–5 ≥ 6	1.0 3.3 (0.9–12.3) 3.0 (0.7–12.2)	Adjusted for age, sex, skin colour and past history of skin cancer

Table 10 (contd)

Reference	Index of exposure	Categories	Odds ratio (95% CI)	Comments
Vitasa et al. (1990)	SCC			
	Cumulative UVB dose to face	Below median	1.0	Proportionate odds ratios; adjusted for age, eye colour, freckling and sunburn reaction
		Above median	2.05 (0.84–5.01)	
		Below 75 percentile	1.0	
		Above 75 percentile	2.53 (1.18–5.40)	
	BCC			
	Cumulative UVB dose to face	Below median	1.0	Proportionate odds ratios; adjusted for age, eye colour, freckling and sunburn reaction
		Above median	0.69 (0.31–1.53)	
		Below 75 percentile	1.0	
		Above 75 percentile	1.11 (0.50–2.44)	

BCC, basal-cell carcinoma; SCC, squamous-cell carcinoma; unless otherwise specified, all analyses are for the two types together

(d) Case–control studies

Design features of the case–control studies of exposure to the sun and the occurrence of nonmelanocytic skin cancer are summarized in Table 11. Most of the studies employed hospital- or clinic-based controls, which introduces potential for selection bias. The results are summarized in Table 12. The methods of analysis and of measurements of exposure to the sun, particularly in the earlier studies, were crude. Neither sensitivity to the sun, usually measured as the ability to tan or propensity to burn, nor pigmentary characteristics (such as skin colour and hair colour), which are likely to be confounding variables, were taken into account in most of the analyses.

The hospital-based study of Lancaster and Nelson (1957) in Sydney, Australia, was primarily a case–control study of melanoma (described in detail on p. 100). It can also be considered to be a case–control study of nonmelanocytic skin cancer, however, because it included two control groups—one of patients with basal-cell carcinoma, squamous-cell carcinoma or solar keratosis and the second of patients with leukaemia or cancer at a site other than the skin. All groups were matched by age and sex. Among males, long duration of occupational exposure to the sun was associated with an increased risk for nonmelanocytic skin cancer or solar keratosis. A summary of total exposure to the sun was devised by assigning scores to a number of factors considered to be related to exposure to the sun. Risk was highest among subjects judged to have excessive exposure to the sun. [The Working Group noted that the proportion of cases who had a solar keratosis is not stated, that no account was taken of matching in the analyses, and that the effect of exposure to the sun was not adjusted for sensitivity to the sun.]

Gellin *et al.* (1965) conducted a study in a single hospital in New York, USA, on 861 patients with basal-cell carcinoma and 1938 non-cancer dermatological patients attending the same clinic. Since 95% of cases and 43% of controls were 40 years old and over, the study was limited to these patients, resulting in 771 cases and 783 controls. The skin cancer patients spent more time outdoors per day than did control patients and were significantly more likely than controls to have light hair, fair complexion, blue eyes and an inability to tan. [The Working Group noted that the analyses were not adjusted for age, sex or sensitivity to the sun, and that confounding by age is likely because controls were younger than cases.]

Urbach *et al.* (1974) conducted a hospital-based study in Philadelphia, USA, and compared exposure to the sun of 392 patients with histologically confirmed basal-cell carcinoma, 59 patients with histologically confirmed squamous-cell carcinoma and 281 outpatients receiving treatment for a skin disease other than cancer. Controls were matched to cases by age and sex. Among male patients, those with basal-cell or squamous-cell carcinoma had more cumulative hours of exposure than did controls. Skin cancer patients also reported more sunburns. [The Working Group noted that the analyses were not adjusted for ability to tan, age or sex (apart from the sex-specific analysis).]

Vitaliano (1978) subsequently reanalysed the data of Urbach *et al.* (1974) and showed that, after adjustment for complexion (dark *versus* pale), ability to tan and age ($< 60, \geq 60$), the cumulative time spent outdoors was related to both types of nonmelanocytic skin cancer. For basal-cell carcinoma, the odds ratio for $\geq 30\,000$ h of exposure relative to $< 10\,000$ h was 3.19; for squamous-cell carcinoma it was 22.8. [The Working Group noted that confidence intervals were not given. Part of the apparently stronger effect for squamous-cell carcinoma could be due to confounding by age: the controls were matched by age to the basal-cell carcinoma cases, who were younger than the squamous-cell carcinoma cases.]

A hospital-based case–control study was conducted in Montréal, Canada (Aubry & MacGibbon, 1985), in which patients with histologically confirmed squamous-cell carcinoma were identified in hospitals in 1977–78. Two patients with other conditions were matched as controls to each case by age, sex and hospital. Information on exposure to the sun was obtained from a postal questionnaire. Among 306 eligible cases, 94 (31%) replied, as did 186 (30%) of the eligible controls; 92 cases and 174 controls completed the questionnaire. Most of the controls who replied had been seen for seborrheic keratoses (61%) or intradermal naevi (16%). Scores for nonoccupational and occupational exposures were estimated, and the two scores were divided into thirds for analysis, which was based on logistic regression. The odds ratios, adjusted for each other and for host factors, were 1.08 and 1.64 for the middle and upper thirds of occupational exposure and 1.23 and 1.58 for the same levels of nonoccupational exposure, respectively. [The Working Group noted the low response rate and that the complexity of the recreational exposure to sun indices and the nature of the control group make the results difficult to interpret.]

O'Loughlin et al. (1985) conducted a case–control study in a hospital in Dublin, Ireland. Patients with histologically confirmed nonmelanocytic skin cancer [types not separated] were compared with age- and sex-matched patients who had cancers of other organs. There was no statistically significant difference between cases and controls in eight measures of exposure to the sun summarized in a single index of exposure and either type of nonmelanocytic skin cancer. [The Working Group noted that the measures of exposure to the sun were crude and likely to be subject to considerable misclassification. No adjustment was made for sensitivity to the sun.]

Herity et al. (1989) conducted a case–control study in the same hospital in Dublin of 396 histologically confirmed nonmelanocytic skin cancers in 1984–85. An equal number of age- and sex-matched patients with other cancers, attending the same hospital, were used as controls. More cases than controls lived in rural areas ($p = 0.007$), and cases reported more frequently spending more than 30 h outdoors per week, but the difference was not significant. For other indices of exposure to the sun, there was little difference between cases and controls. [The Working Group noted that results were not adjusted for reaction to sunlight.]

In a case–control study (reported as an abstract) conducted in 1983–84 in Alberta, Canada (Fincham & Hill, 1989), 225 men with basal-cell carcinoma and 181 men with squamous-cell carcinoma were compared with 406 age-matched male controls. Sunburn in adult life gave an odds ratio of 2.33 ($p < 0.05$) for all nonmelanocytic skin cancer; for basal-cell carcinoma, childhood sunburn gave an odds ratio of 2.48 ($p < 0.05$) and peeling an odds ratio of 1.85 ($p < 0.05$).

A population-based case–control study was conducted in Saskatchewan, Canada (Hogan et al., 1989), which included all patients diagnosed with basal-cell carcinoma in the Province in 1983. Two controls, matched by year of birth, sex and municipality of residence, were selected for each case from a universal Provincial health insurance plan. Replies to mailed questionnaires were received from 55.5% of the cases and 43.7% of the controls. A number of measures of exposure to the sun were associated with incidence of basal-cell carcinoma. In a stepwise logistic regression analysis, occupation as a farmer, history of severe sunburn and working outdoors for more than 3 h per day in winter were independently associated with basal-cell carcinoma, after adjustment for freckles in childhood, family history of skin cancer, 'Celtic' mother, skin colour and hair colour. [The Working Group noted that the measures of exposure were crude and that the estimates do not appear to have been adjusted for the matching variables. The low response rate makes interpretation of the results difficult.]

On the basis of a population-based survey in Western Australia in 1987 of skin cancer among residents aged 40–64 years of age (Kricker et al., 1990), Kricker et al. (1991a) conducted a case–control study of 226 confirmed cases of basal-cell carcinoma and 45 of squamous-cell carcinoma; two sets of 1015 controls with no lesions, who had completed an interview, were available for each type of cancer. The response rate among those eligible to participate was identical for cases and controls: 89%. Separate analyses were undertaken for basal-cell carcinoma and squamous-cell carcinoma using unconditional logistic regression analysis. Risks for both cancers were higher in native-born Australians than in migrants, and the risk for basal-cell carcinoma decreased with increasing age at arrival in Australia. Only four of the subjects with squamous-cell carcinoma had been born outside Australia—an insufficient number to examine the effects of age at arrival. Indicators of sun damage to the skin (facial telangiectasia, solar elastosis of the neck, facial solar lentigines and number of solar keratoses), assessed by dermatologists during the prevalence survey, were examined in models adjusted for age, sex, ethnicity and migrant status and including all other sun damage indicators except solar keratoses, which were considered to be preneoplastic lesions and thus inappropriate for inclusion in models concerned with etiology. Cutaneous microtopography, an objective measure of actinic skin damage, graded without knowledge of the person's skin cancer status, and solar elastosis of the neck had significant residual effects for basal-cell carcinoma, while solar elastosis and facial telangiectasia had significant residual effects for squamous-cell carcinoma. The independently significant indicators of sun damage were analysed in models which included adjustment for age, sex, ethnicity and migrant status as well as measures of sun sensitivity. Solar elastosis of the neck remained an independent predictor of risk of basal-cell carcinoma (odds ratios, $> 1.50; p = 0.003$) and squamous-cell carcinoma (odds ratios, $> 2.00; p = 0.04$).

A subsequent analysis of individual sun exposure was published as an abstract (Kricker et al., 1991b). A positive association was found between nonmelanocytic skin cancer and life-time potential for exposure to the sun, but no evidence of increasing risk for either basal-cell carcinoma or squamous-cell carcinoma with increasing total hours of actual exposure to the sun as recalled by subjects. Risk for basal-cell carcinoma on the trunk was increased substantially in association with maximal exposure of the trunk to the sun, but there was no consistent pattern of association of site-specific basal-cell or squamous-cell carcinoma with exposure of the head and neck or limbs. Neither basal-cell nor squamous-cell carcinoma showed evidence of an association with sun exposure on working days; however, there was persuasive evidence of increased risk for both types of skin cancer with intermediate and high levels of accumulated exposure to the sun on non-working days. Moreover, there was evidence of an association, stronger for basal-cell carcinoma than for squamous-cell carcinoma, with a measure of intermittent exposure to the sun.

Gafà *et al.* (1991) conducted a case–control study of nonmelanocytic skin cancer in Sicily, Italy, in which 133 cases identified from a population-based registry (response rate, 94%) were compared with 266 sex- and age-matched controls. For each case, one control was selected randomly from among patients with non-neoplastic diseases at the same hospital as the case, and a second control was selected randomly from among friends or relatives of the case. After adjustment for family history of skin cancer, 'cancer-related cutaneous disease', skin colour and skin reaction to sunlight, sun exposure for at least 6 h per day and residence for at least 10 years at more than 400 m above sea level were significantly related to risk for nonmelanocytic skin cancer. In crude analyses in which the two types of cancer were separated, sun exposure for at least 6 h per day without a hat was strongly associated with risk for squamous-cell carcinoma [site unspecified] (odds ratio, 6.4; 95% CI, 1.9–21.1) but not for basal-cell carcinoma (1.4, 0.7–2.6). [The Working Group noted that the nature of the control group, the assessment of exposure and the failure to account for age in the analysis make the results difficult to interpret. The crude analysis of the type-specific results, the lack of data on the site of the tumours and the small numbers may explain the different results for the two types.]

(e) Cohort studies (Tables 13 and 14)

In a study in Chicago, IL (USA), Robinson (1987) investigated the incidence of second nonmelanocytic skin cancer among a group of 1000 patients who had had basal-cell carcinoma. Among 978 who were followed for five years after the initial diagnosis, 22% developed a second basal-cell carcinoma at the end of the first year and 36% within five years. There was no significant correlation between developing a second cancer and frequent exposure through sunbathing or outdoor leisure activities, work or currently living in an area with heavy exposure to the sun, or according to estimated number of hours of daily exposure to the sun. Among those with skin types I and II (always burn easily and never or minimally tan) who reported frequent sun exposure, there was an increased risk of second cancer ($p < 0.03$). [The Working Group noted that the methods of assessing exposure and the methods of analysis were not described, and that no numbers were reported. Risk factors for second cancers might not be the same as for the first.]

Marks *et al.* (1989) conducted a longitudinal series of examinations of the head, neck, forearms and hands of a population in Maryborough, north-central Victoria, Australia, for one week annually between 1982 and 1986. The incidence rates of squamous-cell and basal-cell carcinoma were higher in outdoor workers than in indoor workers. In an analysis of the two types combined, occupation was not significantly associated after adjustment for age, sex and reaction to sunlight ($p = 0.09$). [The Working Group noted that no account was taken of lesions that might have been removed between surveys.]

Hunter *et al.* (1990) conducted a study of basal-cell carcinoma in a cohort of female nurses in the USA. A total of 771 cases were identified from responses to follow-up questionnaires sent to the women two and four years after the initial exposure questionnaire was given. In a sample of 29 women, the diagnosis was confirmed for 28; confirmation of the diagnosis was not obtained routinely. Residents of California and Florida had the highest incidence rates. There was a trend of increasing incidence with increasing number of sunburns. With respect to time spent outdoors during the summer, nurses who spent more than

Table 11. Design features of case–control studies of sun exposure and nonmelanocytic skin cancer

Reference	Place	Period of diagnosis	Cases		Controls	
			No.	Source	No.	Source
Lancaster & Nelson (1957)	Sydney, Australia	Unknown	173 BCC, SCC or solar keratosis	Major hospitals	173	Other cancers, same hospitals
Gellin et al. (1965)	New York, USA	1955–59	771 BCC ≥ 40 years old	One skin hospital	783 ≥ 40	Other diagnoses, same skin clinic
Urbach et al. (1974)	Philadelphia, USA	1967–69	392 BCC 59 SCC	One skin and cancer clinic	281	Other diagnoses, same clinic
Aubry & MacGibbon (1985)	Montréal, Canada	1977–78	92 SCC	12 hospitals	174	Skin conditions, same hospitals
O'Loughlin et al. (1985)	Dublin, Ireland	Unknown	63 SCC 58 BCC	One hospital	121	Other cancers, same hospital
Herity et al. (1989)	Dublin, Ireland	1984–85	396 BCC and SCC	One hospital	396	Other cancers, same hospital
Hogan et al. (1989)	Saskatchewan, Canada	1983	538 BCC	Population	738	Population
Kricker et al. (1991a)	Geraldton, Australia	1987	226 BCC 45 SCC	Population	1015 1015	Population
Gafà et al. (1991)	Ragusa, Sicily, Italy	1987–88	133 BCC and SCC	Cancer registry	133 133	Non-neoplastic diseases, same hospital; friends or relatives

BCC, basal-cell carcinoma; SCC, squamous-cell carcinoma

Table 12. Summary of results of case–control studies of nonmelanocytic skin cancer

Reference	Exposure	Categories	Odds ratio (95% CI)	Comments
Lancaster & Nelson (1957)	Years of occupational exposure	< 5 5–10 > 10	1.0 [1.9] [4.2]	[$p < 0.001$, trend; p and odds ratio calculated from raw data]
	Total sun exposure	Minimal Moderate Excessive	1.0 [1.8] [2.4]	[$p = 0.13$; p and odds ratio calculated from raw data]
Gellin et al. (1965)	Hours per day outdoors	0–2 3–5 ≥ 6	1.0 [4.9 (3.8–6.3)] [7.7 (5.6–10.6)]	BCC [$p < 0.001$]
Urbach et al. (1974)	Cumulative hours (× 1000)	< 30 30–50 > 50 < 30 30–50 > 50	1.0 [3.5 (2.0–6.6)] [9.3 (3.2–37.4)] 1.0 [4.0 (1.7–9.6)] [11.1 (2.8–53.6)]	BCC SCC
Aubry & MacGibbon (1985)	Non-occupational exposure score	Low Medium High	1.0 1.23 1.58	SCC [$p = 0.07$] for continuous variable, adjusted for occupation and host factors
	Occupational score	Low Medium High	1.0 1.08 1.64	SCC [$p = 0.02$] for continuous variable, adjusted for non-occupational score and host factors
	Use of sunlamps	Never Ever	1.0 13.4 (1.38–130.48)	SCC [$p = 0.008$], adjusted for sun exposure and host factors
O'Loughlin et al. (1985)	Outdoor occupation	No Yes	1.0 [1.5]	Not significant (McNemar's test) [odds ratio calculated from raw data ignoring matching]
	Hours per week outdoors	< 10 ≥ 10	1.0 [1.4]	Not significant
	Sunbathing > 4 h per day on vacations	No Yes	1.0 [1.0]	Not significant
Herity et al. (1989)	Living in rural area > 30 h outdoors/week		[1.4] [1.1]	$p = 0.007$ $p = 0.7$

Table 12 (contd)

Reference	Exposure	Categories	Odds ratio (95% CI)	Comments
Hogan et al. (1989)	Farmer	No Yes	1.0 1.29 [1.12–1.46]	BCC, adjusted for each other, plus freckles, family history of skin cancer, Celtic mother, skin colour, hair colour
	Severe sunburn	No Yes	1.0 1.19 [1.04–1.35]	BCC
	Working outdoors > 3 h per day in winter	No Yes	1.0 1.13 [1.01–1.27]	BCC
Kricker et al. (1991a)	*BCC* Age at migration (years)	Australian born < 10 > 10	1.0 1.37 (0.55–3.42) 0.32 (0.18–0.59)	$p < 0.001$, adjusted for other variables below and for ethnicity, ability to tan, freckling as a child and number of moles on back
	Solar elastosis of the neck	None Mild Moderate Severe	1.00 1.85 (0.80–4.26) 2.75 (1.16–6.50) 3.96 (1.58–9.93)	$p = 0.03$, comments as above
	Cutaneous microtopography	Grades 1–3 Grade 4 Grade 5 Grade 6	1.0 2.01 (1.00–4.07) 2.42 (1.17–5.01) 2.15 (0.99–4.70)	$p = 0.10$, comments as above
	SCC Migrant to Australia	No Yes	1.0 0.46 (0.15–1.38)	$p = 0.13$, adjusted for variables below plus ability to tan, skin colour, freckling as a child
	Permanent colour difference between neck and adjacent skin	No Yes	1.0 2.58 (1.03–6.47)	$p = 0.03$, comments as above
	Telangiectasia of face	None/mild Moderate Severe	1.0 2.22 (1.06–4.67) 1.88 (0.72–4.90)	$p = 0.10$, comments as above
	Solar elastosis of the neck	None/mild Moderate Severe	1.00 2.31 (1.00–5.34) 3.33 (1.23–9.04)	$p = 0.04$, comments as above

Table 12 (contd)

Reference	Exposure	Categories	Odds ratio (95% CI)	Comments[a]
Gafà et al. (1981)	Residence > 400 m above sea level	No Yes	1.0 2.0 (1.2–3.2)	Adjusted for family history of skin cancer, cutaneous-related conditions, skin colour, skin reaction to sunlight and sun exposure
	Sun exposure ≥ 6 h/day	No Yes	1.0 1.9 (1.2–3.1)	Adjusted for family history of skin cancer, cutaneous-related conditions, skin colour, skin reaction to sunlight and residence > 400 m above sea level

BCC, basal-cell carcinoma; SCC, squamous-cell carcinoma; unless otherwise specified, analyses are for the two types together

8 h per week outside and who used sunscreens had the highest incidence rates. The rates in women who spent the least time outdoors were similar to those who spent more time outdoors and did not use sunscreens. [The Working Group noted that the high incidence rate in nurses using sunscreens, despite control for reaction to sunlight, might be due partly to confounding.]

Table 13. Design features of cohort studies of sun exposure and nonmelanocytic skin cancer

Reference	Place	Period of diagnosis	Population	Sample size	Response rate	Cases	Histological confirmation
Robinson (1987)	Chicago, IL, USA	Not stated	Patients with previous BCC	1 000	98%	BCC, approx. 350	Not stated
Mark et al. (1989)	Maryborough, Australia	1982–86	Population-based	1 981	74%	35 SCC; 113 BCC on light-exposed surfaces only	Yes
Hunter et al. (1990)	USA	1980–84	Female nurses	73 366	74%	771 BCC (self-reported)	Not routinely [records of 28 out of sample of 29 confirmed]

BCC, no. of people with basal-cell carcinoma; SCC, no. of people with squamous-cell carcinoma

(f) Collation of results

The results discussed in this section come from cross-sectional studies by Holman *et al.* (1984a), Engel *et al.* (1988), Green *et al.* (1988a) and Vitasa *et al.* (1990), a case–control study by Kricker *et al.* (1991a) and cohort studies by Marks *et al.* (1989) and Hunter *et al.* (1990), all of which included information pertinent to the association between nonmelanocytic skin cancer and different aspects of sun exposure. Other studies described individually were not considered to provide useful information because of various methodological deficiencies. No data were available on short periods of residence and intermittent exposure, issues which are addressed for melanoma of the skin.

(i) *Total sun exposure: potential exposure by place of residence*

Consistent with descriptive data in a case–control study, migrants to Australia had a lower risk for squamous-cell carcinoma than did native-born Australians, after adjustment for host factors related to risk for nonmelanocytic skin tumours. Late age at arrival in Australia was associated with a lower risk for basal-cell carcinoma (Kricker *et al.*, 1991a).

(ii) *Biological responses to total sun exposure*

Cross-sectional studies and a case–control study are consistent in showing a strong relationship between cutaneous indicators of sun damage and both types of nonmelanocytic skin cancer. In most studies, the indicators of damage and diagnoses of skin cancer were made by the same examiner, but cutaneous microtopography, graded without knowledge of outcome, also showed strong associations.

Table 14. Summary of results of cohort studies of nonmelanocytic skin cancer

Reference	Exposure	Categories	RR (95% CI)	Comments
Marks et al. (1989)	Occupation	*BCC*		
		Indoors	1.0	Adjusted for age, $p = 0.03$
		Outdoors	1.6	
		SCC		
		Indoors	1.0	Adjusted for age, $p = 0.109$
		Outdoors	1.7	
Hunter et al. (1990)	Severe sunburns on face or arms	None	1.0	BCC
		1-2	1.40 (1.13-1.75)	Adjusted for age; p (trend) = 0.001
		3-5	1.78 (1.42-2.25)	
		≥ 6	2.91 (2.37-3.58)	
	Severe sunburns on face or arms	None	1.0	Adjusted for age, time period, region, time spent outdoors, sunscreen habit, hair colour, childhood tendency to sunburn; p (trend) < 0.001
		1-2	1.18 (0.94-1.48)	
		3-5	1.34 (1.05-1.71)	
		≥ 6	1.90 (1.50-2.40)	
	Time spent outdoors during summer (h/week)	≥ 8 (sunscreen)	1.0	Adjusted for age
		≥ 8 (no sunscreen)	0.59 (0.50-0.69)	
		< 8	0.71 (0.58-0.88)	
	Time spent outdoors during summer (h/week)	≥ 8 (sunscreen)	1.0	Adjusted for age, time period, region, number of sunburns, hair colour, childhood tendency to sunburn
		≥ 8 (no sunscreen)	0.70 (0.60-0.82)	
		< 8	0.73 (0.59-0.90)	

[a]BCC, basal-cell carcinoma; SCC, squamous-cell carcinoma

(iii) *Total sun exposure assessed by questionnaire*

No effect of time spent outdoors during summer was seen in a cohort study of basal-cell carcinoma (Hunter *et al.*, 1990). In a cross-sectional study of fishermen, cumulative exposure to UVB radiation was positively associated with the occurrence of squamous-cell carcinoma but not of basal-cell carcinoma (Vitasa *et al.*, 1990). The different results may be attributable in part to small numbers and incomplete histopathological confirmation of diagnoses.

(iv) *Occupational exposure*

In two studies from Australia, outdoor occupation was not significantly associated with the prevalence of the two types of carcinoma combined (Green *et al.*, 1988a) or with the incidence of squamous-cell carcinomas (Marks *et al.*, 1989).

(v) *Sunburn*

A cohort study of basal-cell carcinoma in the USA showed a trend of increasing risk with increasing number of sunburns after adjustment for various factors, including tendency to sunburn (Hunter *et al.*, 1990). Number of sunburns showed a nonsignificant positive association with risks for basal-cell and squamous-cell carcinoma of the skin after adjustment for various constitutional variables, including propensity to burn (Green *et al.*, 1988a).

2.1.2 *Cancer of the lip*

Assessment of the carcinogenicity of solar radiation for the lip is complicated by the fact that carcinoma at this site is actually diagnosed as a mixture of cancers of the external lip and cancers of the buccal membranes (oral cavity). Use of alcohol and tobacco are known causes of the latter tumours (IARC, 1985, 1986b, 1988).

While there are wide variations in the apparent incidence of cancer of the lip with latitude, evaluation of the association is difficult because of inconsistency in the definitions of the boundaries of the lip. 'Cancer of the lip' is defined as cancer of the vermilion border and adjacent mucous membranes and thus excludes cancers of the skin of the lip (WHO, 1977). Most are squamous-cell carcinomas and are located on the lower lip (Keller, 1970; Lindqvist, 1979), which is more heavily exposed to sunlight than is the upper lip (Urbach *et al.*, 1966).

In general, case reports were not considered, because of the availability of more informative data. One case report from Nigeria described the occurrence of two lip tumours in albinos (Onuigbo, 1978).

(*a*) *Descriptive studies*

The incidence of lip cancer is 4–10 times higher in men than in women in most white populations, and higher in whites than in populations of darker skin complexions living in the same geographical areas (Muir *et al.*, 1987).

(i) *Geographical variation*

The incidence of lip cancer is higher in rural than in urban areas, in particular among men (Doll, 1991).

Mortality from and incidence of lip cancer are substantially lower in migrants to Australia than in native-born Australians (Armstrong *et al.*, 1983; McCredie & Coates,

1989). Groups of migrants to Israel all show lower risks for lip cancer than the locally born population (Steinitz *et al.*, 1989).

(ii) *Occupation*

As reviewed by Clemmesen (1965), several observations during the nineteenth century pointed to an increased risk of lip cancer among people in outdoor occupations, in particular farmers and farm labourers. In England and Wales, increased risks for lip cancer were reported among agricultural labourers, fishermen, other dock workers and railwaymen employed outdoors (Young & Russell, 1926). Atkin *et al.* (1949) studied the occupations of 1537 men in England and Wales who died from lip cancer between 1911 and 1944. They reported that mortality from cancer of the lip was 13 times higher among men employed in agriculture than in men with professional jobs. Excess risks for lip cancer have also been observed in farmers in western Canada (Gallagher *et al.*, 1984) and in Denmark (Olsen & Jensen, 1987; Lynge & Thygesen, 1990).

(b) *Case–control studies*

Keller (1970) compared 301 men with lip cancer admitted to veterans' hospitals in the USA between 1958 and 1962 with two groups of white age-matched controls admitted to the same hospitals, comprising 301 oral cancer controls and 265 general controls. Altogether, 59.9% of the lip cancer cases, 37.1% of the cancer controls and 40.6% of the general controls had been born in the south of the USA. Farming was recorded as the occupation of 27% of the lip cancer cases but of only 8% of cancer controls and 4% of the general controls [crude odds ratios, 4.0 and 8.4, respectively]. Any type of outdoor work was recorded for 39% of cases of lip cancer, for 20% of cancer controls and for 12% of the general controls [crude odds ratios, 2.6 and 4.8, respectively]. Risk estimates were not adjusted for smoking, another risk factor identified in the study.

Spitzer *et al.* (1975) obtained information by personal interview on 339 men with squamous-cell carcinoma of the lip registered with the Newfoundland (Canada) Cancer Registry between 1961 and 1971 and 199 male controls chosen from the electoral register, matched for age and geographical location in nine census divisions; the overall response rate was 93%. An association was found between lip cancer and outdoor work (odds ratio, 1.52; $p < 0.05$); an odds ratio of 1.50 ($p < 0.05$) was found for occupation as a fisherman for at least eight full seasons, after adjustment for outdoor work, pipe smoking and age. No positive association was found for specific fishing activities, such as use of mouth as a third hand or of cast nets.

Lindqvist (1979) obtained information by mailed questionnaires from 171 cases (149 men, 22 women; 74% response rate) of epidermoid carcinoma of the lip registered with the Finnish Cancer Registry in 1972–73 and from a control group of 124 patients (56 men, 68 women; 77% response rate) registered with squamous-cell carcinoma of the skin of the head and neck. Risk estimates were adjusted for age. Odds ratios for men working outdoors ranged from 2.2 to 3.2 according to the calendar period during which the subjects had worked outdoors. The odds ratio was significantly increased only for those who both worked outdoors and smoked. [The Working Group noted that the choice of head and neck skin cancer patients as controls would lead to an underestimate of the odds ratio for outdoor work.]

Dardanoni et al. (1984) obtained information by personal interviews from 53 men with lip cancer registered in the Ragusa Cancer Registry in Italy and from 106 male controls matched for age and municipality of residence and admitted to the same hospitals for non-neoplastic diseases. An association was found between lip cancer and working or spending at least 6 h each day outdoors (odds ratio, 4.9; $p < 0.001$). After control for socio-economic level, the odds ratio was 1.7 ($p < 0.001$). [The Working Group noted that the latter p value is inconsistent with the number of subjects.]

2.1.3 Malignant melanoma of the skin

Melanoma of the skin is divided into three major histological types. The majority of melanomas in white-skinned populations (of European origin) are superficial spreading and nodular melanomas. Lentigo maligna melanoma—also known as Hutchinson's melanotic freckle—occurs later in life than the other types, and more specifically on exposed sites; however, the body site and evidence of sun damage in surrounding skin may influence its pathological classification (McGovern et al., 1980). Acral lentiginous melanoma has not been studied epidemiologically; it is rare in white-skinned populations, although it comprises a substantial proportion of melanomas in Japan (Elwood, 1989a).

(a) Case reports

In general, case reports were not considered, owing to the availability of more informative data.

In a survey of 830 cases of xeroderma pigmentosum located through published case reports (Kraemer et al., 1987), melanomas were reported in 37 patients (5%). As the median age at last follow-up of these cases was only 19 years, this observation is likely to represent a substantial excess over the number expected, although the exact nature of the study population precludes an accurate comparison. Site was specified for 29 of the 37 cases; 65% of these were on the face, head and neck (normally constantly UVR-exposed sites) as compared with 19.4% on this site among affected members of the US general population. [The Working Group recognized that data collected from previously published case reports are not uniform and may be atypical of a true incidence or prevalence series. Furthermore, no information is available on the relationship between solar exposure and the occurrence of malignant cutaneous melanoma in these patients.]

(b) Descriptive studies

(i) *Sex distribution*

The sex distribution of melanoma, adjusted for age, varies widely between populations. In many, it occurs as often as or more commonly in women than in men (Lee & Storer, 1980; Lee, 1982), in contrast to other types of skin cancer which are uniformly commoner in men (Muir et al., 1987).

(ii) *Age distribution*

Age distributions of melanoma in human populations vary with sex (Lee, 1982). They cannot easily be interpreted because they represent a variable combination of the different patterns of melanomas at different sites as well as a combination of time trends and trends in the experience of birth cohorts.

(iii) *Anatomical distribution*

Melanoma is proportionately commonest on the back and face in men and on the legs in women (Crombie, 1981); however, the incidence of melanoma per unit of body area is similar on fully exposed sites, such as the face, and on partially exposed sites, such as the lower limbs in women and the back in men. The frequency on body sites that are usually covered, such as the buttocks, is much lower (Elwood & Gallagher, 1983).

(iv) *Ethnic origin*

Melanoma is predominantly a disease of white-skinned populations. Rates in dark-skinned populations are much lower, the age-standardized incidence rate in India being 0.2 per 100 000 compared to around 30 in Queensland, Australia. In Los Angeles, USA, rates were less than 1 per 100 000 in Japanese and Chinese subjects and 11-12 in white subjects (Muir *et al.*, 1987; Whelan *et al.*, 1990). The site and histological distribution of melanoma are different in non-white populations and have been little studied epidemiologically. The remainder of this section deals only with melanoma in white populations.

The incidence of melanoma is substantially lower among Hispanics than among other whites in the USA. For example, the incidence among Hispanics in New Mexico is less than 2 per 100 000 person years, but in other whites it is about 11 per 100 000 (Muir *et al.*, 1987). In several case-control studies (described in detail below), subjects with a southern or eastern European background had lower risks than those with northern European or British origins (Elwood *et al.*, 1984; Holman & Armstrong, 1984a).

In a Canadian study (Elwood *et al.*, 1984), people with an eastern or southern European background had a crude odds ratio of 0.5 relative to those with an English background. This effect was not changed appreciably after adjustment for constitutional factors of hair, eye and skin colour and the skin's reaction to sun exposure. In contrast, the effect of ethnic origin observed in Western Australia was substantially reduced after adjustment for pigmentation characteristics (Holman & Armstrong, 1984a).

(v) *Geographical variation*

Armstrong (1984) showed that the relationship between melanoma incidence in Caucasians and latitude of residence decreases from around 35 ° to a minimum around 55 ° and then rises with latitude due to high rates in Scandinavian and Scottish populations. This pattern is likely to be due to both latitudinal and pigmentation factors. Within countries, inverse relationships of incidence or mortality with latitude have been seen in England and Wales (Swerdlow, 1979), Norway (Magnus, 1973), Sweden (Eklund & Malec, 1978) and Finland (Teppo *et al.*, 1978).

In the first comprehensive analysis of the geography of melanoma in whites, Lancaster (1956) noted that mortality from the disease was higher in Australia and South Africa than in the parts of Europe from which their populations originated; that mortality in Australia, New Zealand and the USA increased with proximity to the equator; but that within Europe it was higher in Norway and Sweden in the north than in France and Italy in the south. These patterns are also evident in more recent data (Armstrong, 1984).

Geographical variation in relationship to ambient UV irradiation levels: Several studies have compared melanoma incidence and mortality rates in different areas of North America to estimated or measured levels of ambient UVR, and Elwood (1989b) estimated the change in rate for a 10% change in UVR level (Table 15). [The Working Group noted that these studies did not assess any other component of the solar spectrum.]

Elwood et al. (1974) showed, using mortality data for US states and Canadian provinces, that the correlation coefficients with latitude were 0.79 for men and 0.72 for women. A variation in latitude of 2°, which is equivalent to 138 miles, was associated with a change in death rates from melanoma of about 10%. Annual UV flux at erythema-producing wavelengths was calculated from information on latitude and meteorological data on cloud cover. This calculated index of exposure was very strongly correlated with latitude (correlation coefficient, 0.89), so melanoma mortality rates were strongly related to this index; a 10% increase in received UVR dosage would be expected to give an increase of 3.7–4.5% in the death rate from melanoma at latitude 50°, and 6.8–10.3% at latitude 30° (Table 15). These values were somewhat higher for men than for women; for example, 4.4% in men compared with 3.0% in women at latitude 50° using the exponential model.

Fears et al. (1976) related melanoma incidence to latitude and to a calculated measure of UVR. Their data cover a slightly narrower range of latitude, and they calculated that a 10% increase in UVR would cause an increase in melanoma mortality of 7–12%, the higher figure applying to more southerly latitudes, which already have higher rates. Incidence rates vary more rapidly with latitude than do mortality rates, and therefore they predicted that a 10% increase in UVR would be likely to give a 14–24% increase in the incidence of melanoma (see Table 15).

Estimates using calculated UVR levels: Fears et al. (1977) used measurements from Robertson–Berger meters for four areas and a power model, in which the calculated percentage changes are not dependent upon the initial latitude. These calculations showed considerably stronger effects, with an estimated 25% increase in incidence for a 10% increase in solar UVR (see Table 15).

Scotto and Fears (1987) used annual UVR counts from Robertson–Berger meters in seven areas of the USA (Detroit, Seattle, Iowa, Utah, San Francisco, Atlanta and New Mexico) and data on melanoma from incidence registries (the Surveillance Epidemiology and End Results system). They fitted a power model and presented analyses by sex and by body site of the melanoma divided into trunk and lower limb *versus* head, neck and upper limb. They obtained data on covariates, including ethnic origin, pigmentation characteristics, hours spent outdoors during weekdays and during weekends and use of suncreens, suntan lotion and protective clothing, from telephone interviews with at least 500 households in each area. Data on the melanoma patients were not available, however. The results predict greater increases for females than for males, unlike the earlier work. The overall effects of a 10% increase in UVR are a 5.5% increase for trunk and lower limb tumours and a 9% increase for head, neck and upper limb tumours, averaged over the two sexes. Adjustment for the various covariates reduces the predicted increases to a 3.5% increase for trunk and lower limb tumours, and 5.5% for head, neck and upper limb tumours (see Table 15).

Pitcher and Longstreth (1991) used data on melanoma mortality over a 30-year period and calculated UV flux on the basis of satellite data from the US National Aeronautics and Space Administration, including measurements of ozone concentrations at high atmospheric conditions. The models fitted are complex, as they are fitted for the two sexes, for three different places covering a range of latitudes, and separately for changes in the annual UV

Table 15. Estimates by Elwood (1989b) of percentage increase in frequency of melanoma among whites with a 10% increase in solar ultraviolet radiation, based on differences with latitude in Canada and the USA

Ultraviolet radiation level derived from[a]	Model	50° latitude		30° latitude		References on which estimates based
		Incidence	Mortality	Incidence	Mortality	
Calculation of erythema-weighted index	Linear		4.5		6.8	Elwood et al. (1974)[b]
	Exponential		3.7		10.3	
Calculation of erythema-weighted index	Exponential	14.0	7.0	23.5	12.0	Fears et al. (1976)[c]
RB meter (1974)	Power	25.0		25.0		Fears et al. (1977)[d]
RB meter (1978–81)	Power					Scotto & Fears (1987)[e]
	Trunk and lower limb					
	Crude	5.5		5.5		
	Adjusted	3.5		3.5		
	Head, neck and upper limb					
	Crude	9.0		9.0		
	Adjusted	5.5		5.5		
	Total					
	Crude	6.7		6.7		
	Adjusted	4.2		4.2		
Calculation of erythema-weighted estimate from NASA including satellite ozone column measurements	Power					Pitcher & Longstreth (1991)[f]
	Annual		3.2		3.2	
	Peak		7.0		7.0	
	Exponential					
	Annual		2.1		4.5	
	Peak		5.8		8.2	

Both sexes (simple average of sex-specific results)

[a]RB, Robertson–Berger; NASA, National Aeronautics and Space Administration
[b]Mortality data, USA and Canada 1950–67 by state/province; 58 areas
[c]Incidence data, Third National Cancer Survey (1969–71) for nine areas; US mortality by state. Calculation based on latitude equivalent to change in ultraviolet radiation
[d]Incidence data, Third National Cancer Survey (1969–71) for four areas
[e]Incidence data, Surveillance Epidemiology and End Results Program for seven areas. Crude results take account only of age; adjusted results are controlled for ethnic origin, hair or skin colour, suntan lotion use and hours spent outdoors; total, for comparison, is based on 67% trunk and lower limb and 33% head, neck and upper limb tumours
[f]Mortality data by US county 1950–79; estimates of changes in mean annual dose and in peak doses (clear day in June); estimates using DNA action spectrum were also made and were 1–8% higher than those shown.

flux and changes in the peak levels in clear summer conditions. Larger effects were again found for males than for females, and a larger effect when using the peak measurements than when using the annual measurements. The overall estimates of the percentage increase in melanoma mortality associated with a 5% decrease in ozone level, on the assumption that this is roughly equivalent to a 10% increase in solar UVR, ranged from 2.1 to 7.0 at 50 °N and from 3.2 to 8.2 at 30 °N (see Table 15).

[The Working Group noted that, despite the sophistication of some of the mathematical models, these results are derived from population-based descriptive data and not from individual measurements and are restricted to North America.]

(vi) *Migration*

The most informative data on risk in migrants come from Australia, New Zealand, Israel and the USA. Native residents of Australia (McCredie & Coates, 1989; Khlat *et al.*, 1992) and New Zealand (Cooke & Fraser, 1985), mostly of British origin, experienced incidence and mortality rates of melanoma roughly twice those of British immigrants. Native Israelis had a risk at least twice that of immigrants to Israel from Europe for at least 30 years after immigration (Steinitz *et al.*, 1989).

The higher incidence in white immigrants to Hawaii from the US mainland compared with white natives has been attributed to a difference in skin colour (Hinds & Kolonel, 1980). Non-Hispanic migrants to Los Angeles County (California, USA) from higher latitudes in the USA are still substantially protected against melanoma of all histological types decades after migration. Similar relative protection is enjoyed by native residents of more northerly US communities in comparison with co-resident migrants from the south-western USA (Mack & Floderus, 1991).

(vii) *Socioeconomic status and occupation*

Melanomas are much commoner in higher socioeconomic groups, as shown in data from the United Kingdom since 1949–51. In the United Kingdom, the distribution of melanoma in married women by social class (categorized by their husbands' social class) is similar to that of men, indicating that this is a social rather than a specific occupational factor (Lee, 1982). In the USA, the risk increases with income for men aged 30–69; at age 70 and above, the trend is reversed, suggesting a role for long-term exposure to the sun (Kirkpatrick *et al.*, 1990). In case–control studies, the effect of socioeconomic status is weakened after adjustment for measures of exposure to the sun (Gallagher *et al.*, 1987; Østerlind *et al.*, 1988b).

Assessment of outdoor exposure on the basis of routine data on job descriptions showed that melanoma is commoner in indoor than in outdoor workers, even within the same socio-economic group (Lee & Strickland, 1980; Lee, 1982). Cutaneous melanoma incidence rates during 1972–76 in New Zealand showed no pattern according to outdoor workplace (Cooke *et al.*, 1984). An analysis of 3991 cases of cutaneous melanoma registered during 1971–78 in England and Wales and of 5003 cases registered during 1961–79 in Sweden suggested an elevated incidence in professional occupations. The incidence among farmers was close to that expected (Vågerö *et al.*, 1990).

Garland *et al.* (1990) reported 176 incident cases of melanoma among US Navy personnel. The rate for indoor occupation was higher than that for outdoor workers.

(c) *Case–control studies*

Elements of each case–control study described below are given in Table 16.

(i) *Australia*

Lancaster and Nelson (1957) carried out a case–control study on 173 patients aged over 14 years treated for malignant melanoma in hospitals in Adelaide, Melbourne and Brisbane, and 173 hospital controls with cancers other than of the skin, matched for sex and age. Information was obtained by interviews [response rate not given], and analysis was done by single factor cross-tabulations only. Unmatched crude odds ratios were calculated by the Working Group. Skin [odds ratio, 1.95 for fair *versus* olive and medium], hair colour [odds ratio, 1.7 for fair and red *versus* black and brown], eye colour [odds ratio, 1.75 for blue and green-grey *versus* brown and hazel] and skin reaction to sunlight [2.9; 95% CI, 1.9–4.5 for red *versus* brown reaction] were significantly associated with risk for malignant melanoma. Among the other factors studied were birth outside Australia [0.8; 0.4–1.6], 10 years' or more occupational exposure to sunlight in males [1.4; 0.7–2.7], sunbathing [1.5; 0.9–2.4] and moderate [1.2; 0.5–3.1] and excessive [2.3; 0.8–6.3] total exposure to the sun compared to minimal exposure. There were only eight cases and 11 controls in the latter category of sun exposure.

Beardmore (1972) studied 468 cases of histologically confirmed malignant melanoma and 468 sex- and age-matched hospital controls (including patients with skin cancer) at one hospital in Brisbane. Information was obtained by interview [response rate and method of evaluation of hair, skin and eye colour not given]. Hair, skin and eye colour and skin reaction to sunlight were not associated with risk for malignant melanoma. Comparison of exposure to sunlight from mainly outdoor occupations to that from mainly indoor occupations resulted in a crude odds ratio of [1.42; 95% CI, 1.03–1.97]; a similar comparison for recreational activities gave a crude odds ratio of [1.03; 0.75–1.42]. Fewer cases than controls had a history of treatment for keratosis and/or skin cancer or currently had keratosis and/or skin cancer [crude odds ratios, 0.51, 0.38–0.69; and 0.16, 0.12–0.22, respectively].

In the Western Australia Melanoma study (Holman & Armstrong, 1984a,b), 511 cases aged 10–79 years and 511 population controls matched for sex, age and area of residence were interviewed at home using a questionnaire based on that of the Western Canada study, which included objective measurements and naevi counts. The study also included a review of pathology slides. Analyses were presented for superficial spreading, nodular and lentigo maligna melanomas and for a fourth, unclassifiable group. Response rates were 76% for cases and 62% for controls, and adjustment was made for chronic and acute skin reaction to sunlight, hair colour, ethnic origin and age at arrival in Australia using a multiple logistic regression model. Hair colour, acute and chronic reaction to sunlight, number of naevi and family history of melanoma were significantly associated with risk; skin and eye colour were significantly associated in a crude analysis only. Duration of residence in Australia was strongly, positively associated with risk for all melanomas and for all sub-types except for unclassifiable melanoma. After control for ethnic origin, the odds ratios for superficial spreading melanoma were 1.2 (95% CI, 0.25–5.5) for people arriving in Australia at age 0–4, 1.7 (0.34–8.0) for those arriving at age 5–9, 0.74 (0.17–3.3) for those arriving at age 10–14, 0.25 (0.05–1.4) for those arriving at age 15–19 years or older (< 30 years) and 0.38

(0.19–0.78) for those arriving at age ≥ 30 years (p for trend, < 0.0001) compared to those born in Australia. A lifetime residential history was used to calculate the mean annual hours of bright sunlight based on place of residence as a measure of potential exposure to the sun. An analysis restricted to native-born Australians showed positive associations for all melanomas and for each subtype except nodular melanoma. An analysis dichotomizing exposure at an annual mean of > 2800 h sunlight at different ages showed that the highest risk ratio for all melanomas and for the superficial spreading subtype were for high exposure at ages 10–24. Cutaneous microtopography was used to measure skin damage; a positive association was found with all melanomas, being strongest for lentigo maligna melanoma.

In a further analysis by individual habits of exposure to the sun (Holman *et al.*, 1986a), no significant association was seen for total outdoor exposure. Analysis by recreational outdoor exposure, expressed as a proportion of total exposure, at ages 10–24 years showed no significant association. For superficial spreading melanoma, analysis by specific activity showed positive associations with boating (p = 0.04) and fishing (p = 0.07) and weaker, nonsignificant associations with swimming and sunbathing at ages 15–24 or 0–9 years before diagnosis. For other types of melanoma, no clear positive association was found; regular swimmers had a lower risk of lentigo maligna melanoma (trend test significant). Occupational exposure was analysed on the basis of whether the site of the melanoma was usually covered by clothing and compared to that of a referent group for whom the site was usually covered: subjects for whom the site was exposed showed a significant positive association. In comparison with the same referent group, patients who had never worked outdoors had significantly increased risks for all melanomas. The type of bathing suit usually worn by females in summer was assessed, and a positive association was found for wearing bikinis or for nude bathing, which was significant for all trunk melanomas and for superficial spreading melanoma on the trunk. When previous sunburns were classified by severity, no significant trend was observed for all melanomas; but there was a positive trend for lentigo maligna melanoma (p = 0.06) and a significant negative association for nodular melanoma.

In the smaller Queensland Melanoma study (Green, 1984; Green *et al.*, 1985a), 183 patients with histologically confirmed melanoma, other than lentigo maligna melanoma or acral lentiginous melanoma, and 183 population controls matched for sex, age and area of residence were interviewed at home using a standardized questionnaire, which included objective measurements and naevi counts. The response rates were 97% and 92%, respectively. Adjustment was made using a multiple logistic regression model. Hair colour, acute sun reactions and naevi were significantly associated with risk. Skin colour, eye colour, chronic sun reaction, freckling and family history of melanoma were significant in a crude analysis only. Hours of occupational and recreational exposure to the sun from 10 years of age across three categories gave risks of 1, 3.2 (95% CI, 0.9–12.4) and 5.3 (0.9–30.8) after adjustment for naevi, hair colour and propensity to sunburn. Average levels of exposure to UVB radiation were also allocated by residential history but showed no association with risk for melanoma. People born in Queensland had moderately higher risks than those who arrived there later in life or who had lived somewhere else at any time. Melanoma patients had more kerotoses or skin cancers on their faces (odds ratio, 2.8; 1.1–7.2). Sunburn (Green *et al.*, 1985a) was defined as pain persisting longer than 48 h, with or without blistering, and was recorded as the number of episodes in each decade. Risk increased with the number of severe sunburns and was 1.9 and 5.0 in the two higher categories on matched analysis, decreasing to 1.5 (0.7–3.2) and 2.4 (1.0–6.1), respectively, when adjusted for naevi and exact age. An additional analysis of 49 cases of lentigo maligna melanoma and 49 controls showed no association with sunburn (Green & O'Rourke, 1985; Green *et al.*, 1986).

In a more detailed review of these data (Green *et al.*, 1986), no association was observed with occupational exposure to the sun. Analyses of recreational hours spent on the beach in the sun were made for lifetime exposures, exposures at 10–19 years of age and exposures in the five years prior to diagnosis; no strong or consistent association was seen in either crude or adjusted analyses. Associations with total accumulated hours of exposure to the sun (calculated by adding occupational and total recreational exposures) showed a positive trend for lifetime exposure and exposure at ages 10–19 (odds ratio, 4.4; 95% CI, 1.8–184.5), but no association was seen for exposure during the previous five years. Analysis of levels of UVR by lifetime residential history showed no major association and no site-specific association.

(ii) *Europe*

In a case–control study of residents of Oslo, Norway (Klepp & Magnus, 1979), 78 malignant melanoma patients over 20 years of age were compared with 131 unmatched hospital controls with other cancers. Both cases and controls with advanced disease were excluded. Information was obtained by questionnaire [response rate not given]. Hair and eye colour were recorded independently by the interviewer and subject but were not associated with risk for the disease, whereas skin reaction to sunlight and freckling were. A nonsignificant odds ratio of [1.5] was found for men working outdoors for more than 3–4 h/day; the odds ratio for taking sunbathing holidays in southern Europe was 2.4 ($p = 0.05$). No significant association was seen with degree of exposure of different body sites, classified from 'as often as possible' to 'hardly ever'.

Adam *et al.* (1981) conducted a population-based case–control study in the United Kingdom of 111 female cases of malignant melanoma aged 15–49 traced from registries and 342 female controls randomly selected from general practitioners' lists and matched for age and marital status. Information was obtained by postal questionnaire; response rates were 66% for cases and 68% for controls. Hair colour and skin reaction to sunlight, but not skin colour, were significantly associated with risk for malignant melanoma. Slightly more cases than controls reported deliberately tanning their legs or trunk, either at home or abroad. No difference was reported in the amount of work, leisure or total time spent outdoors. [The Working Group noted that the study concentrated on oral contraceptive use and that information on exposure to the sun was very limited.]

MacKie and Aitchison (1982) conducted a case–control study in western Scotland of 113 malignant melanoma patients aged 18–76 years and 113 sex- and age-matched hospital controls with conditions not related to the skin. Cases of lentigo maligna melanoma were excluded. Information about exposure to the sun within the previous five years was obtained by questionnaire [response rate not given] and included occupational and recreational exposure (≥ 16 h *versus* < 16 h outdoor exposure per week) and history of severe sunburn, defined as either 'blistering sunburn' or 'erythema persisting for a week or longer'. Other factors included in the multivariate analysis were social class and skin type. A significant negative association was observed for recreational exposure and for occupational exposure to the sun in males. A significant positive association was observed for severe sunburn. No significant difference was observed for the number of continental holidays taken or total number of days spent in sunnier climates.

Sorahan and Grimley (1985) studied 58 patients aged 20–70 years with cutaneous malignant melanoma (other than lentigo maligna melanoma) in two hospitals in the United Kingdom and 182 hospital controls with diseases other than of the skin and 151 unmatched controls from electoral rolls. The response rates were 64% for cases and 60% for each control group. Information was obtained by postal questionnaire, and analyses were adjusted using a multiple logistic regression model. A significant positive association was observed for number of bouts of painful sunburn ever experienced, with an odds ratio reaching 7.0 for five or more bouts compared to none. A significant positive association was also seen with the number of holidays ever spent abroad in a hot climate, reaching 6.5 for 21 holidays or more, compared to none. Both associations were weakened, and the latter became nonsignificant, after adjustment for propensity to sunburn, number of moles and history of sunburn.

In another study in the United Kingdom (Elwood et al., 1986), 83 histologically confirmed cases over 18 years of age and 83 hospital controls (in- and out-patients), matched for sex, age and area of residence, were interviewed at home using a questionnaire which included objective measurements and naevi counts. The responses were validated by replies to a postal questionnaire. The response rates were 74% for cases and 92% for controls. Adjustment was made using a multiple logistic regression model. Skin reaction to sunlight, freckling and naevi were significantly associated with risk. A history of sunburn causing pain for two days or more gave a significant odds ratio of 3.2 (95% CI, 1.7–5.9). Past outdoor occupational exposure showed a significantly reduced odds ratio of 0.2 (0.1–0.9) for the second highest category but a nonsignificant odds ratio of 1.7 (0.3–8.6) for the highest category and no overall trend.

In northern Italy, Cristofolini et al. (1987) compared 103 patients aged 21–79 under treatment for cutaneous malignant melanoma at one hospital with 205 hospital controls with diseases other than skin tumours. Subjects were interviewed [response rate not given] and assessed by a dermatologist. Adjustment was made using a multiple logistic regression model. Hair and skin colour and family history were significantly associated with risk, but eye colour, freckling and number of naevi were not. A history of frequent sunburn as an adult gave an odds ratio of 1.2 (95% CI, 0.7–2.1) and that of severe sunburn in early life an odds ratio of 0.7 (0.4–1.2). Heavy or frequent exposure to sunlight during the previous 20 years, categorized as yes or no, gave a significantly reduced odds ratio of 0.6 (0.4–0.95). Outdoor compared to indoor occupation gave a nonsignificant odds ratio of 0.9 (0.5–1.7), and a history of carcinoma of the skin gave a risk ratio of 0.4 (0.02–2.9), based on small numbers. Melanoma at exposed sites showed positive associations with heavy sun exposure (1.44; 0.8–2.8) and outdoor occupation (1.8; 0.9–3.7), while melanoma at normally unexposed sites showed a significant negative association with heavy exposure to the sun (odds ratio, 0.25; 95% CI, 0.13–0.47).

In a study of melanoma in eastern Denmark (Østerlind et al., 1988b,c; Østerlind, 1990), 474 cases of melanoma, excluding lentigo maligna melanoma patients, aged 20–79 were compared with 926 population controls and matched for sex and age. Subjects were interviewed at home using a questionnaire which included objective measurements and naevi counts, and adjustment was made using a multiple logistic regression model. Response rates were 92% for cases and 82% for controls. The number of sunburns (defined as those causing pain for two days or longer) before age 15, from age 15 to 24 and over the previous 10 years were all significantly associated with risk: crude odds ratios for the maximal categories, 3.7 (95% CI, 2.3–6.1), 2.4 (1.6–3.6) and 3.0 (1.6–5.4), respectively. Adjustment for sex and host factors, including naevi, freckles and hair colour, reduced the risk ratios, but they remained significant. Adjustment for sunburns before age 15 rendered the associations with later sunburn weak and nonsignificant. Joint analysis of sunburns and naevi suggested independent, additive risks. Significantly increased risks were seen with residence near the coast before age 15 or for more than 30 years. Specific recreational activities were investigated and categorized by the number of years of regular participation, adjusted for sex and host factors, including number of naevi, and for other activities. Significant positive associations were observed with sunbathing, boating, winter skiing and swimming, the latter becoming nonsignificant after adjustment. Regular participation in gardening, ball games, golf, horseback riding or hiking was not associated with risk for melanoma. A positive trend was seen with vacations spent in beach resorts in southern Europe (odds ratio, 1.7; 95% CI, 1.2–2.4), which was weakened after adjustment for sunbathing and sunburn (1.4; 1.0–2.1). Socioeconomic status showed a strongly positive association in men, which became nonsignificant when adjusted for sunburn and recreational exposure to the sun. Occupational exposure outdoors for at least six months was associated with a significantly reduced odds ratio of 0.7 (0.5–0.9) in men; the protective effect was most pronounced in men who started working outside at an early age and continued for at least 10 years. No association was seen with skin grading categories defined by microtopography.

In a study in northern Italy (Zanetti *et al.*, 1988), 208 cases of histologically confirmed malignant melanoma were identified from the regional tumour registry and were compared with 416 controls chosen from the National Social Service Registry. Response rates were 87% for cases and 68% for controls. An increased risk was observed with light hair colour, tendency to burn and a history of sunburn in childhood. No significant effect of region of origin was observed. Exposure to the sun was assessed by activity: for outdoor work, a nonsignificant increased risk was seen with the maximal duration of exposure ($\geq$ 33 years) in men, but the overall trend was nonsignificant. Outdoor sports, assessed by years of participation, showed an increased risk at the maximal level in men and women (significant for men). A significantly increased risk was found for men participating in sports categorized as involving the greatest exposure to the sun. A nonsignificantly increasing trend in men was observed for total number of weeks' holiday, but little effect was seen in women; a significant positive trend was observed in men, but not for women, for the number of weeks spent at the seaside in childhood. Similar exposure in adult years resulted in a nonsignificant positive trend.

Garbe *et al.* (1989) studied 200 malignant melanoma patients at a dermatological follow-up clinic in Berlin, Germany, in 1987 and 200 controls from the same clinic who had any other skin disease (response rate, 90%). Subjects of non-German origin were excluded, as were those seeking consultation for pigmented naevi or who had been treated previously by UVR (10%). Occupational exposure to the sun, assessed as none, sometimes or nearly all the time, showed a strongly increased risk up to an odds ratio of 5.5 (1.2–25.3). No significant relationship was found with duration of leisure-time exposure to the sun or number of sunburns. [The Working Group noted that little detail was given about exposure and that the control group consisted of patients with other skin disease.]

Weiss et al. (1990) studied 1079 cases of malignant melanoma reported to the German Dermatological Society Registries in 1984–87 and 778 hospital controls from the same clinics. Positive associations were seen with occupational exposure to the sun, which increased with the number of years of exposure. No association was seen with exposure to the sun during leisure time or with sunbathing. [The Working Group noted that this study appears to overlap with that of Garbe et al. (1989) and that the data were presented with relative risks but with no test of significance.]

Beitner et al. (1990) studied 523 incident cases of malignant melanoma seen at a hospital in Stockholm, Sweden (representing 64% of all cases registered in Stockholm County), and 505 controls selected from the population register for Stockholm County. Cases completed a questionnaire while waiting at the clinic, and controls received the questionnaire by mail (response rates, 99.6% and 96.2%, respectively). A significant positive effect was seen for the number of sunbathing sessions each summer, with a history of erythema after sunbathing and with sunbathing vacations abroad. Residence in countries around the Mediterranean or in a sub-tropical or tropical climates for more than one year during the previous 10 years gave a significant odds ratio of 1.9 [95% CI, 1.0–3.6]. There was no increase in risk with sunbathing during winter vacations at high altitudes. Outdoor workers had a significantly reduced risk of 0.6 (0.4–1.0) after adjustment for age, sex and hair colour.

Elwood et al. (1990) studied 195 cases of superficial spreading or nodular melanoma in people aged 20–79 from five pathology laboratories in the United Kingdom and 195 controls chosen from among all in- and out-patients in the region. Cases and controls underwent an interview and a limited examination by an interviewer in their homes (participation rate—cases and controls, 73%; voluntary response rate—cases, 91%; controls, 78%). Risk was significantly increased with sunburn at age 8–12 (odds ratio, 3.6; 1.4–11.2), but no significant increase was observed with sunburn at age 18–22 or with sunburn received 18–20 or five years prior to diagnosis. No other sun exposure variable was reported.

Grob et al. (1990) compared 207 consecutive white patients, 18–81 years old, with histologically confirmed invasive melanoma (at least level 2; lentigo melanoma and acral lentiginous melanoma excluded) seen in one dermatology clinic in Marseilles, France, with 295 controls. Controls under 65 years of age were chosen from among subjects interviewed after reportedly random selection and examined at a public health centre; those over 65 were chosen from among out-patients with non-cancer and non-dermatological conditions. Patients and controls were examined and interviewed by the same dermatologist. Multiple logistic model analysis was used. The risk for melanoma was increased significantly in association with annual outdoor leisure exposure during the previous two years (odds ratio, 8.4; 95% CI, 3.6–19.7), outdoor occupation (6.0; 2.1–17.4) and total lifetime sun exposure (odds ratio for maximum category, 3.4; 1.6–7.1). There was a nonsignificant association with sunburns in recent years (1.7; 0.63–4.6) after adjustment for number of naevi, maximal depth of suntan, hair colour, social level, complexion and age. [The Working Group found the study difficult to interpret because of the nature of the control group and the relative recency of measurements of exposure to the sun.]

In a report designed to produce a risk prediction model, MacKie *et al.* (1989) studied 280 cases of invasive cutaneous malignant melanoma (level 2 or deeper) from Scottish melanoma registries. Controls were 280 hospital patients with non-dermatological diseases. Response rates were 76% for cases and unknown for controls. An increased risk was observed for history of severe sunburn (adjusted odds ratio, 7.6 (95% CI, 1.8–32.0) for men and 2.3 (0.9–5.6) for women). A significant positive association for tropical residence was noted for men, which became nonsignificant after adjustment. [The Working Group noted that, apart from tropical residence, no data were presented on exposure to the sun.]

(iii) *North America*

Gellin *et al.* (1969) studied 79 patients, aged 30–79, with histologically confirmed malignant melanoma at one hospital in New York, USA, and compared them with 1037 hospital controls with skin conditions other than cancer. Information was obtained by interview and examination [response rate not given]. The odds ratios for duration of daily outdoor activity were [2.8 (95% CI, 1.3–5.8)] for 6 h or more and [4.1 (2.5–6.8)] for 3–5 h, compared to 0–2 h. [The Working Group noted that the controls had skin diseases.]

Paffenbarger *et al.* (1978) reported on cases found by follow-up of subjects first examined when entering Harvard University in 1916–50 and the University of Pennsylvania in 1931–40. Out of a total of 50 000 male subjects and 1.71 million person-years of observation, 45 deaths from melanoma were observed and each compared to four controls born in the same year, who were classmates and who had survived as long as the case subjects. Of the many factors investigated, only outside remunerative work was associated with a significant risk for melanoma (odds ratio, 3.9; $p = 0.01$). Within the cohort, students from New England had a 50% lower risk for melanoma than other students, presumably owing to more northerly residence.

Lew *et al.* (1983) carried out a study in Massachusetts on 111 cases of cutaneous malignant melanoma, aged 23–81, followed at one hospital and 107 controls who were friends of cases, matched by age and sex. Information was obtained by interview at the clinic; response rates were 99% for cases and 90% for controls, and analysis was made using a logistic regression model. Cases showed poorer tanning ability, and a significant association was observed with blistering sunburn during adolescence (odds ratio, 2.1; 95% CI, 1.2–3.6) and with 30 days or more vacation in sunny, warm places during childhood (2.5; 1.1–5.8). The association with history of sunburn persisted after controlling for tanning ability. [The Working Group noted that the nature of the controls and the simplicity of the analyses presented make interpretation of the results difficult.]

Rigel *et al.* (1983) analysed data on 114 melanoma patients (out of a total of 328) seen in a referral centre in New York between 1978 and 1981, and on 228 controls who were staff and patients at the centre. Significantly increased risks were seen with > 2 h per day sun exposure 11–20 years previously (odds ratio, 2.5; $p = 0.005$) and outdoor *versus* indoor recreation (2.4; $p = 0.01$). [The Working Group noted that the selection of subjects and the nature of the control group make these results difficult to interpret.]

In the Western Canada Melanoma case-control study (Elwood *et al.*, 1984, 1985a,b), carried out in four Canadian provinces, 595 cases of malignant melanoma, aged 20-79, and 595 population controls, matched for sex, age and province of residence, were questioned by trained interviewers at their homes (response rates: cases, 83%; controls, 48-59%). Cases of lentigo maligna melanoma and acral lentiginous melanoma were excluded. Analyses were made using a multiple logistic regression model. Significant positive associations were found after adjustment for host factors and ethnic origin for frequent recreational (odds ratio, 1.7; 95% CI, 1.1-2.7) and holiday exposure (1.5; 1.0-2.3) and with the number of sunny vacations per decade (1.7; 1.2-2.3). No overall trend was observed for occupational exposure, but a significantly increased risk was associated with moderate occupational exposure, defined as seasonal or short-term occupational exposure. Maximal occupational exposure was associated with a significantly reduced odds ratio in men (0.5 [CI not given]) but not in women (1.5 [CI not given]). Analysis of total annual exposure to the sun from all sources showed no overall trend (odds ratio, 1.0-1.6 in various categories above the minimal exposure referent group). Severe or frequent sunburn in childhood resulted in a nonsignificant odds ratio of 1.3, after adjustment for host factors and sun sensitivity. From variables relating to sunburn on vacation and the usual degree of suntan in winter and summer, positive associations were observed for increasing sunburn and with decreasing usual tan. Cross-tabulation of sunburn with tendency to sunburn (skin type) did not change the significant positive effect of tendency to burn, but the odds ratio for sunburn fell from 1.8 in the maximal category to 1.4 ($p > 0.2$) after adjustment for sun reaction. Similarly, cross-tabulation of usual degree of suntan against skin type gave little difference in the positive association with reaction to the sun, but a weakening of the association with usual degree of suntan was seen which became nonsignificant. A multivariate analysis including history of sunburn, usual degree of suntan, skin type and host factors showed significance for the two latter factors, nonsignificant positive effects of holiday sunburn and a significant negative effect of usual degree of suntan. These results are interpreted as showing a primary association with tendency to burn easily or to tan poorly rather than with history of either sunburn or suntan. For men, a significant negative association was seen with outdoor occupation, but this weakened and became nonsignificant when adjusted for recorded exposure to the sun. Similarly, the crude odds ratio for upper compared to lower socioeconomic groups was 3.8 (2.0-7.4) but was reduced to 2.3 (1.0-5.1) after adjustment for host factors and for occupational, recreational and holiday sun exposure (Gallagher *et al.*, 1987).

Elwood *et al.* (1987) made an analysis separating superficial spreading melanoma, nodular melanoma and lentigo maligna melanoma in the western Canada study, based on 415, 128 and 56 cases, respectively. Recreational exposure, holiday exposure and the number of sunny vacations per decade were positively and significantly (trends) associated with superficial spreading melanoma (odds ratios, 1.4, 2.0 and 2.2; 95% CI, 1.0-2.0, 1.4-2.9 and 1.5-3.3, respectively); recreational exposure was also positively associated with nodular melanoma (2.4; 1.3-4.5), but neither holiday exposure nor the number of sunny vacations showed an association. None of these measures of intermittent exposure was significantly associated with lentigo maligna melanoma. Occupational exposure showed no significant association with any of the three types. History of sunburn showed positive but nonsignificant associations with superficial spreading and lentigo maligna melanomas but not with nodular melanoma.

Brown et al. (1984) identified 120 men who had been aged 18–31 during the Second World War from among 1067 patients seen at a melanoma clinic in New York City in 1972–80 and sent them questionnaires (response rate, 74%). Controls were 65 age-matched subjects attending the same dermatology department with skin diseases other than melanoma [response rate unknown]. Within the total of 74 cases and 49 controls who had been in the armed services, the odds ratio for service in the tropics as compared to service in the USA or Europe was [7.7; 95% CI, 2.5–23.6].

In a hospital-based study in Buffalo, NY, USA (Graham et al., 1985), 404 cases of cutaneous malignant melanoma referred to the Roswell Park Memorial Institute, aged from under 30 to over 65, were compared with 521 controls with other neoplasms at the same institute, using questionnaires completed on admission. There was a weak negative trend with total number of hours of exposure to the sun, which was significant in men; a similar trend was observed for average annual exposure to the sun. Occupational exposure to the sun gave a nonsignificant reduction in risk in men in the highest exposure group after adjustment for tendency to burn. Multivariate analysis showed a negative association with cumulative exposure to the sun, which was significant in men when adjusted for tendency to burn, freckling and light complexion. Results specific to recreational or holiday exposure to the sun were not presented.

Dubin et al. (1986) compared 1103 cases of melanoma seen at the New York University Medical Center from 1972 to 1982 (mostly in 1977–79) to 585 controls interviewed in 1979–82 at the skin clinic for conditions excluding cancer. Both cases and controls were interviewed by physicians; response rates were 98% for cases and 78% for controls. In order to complete the data on risk factors, a postal questionnaire was sent requesting information on exposures to fluorescent lights and to the sun and on skin colour (response rates, 45% of cases and 30% of controls). Mostly outdoor compared to mostly indoor work gave an odds ratio of 2.5 (95% CI, 1.4–4.4) and mostly outdoor compared with mostly indoor recreation gave an odds ratio of 1.7 (1.2–2.3), although mixed indoor and outdoor recreation gave a significantly reduced risk of 0.6 (0.5–0.8). Overall exposure to the sun (three categories) showed no trend. A history of the presence of solar keratosis gave a significant risk ratio of 5.0 (2.3–10.5). Quantitative total sun exposure was assessed for 623 cases and all 585 controls: there was no significant trend with total hours of exposure to the sun per day 0–5, 6–10 or 11–20 years before diagnosis. [The Working Group noted that the cases and controls were not interviewed over the same period.]

In a study based on a subset of the above (Dubin et al., 1989), 289 cases and 527 controls were interviewed using the same method (response rates, 100% of eligible cases; 70% of controls [19% of potential controls were excluded because of diagnosis of a lesion known to be caused by exposure to the sun]). Mostly outdoor occupation gave a nonsignificant elevated risk. Mostly outdoor recreation was associated with a significantly elevated risk in light tanners but a nonsignificant elevated risk in dark tanners (interaction nonsignificant). Overall exposure to the sun was associated with significantly increased risks in all groups. A history of sunburn was associated with a significantly increased risk in light tanners and in all subjects but had a nonsignificant protective effect in dark tanners (interaction significant).

When analysed by age group, a history of sunburn gave a positive association at age 20–39, a weak association at 40–59 and a negative association at 60 or over (interaction significant). Prior skin cancer or solar keratosis had a significant effect, which was stronger in men than in women (interaction nonsignificant).

In a study in San Francisco, Holly *et al.* (1987) compared 121 patients with nodular or superficial spreading melanoma at a university melanoma clinic with 139 controls from a medical screening clinic or from an orthopaedic clinic at the same centre. Response rates were 'over 95%'. Sunburn score, based on the number of blistering sunburns during school and young adult years, showed a significant odds ratio of 3.8 (95% CI, 1.4–10.4) after controlling for naevi, hair colour and previous skin cancers. A positive association was seen with previous skin cancer (3.8; 1.2–12.4).

Weinstock *et al.* (1989) reported a case–control study within a cohort of US nurses (see Hunter *et al.*, 1990, p. 86). Data on 130 cases and 300 controls (response rates to post-diagnosis questionnaire, 85% and 81%, respectively) were analysed using multivariate models. Following adjustment for skin sensitivity, significant positive effects were seen for sunburn at ages 15–20 (odds ratio, 2.2; 95% CI, 1.2–3.8), but not at age ≥ 30 (1.3; 0.7–2.3), and for residence at a southern latitude at age 15–20 (2.2; 1.1–4.2), but not at age ≥ 30 (1.6; 0.9–2.8). No direct recording of exposure to the sun was reported.

A further analysis (Weinstock *et al.*, 1991a) assessed the use of swimsuits in these subjects. There was a significant positive association of melanoma risk with the frequency of use of swimsuits of any type in sun-sensitive women (odds ratio, 6.4; 95% CI, 1.7–23.8) but not in sun-resistant women (0.3; 0.1–1.0). After controlling for type of swimsuit and sensitivity factors, melanoma risk was increased with increasing hours per day of outdoor swimsuit use (any type) after age 30, but no association was seen with intensity of exposure or with the number of winter vacations in warm and sunny locations. The use at age 15–20 of a bikini compared to high backline, one-piece swimsuits, gave an odds ratio for all melanomas of 1.9 (1.0–3.7) and for trunk melanoma specifically of 0.8 (0.3–2.6); the risks were 3.5 [CI not given] among sun-sensitive women and 1.3 [CI not given] among less sun-sensitive women, but the interaction was not significant.

In a case–control study of patients attending a pigmented lesion clinic in Boston, USA (Weinstock *et al.*, 1991b), 186 had cutaneous melanoma; the 239 controls had other dermatological diagnoses, the most frequent of which were common naevus and solar keratosis. Data were obtained from medical records and from a self-administered questionnaire completed before clinical examination and were analysed by a multivariate method. Significantly increased risks for melanoma were associated with lack of tan after repeated exposures as a teenager (odds ratio, 2.3; 95% CI, 1.0–4.9). A nonsignificant trend towards increased risk was observed for residence in southerly areas. [The Working Group noted that the paper dealt primarily with dysplastic naevi and the results on melanoma are not given in detail, and that the controls also had dermatological conditions.]

Table 16. Case–control studies of melanoma in which exposure to the sun and/or artificial ultraviolet radiation was assessed

Place	Period of diagnosis	No. of cases	Source of cases	Melanoma type	No. of controls	Type of control	Reference
Australia							
East Australia	NS	173	3 hospitals	All types	173	Other cancers	Lancaster & Nelson (1957)
Queensland, Australia	1963–69	468	1 hospital	All types	468	Hospital patients, including skin cancers	Beardmore (1972)
Western Australia	1980–81	511	Population	All types	511	Population	Holman & Armstrong (1984a,b)
Queensland, Australia	1979–80	183	Population	No LMM	183	Population	Green (1984); Green et al. (1985a)
Europe							
Oslo, Norway	1974–75	78	1 hospital	All types	131	Other cancers, same hospital	Klepp & Magnus (1979)
United Kingdom	1971–76	111	Population	All types	342	General practice lists	Adam et al. (1981)
Western Scotland	1978–80	113	Hospital	No LMM	113	Hospital, non-skin	MacKie & Aitchison (1982)
Birmingham, UK	1980–82	58	2 hospitals	No LMM	333	Hospital and population	Sorahan & Grimley (1985)
Nottingham, UK	1981–84	83	Population (2 hospitals)	All types	83	Matched hospital	Elwood et al. (1986)
Trento, Italy	1983–85	103	1 hospital	All types	205	Hospital	Cristofolini et al. (1987)
East Denmark	1982–85	474	Population	No LMM	926	Matched population	Østerlind et al. (1988a,b); Østerlind (1990)
Turin, Italy	1984–86	208	Population	All types	416	Population	Zanetti et al. (1988)
Berlin, Germany	1987	200	1 hospital	All types	200	Skin clinic patients	Garbe et al. (1989)

Table 16 (contd)

Place	Period of diagnosis	No. of cases	Source of cases	Melanoma type	No. of controls	Type of control	Reference
Scotland	1987	280	Population	Invasive MM at least type 2	280	Hospital, excluding skin	MacKie et al. (1989)
Germany	1984–87	1079	6 dermatology clinics	All types	778	Skin clinic patients	Weiss et al. (1990)
Stockholm, Sweden	1978–83	523	1 hospital	All types	505	Matched population	Beitner et al. (1990)
Midlands, UK	1984–86	195	Population	SSM and NM	195	Hospital in-/out-patients	Elwood et al. (1990)
Southeast France	1986–88	207	Hospital	Invasive, all types	295	Health centre	Grob et al. (1990)
North America							
New York, USA	1955–67	79	1 hospital	All types	1037	Other skin diseases, non-cancer	Gellin et al. (1969)
Boston, MA, USA Philadelphia, PA, USA	NS	45	Cohort of university alumni	All types	180	Classmates	Paffenbarger et al. (1978)
Boston, MA, USA	1978–79	111	1 hospital	All types	107	Friends of cases	Lew et al. (1983)
New York, USA	1978–81	114	1 hospital	All types	228	Patients and staff	Rigel et al. (1983)
New York, USA	1972–80	74	1 melanoma clinic	All types	49	Skin clinic patients	Brown et al. (1984)
Western Canada	1979–81	595	Population	SSM, NM or UCM	595	Population	Elwood et al. (1984, 1985a,b)
Buffalo, NY, USA	1974–80	404	Hospital patients	All types	521	Cancer patients	Graham et al. (1985)
New York, USA	1972–82	1103	3 hospitals	All types	585	Skin clinic patients	Dubin et al. (1986)
Western Canada	1979–81	415 128 56	Population	SSM NM LMM	415 128 56	Population	Elwood et al. (1987)
San Francisco, CA, USA	1984–85	121	1 melanoma clinic	NM and SSM	139	Clinic patients	Holly et al. (1987)

Table 16 (contd)

Place	Period of diagnosis	No. of cases	Source of cases	Melanoma type	No. of controls	Type of control	Reference
New York, USA	1979–82	289	3 hospitals	All types	527	Non-cancer skin patients	Dubin et al. (1989)
USA	1976–84	130	Nurses cohort	AM excluded	300	Nurses cohort	Weinstock et al. (1989)
Boston, MA, USA	1982–85	186	1 hospital	All types	239	Skin clinic patients	Weinstock et al. (1991b)

NS, not specified; SMM, superficial spreading melanoma; NM, nodular melanoma; UCM, unclassifiable melanoma; LMM, lentigo maligna melanoma (or Hutchinson's melanotic freckle); AM, acral lentiginous melanoma

(d) *Collation of results*

The studies summarized above show that a range of host characteristics are related to melanoma risk, including ethnic origin, skin, hair and eye pigmentation, and, importantly, a tendency to sunburn or suntan, often expressed clinically as skin type. These factors can be assumed to reflect genetic sensitivity to cutaneous effects of sun exposure and, in addition to the indirect evidence of a role of exposure to the sun in melanoma that they provide, should be considered as confounders in a relationship between sun exposure and melanoma. The numbers of acquired benign naevi and of dysplastic naevi have been shown to be very strong risk factors for melanoma in several studies; the density of freckling on the skin has also been shown to be a risk factor. Because there is evidence that these outcomes are themselves related to sun exposure, and in the case of naevi may be intermediate steps in the genesis of melanoma, they should not be considered confounding factors (Armstrong, 1988). Most of the studies relied on a wide range of questions to assess different aspects of sun exposure. Armstrong (1988) developed a useful classification of such questions, dividing them into those that assess potential exposure, such as place of residence and time of migration, those that record actual exposure and those that record response to exposure, such as questions on sunburn and suntanning.

(i) *Total sun exposure: potential exposure by place of residence* (Table 17)

Consistent with the descriptive studies, Holman and Armstrong (1984b) showed that the risk in migrants arriving in Australia before age 10 (odds ratio, 0.89; 95% CI, 0.44–1.80) is as high as that of the Australian born (1.00), and the risk in those arriving at age 10 or above is much less (0.34; 0.16–0.72 for age 10–29; 0.30; 0.08–1.13 for age $\geq$ 30). These data are an improvement on descriptive data as they allow control for ethnic background and pigmentation. In the same study, an association was seen with annual hours of bright sunlight averaged over all places of residence.

In the USA, two case–control studies (Graham *et al.*, 1985; Weinstock *et al.*, 1989) showed increased risks for people who had lived at southerly latitudes.

Increased risks in people who have lived near the coast were seen in Denmark (Østerlind *et al.*, 1988b) and in Queensland, Australia (Green & Siskind, 1983). It was assumed in the Danish study that coastal residence would involve more exposure to the sun. In Queensland, living near the coast is not related to annual ambient UVR, which varies with latitude, so that peak summer UV irradiance is higher in the interior than on the coast (Green & Siskind, 1983). The observations are thus due either to different behavioural patterns with geographical location or to differences in exposure to UVR.

(ii) *Biological response to total sun exposure*

It has been assumed that a history of nonmelanocytic skin cancer, solar keratoses, actinic tumours or changes on cutaneous microtopography are all indicators of cumulative sun damage. Positive associations are seen with these measures in studies in Australia and in the USA, although Østerlind *et al.* (1988b) in Denmark saw no relationship with microtopographical change (Table 17).

Table 17. Results of case-control studies on melanoma: place of residence, biological markers

Place	Direction of association	OR[a]	95% CI	p value	Measurement of exposure	Reference
Potential exposure by place of residence						
Australia	Up	5			Residence near coast; mortality rate/100 000 (incidence rate/100 000, 37)	Green & Siskind (1983)
Australia	Down	0.3	(0.1–1.1)	< 0.001	Age at arrival in Australia; OR given for age ≥ 30 years; p value for trend	Holman & Armstrong (1984b)
Australia	Up	2.8	(1.8–4.8)	< 0.001	Mean annual hours of bright sunlight at places of residence; p for trend	Holman & Armstrong (1984b)
USA	Up	1.4	(0.9–2.0)	> 0.05	Ever resided below 40 °N latitude	Graham et al. (1985)[b]
Australia	Down	0.3	(0.1–1.4)	> 0.05	Length of residence in Australia; risk associated with migration to Australia	Green et al. (1986)
Denmark	Up	1.7	(1.1–2.7)	0.006	Residence near coast; crude OR	Østerlind et al. (1988b)
USA	Up	2.2	(1.1–4.2)	0.02	Residence in southerly latitude at age 15–20, OR for 12.6 °	Weinstock et al. (1989)
Biological markers of cumulative sun exposure						
Australia	Up	2.7	(1.4–5.0)	0.003	Cutaneous microtopography; p for trend	Holman & Armstrong (1984b)
Australia	Up	3.7	(2.1–6.6)	< 0.001	History of nonmelanocytic skin cancer	Holman & Armstrong (1984b)
Australia	Up	3.6	(1.8–7.3)	< 0.001	Actinic tumours on face	Dubin et al. (1986)
USA	Up	5.0	(2.3–10.5)	< 0.01	History of solar keratosis	Green & O'Rourke (1985)
USA	Up	3.8	(1.2–12.4)	0.03	History of nonmelanocytic skin cancer, adjusted	Holly et al. (1987)
Denmark	Flat	1.1	(0.7–1.8)	> 0.05	Cutaneous microtopography; crude OR	Østerlind et al. (1988b)

[a]Odds ratio for maximal category
[b]Results calculated by Armstrong (1988)

(iii) *Total sun exposure assessed by questionnaire*

The results of studies in which total sun exposure was assessed using questionnaires, either over lifetime or at different periods of life, have been mixed (Table 18). Positive associations were seen by Green (1984) in Queensland, Australia; no consistent overall association was seen in western Canada, and in Western Australia the association was negative. The results of the other studies are similarly mixed. This inconsistency, in contrast to the results noted above by place of residence and by biological response, could be due either to the difficulty of assessing total sun exposure by questionnaires (Armstrong, 1988) or to different effects of differing patterns of exposure to the sun.

(iv) *Short periods of residence implying high potential exposure*

Several case–control studies have reported, usually as incidental findings, that subjects who have had a short period of residence in tropical or sub-tropical environments have an increased risk for melanoma (Table 19).

(v) *Occupational exposure*

Regular outdoor occupational exposure is probably the most convenient measure of relatively constant sun exposure and has been assessed with differing degrees of detail, from simple questions on ever/never or a basic amount of outdoor exposure, to detailed assessments involving assessments of clothing habits, geographical location of work and so on. The results appear to be inconsistent (Table 20). The more detailed studies, however, show more consistency, with a significant negative association, particularly in men, who constitute most of the highly exposed subjects (Table 21).

An overall irregular pattern was seen in western Canada, probably because individuals with relatively little occupational exposure are those who perform outdoor work seasonally or for short periods, often in early life, so that this exposure may be an indication of intermittent rather than constant exposure (Elwood *et al.*, 1985b). Such results are consistent with the effects of a short period of residence in a sunny place, as reviewed earlier. Paffenbarger *et al.* (1978) also showed that students who recorded outdoor work before college [presumably summer employment] had a significantly increased risk of melanoma in later life.

(vi) *Intermittent exposure*

To assess the effects of intermittent exposure, investigators have asked questions about specific activities that would be likely to represent relatively severe intermittent exposure, such as sunbathing, or asked particularly about holidays in sunny places, or used more complex questionnaires to attempt to assess total intermittent exposure through recreational or holiday activities. Most of these studies show positive associations, but few show large effects (Table 22).

In general, the more detailed studies show reasonably consistent positive results. For example, in western Canada, significant positive associations were seen with recreational and holiday sun exposures in activities involving reasonably intense sun exposure, such as beach activities (Elwood *et al.*, 1985b). In Denmark, rather similar relative risks of 1.5–1.9 were seen with regular participation in activities such as sunbathing, boating, skiing, swimming and vacations in sunny places (Østerlind *et al.*, 1988b). Significant positive associations with sunbathing were seen in the Swedish study of Beitner *et al.* (1990). In the study of Zanetti *et al.*

Table 18. Results of case–control studies on melanoma: total sun exposure assessed by questionnaire

Place	Direction of association	OR[a]	95% CI	p value	Measurement of exposure	Reference
USA	Up	2.5	NA	< 0.001	Sun exposure 2 h/day, 11–20 years previously	Rigel et al. (1983)
Australia	Up	5.3	0.9–30.8	NA	Total sun exposure throughout life > 50 000 h, adjusted	Green (1984)
Canada	Weakly up	1.2	0.7–2.0	> 0.1	Hours of sun exposure per year, p for trend	Elwood et al. (1985b)
USA	Down	0.6	0.4–0.9	< 0.05	Total sun exposure throughout life	Graham et al. (1985)[b]
USA	Weakly up	1.1	0.6–2.1	> 0.05	Hours of sun exposure 0–5 years previously, > 5 h/day	Dubin et al. (1986)
USA	Down	0.85	0.5–1.4	> 0.05	Hours of sun exposure 11–20 years previously, > 5 h/day	Dubin et al. (1986)
USA	Weakly up	1.1	0.8–1.6	> 0.05	Lifetime sun exposure	Dubin et al. (1986)
Australia	Down	0.7	0.4–1.1	0.13	Mean total outdoor hours/week in summer, > 23 h/week; p for trend	Holman et al. (1986a)
Italy	Down	0.7	0.4–1.1	> 0.05	Heavy or frequent exposure in previous 20 years	Cristofolini et al. (1987)
France	Up	3.4	1.6–7.1	< 0.05	Total lifetime outdoor sun exposure, adjusted	Grob et al. (1990)

[a]Odds ratio for maximal category
[b]Results calculated by Armstrong (1988)

Table 19. Evidence of melanoma risk with short periods of residence implying high potential exposure

Place	Direction of association	Odds ratio	95% CI	p value	Measurement of exposure	Reference
USA	Up	[7.7	2.5–23.6]	0.0002	US service: tropics *versus* USA/Europe	Brown *et al.* (1984)
UK	Up	1.8	0.6–5.1	> 0.05	≥ 1 year living in tropics, subtropics	Elwood (1986)
Scotland	Up	2.6 (males) 1.8 (females)	1.3–5.4 0.8–4.0	< 0.05 > 0.05	> 5 years living in tropics, subtropics; crude OR	MacKie *et al.* (1989)
Sweden	Up	1.9	1.0–3.6	< 0.05	Living in Mediterranean, tropics, subtropics > 1 year in last 10 years	Beitner *et al.* (1990)

Table 20. Results of case–control studies on melanoma: occupational exposure

Place	Direction of association	OR[a]	95% CI	p value	Measurement of exposure	Reference
USA	Up	3.9	NR	0.01	Outdoor work recorded at college medical examination; prospective	Paffenbarger et al. (1978)
Norway	Up	1.4	0.6-3.5	0.37	At least 3-4 h of outdoor work a day	Klepp & Magnus (1979)[b]
Scotland	Down	0.5	0.2-1.2	> 0.05	Hours of outdoor occupation a week	MacKie & Aitchison (1982)[b]
USA	Up	1.2	NR	> 0.05	Outdoor occupation versus indoor	Rigel et al. (1983)
Canada	Irregular	0.9	0.6-1.5	< 0.01	Hours of outdoor occupation a week in summer	Elwood et al. (1985b)
USA	Down	0.7	0.3-1.3	> 0.05	Lifetime hours of outdoor occupation	Graham et al. (1985)
USA	Up	2.5	1.4-4.4	< 0.05	Mostly outdoors; multiple logistic OR = 2.4, $p < 0.05$	Dubin et al. (1986)
UK	Irregular	1.7	0.3-8.6	0.5	Lifetime hours of outdoor occupation	Elwood et al. (1986)
Australia	Down	0.5	NR	0.04	Mean hours of outdoor occupation a week in summer	Holman et al. (1986a)
Denmark	Down	0.7	0.5-0.9	< 0.05	Outdoor occupation versus indoor	Østerlind et al. (1988b)
Italy	Irregular	2.1	0.6-6.8	0.32	Outdoor occupation	Zanetti et al. (1988)
Germany	Up	5.5	1.2-25.3	< 0.05	Outdoor occupation; adjusted OR = 11.6 (2.1-63.3)	Garbe et al. (1989)
Sweden	Down	0.6	0.4-1.0	NR	Outdoor occupation, yes/no	Beitner et al. (1990)
France	Up	6.0	2.1-17.4	< 0.05	Outdoor occupation versus indoor	Grob et al. (1990)

NR, not reported
[a]Odds ratio for maximal category
[b]Calculated by Armstrong (1988)

Table 21. Results of case–control studies on different types of melanoma and occupational exposure

Place	Type of melanoma	Odds ratio	95% CI	p value	Measurement of exposure	Reference
Canada	Excluding LMM and ALM	0.5	[0.3–1.0]	NR	> 32 h outdoor occupation a week in summer (men)	Elwood et al. (1985b)
Queensland, Australia	Excluding LMM and ALM	No association			Outdoor occupation	Green et al. (1986)
Western Australia	SSM	0.5	NR	0.04 for trend	Top quartile, hours of outdoor occupation a week in summer	Holman et al. (1986a)
Denmark	Excluding LMM and ALM	0.7	0.5–0.9	< 0.05	Outdoor occupation (men)	Østerlind et al. (1988b)

LMM, lentigo maligna melanoma; ALM, acral lentiginous melanoma; SSM, superficial spreading melanoma; NR, not reported

Table 22. Results of case–control studies on melanoma: intermittent exposure

Place	Direction of association	OR[a]	95% CI	p value	Measurement of exposure	Reference
Norway	Up	2.4	1.0–5.8	0.06	Sunbathing holidays in southern Europe in previous 5 years	Klepp & Magnus (1979)[b]
UK	Up	1.5	0.9–2.5	0.16	Spent some time deliberately tanning their legs	Adam et al. (1981)[b]
	Up	1.6	1.0–2.5	0.05	Spent some time deliberately tanning their trunk	
Scotland	Down	0.4	0.2–0.9	< 0.05	Hours a week in outdoor recreation	Mackie & Aitchison (1982)[b]
USA	Up	2.5	1.1–5.8	< 0.05	Days of vacation in a sunny warm place in childhood	Lew et al. (1983)
USA	Up	2.4	NR	0.01	Outdoor versus indoor recreation	Rigel et al. (1983)
Canada	Up	1.7	1.1–2.7	< 0.01	Hours of high exposure in recreational activities per week in summer	Elwood et al. (1985b)
	Up	1.5	1.0–2.3	< 0.01	Hours of high and moderate exposure in recreational activities per day in summer vacations	
	Up	1.7	1.2–2.3	< 0.001	Number of sunny vacations per decade	
UK	Up	5	NR	> 0.05	Number of holidays abroad in hot climate; adjusted	Sorahan & Grimley (1985)
USA	Irregular	1.7	1.2–2.2	< 0.01	Recreation type; multiple logistic OR, 1.0	Dubin et al. (1986)
Australia	Irregular	1.9	0.5–7.4	0.62	Recreational hours spent in sun on beach over whole life; crude RR	Green et al. (1986)
Australia	Up	1.3	0.9–1.9	0.25	Proportion of recreational outdoor exposure in summer at 10–24 years of age; p for trend	Holman et al. (1986a)
	Up	2.4	1.1–5.4	0.04	Boating in summer; p for trend	
	Up	2.7	1.2–6.4	0.07	Fishing in summer; p for trend	
	Irregular	1.1	0.7–1.8	0.66	Swimming in summer; p for trend	
	Up	1.3	0.8–2.2	0.26	Sunbathing in summer at 15–24 years of age; p for trend	
Denmark	Up	1.9	1.3–2.9	0.004	Sunbathing; crude RR; p for trend	Østerlind et al. (1988b)
	Up	1.7	1.1–2.8	0.012	Boating; crude RR; p for trend	
	Up	1.5	0.9–2.4	0.006	Skiing; crude RR; p for trend	
	Up	1.5	1.2–2.0	0.004	Swimming (outdoors); crude RR; p for trend	
	Up	1.7	1.2–2.4	< 0.01	Vacations in sunny resorts; crude RR; p for trend	

Table 22 (contd)

Place	Direction of association	OR	95% CI	p value	Measurement of exposure	Reference
Italy	Irregular	2.6	1.0–6.9	0.003	Years of outdoor sport (men); p for trend	Zanetti et al. (1988)
	Up	3.8	1.1–13.0	NR	High-exposure sports (men)	
	Irregular	1.9	0.6–5.8	0.27	Total weeks' vacation (men); p for trend	
	Up	3.7	1.4–9.7	0.001	Weeks' vacation near sea; early life (men); p for trend	
	Up	1.6	0.7–3.6	0.77	Weeks' vacation near sea; adult life (men); p for trend	
	Irregular	2.1	0.6–7.9	0.37	Years of outdoor sport (women); p for trend	
	Up	2.3	0.6–9.1	NR	High-exposure sports (women)	
	Irregular	1.1	0.5–2.4	0.56	Total weeks' vacation (women); p for trend	
	Up	1.2	0.6–2.5	0.56	Weeks' vacation near sea; early life (women); p for trend	
	Up	1.5	0.9–2.7	0.16	Weeks' vacation near sea; adult life (women); p for trend	
Germany	No association	NR	NR	NR	Free-time sun exposure	Garbe et al. (1989)
Sweden	Up	1.8	1.2–2.6	< 0.05	Number of sunbaths per summer	Beitner et al. (1990)
	Up	2.4	1.5–3.8	< 0.05	Sunbathing vacations abroad	
France	Up	8.4	3.6–19.7	< 0.05	Outdoor leisure exposure	Grob et al. (1990)

NR, not reported
[a]Odds ratio for maximal category
[b]Calculated by Armstrong (1988)

(1988) in Turin, Italy, positive associations were seen with doing an outdoor sport for many years and with number of weeks of holidays spent near the sea. These consistently positive associations contrast with the less consistent pattern seen in Australia. In Western Australia, stronger associations are seen with boating and fishing than with swimming and sunbathing, which would be expected to involve more intense exposure to the sun, and only a weak association was seen with the proportion of outdoor time spent on recreational activities in teenage and early adult years (Holman *et al.*, 1986a). In Queensland, Green *et al.* (1986) found only irregular associations with recreational hours spent at the beach or in other activities with intense exposure to the sun. This finding might be consistent with the concept that, in a sunny environment, recreational activities may involve sufficient frequency or intensity of sun exposure to result in a constant rather than an intermittent dose pattern.

(vii) *Sunburn*

Most of the studies show positive associations between risk for melanoma and a history of sunburn (Table 23). The questionnaires usually defined very severe sunburn as a burn that causes pain lasting for at least two days or blistering. The greater consistency of this relationship compared to that with intermittent exposure may indicate a specific association with sunburn *per se* or that sunburn is simply a more easily remembered measure of intermittent and/or intense exposure to the sun.

A history of sunburn indicates both unusually intense exposure and skin sensitivity, and therefore studies which assess sunburn while controlling for sensitivity through a separate question on tendency to burn are important. Both the western Canada and Western Australia studies when analysed in this way show that the association is primarily with tendency to burn rather than with a history of sunburn (Elwood *et al.*, 1985a; Holman *et al.*, 1986a). The studies in Queensland, Denmark and Scotland, however, show strong associations with sunburn history even after controlling for tendency to burn and other measures of skin sensitivity.

Because sensitivity to the sun and sunburn are likely to be highly correlated and both are likely to be measured with a degree of error, it is difficult to distinguish their effects. Similarly, sunburn is likely to be confounded with intermittent exposure of a less intense nature, from which it cannot readily be distinguished because of measurement error (Armstrong, 1988).

The study in England by Elwood *et al.* (1990) assessed sunburn at different ages and showed the strongest association with sunburn at ages 8–12; a stronger association with sunburns at young age was also seen by Weinstock *et al.* (1989) and by Østerlind *et al.* (1988b).

2.1.4 *Malignant melanoma of the eye*

(a) *Case reports*

In general, case reports were not considered, owing to the availability of more informative data.

Kraemer *et al.* (1987) reported on 830 cases of xeroderma pigmentosum, with a median age of 12 years at last observation, located through a survey of published case reports. Ocular abnormalities were found in 328 of 337 patients on whom information was available. Of these, 88 were reported to have some form of ocular neoplasm, mostly in the limbus, cornea and conjunctiva. Five of these patients were reported as having ocular melanoma; only one

Table 23. Results of case-control studies on melanoma: history of sunburn

Place	Direction of association	OR[a]	95% CI	p value	Measurement of exposure	Reference
Scotland	Up	2.8	1.1-7.4	< 0.05	Blistering sunburn or erythema persisting > 1 week	MacKie & Aitchison (1982)
USA	Up	2.1	1.2-3.6	< 0.05	Blistering sunburn during adolescence (yes/no)	Lew et al. (1983)
Canada	Up	1.8	1.1-3.0	< 0.01	Vacation sunburn score	Elwood et al. (1985a)[b]
Australia	Up	2.4	1.0-6.1	< 0.05	Number of severe sunburns throughout life	Green et al. (1985a)
UK	Up	4.2	NR	< 0.01	Bouts of painful sunburn; adjusted	Sorahan & Grimley (1985)
Canada	Up	3.2	1.7-5.9	< 0.001	Sunburn causing pain for ≥ 2 days	Elwood et al. (1986)[c]
Australia	Irregular	0.9	0.5-1.5	0.43	Sunburn causing pain for ≥ 2 days, during last 10 years	Holman et al. (1986a)[b]
	Up	1.2	0.6-2.3	0.1	Sunburn causing pain for ≥ 2 days, < 10 years of age	
	Up	1.7	1.0-2.9	0.003	Blistering sunburn	
Italy	Down	0.7	0.4-1.2	> 0.05	Severe sunburn in adolescence or early adult life (yes/no)	Cristofolini et al. (1987)
	Up	1.2	0.7-2.1	> 0.05	Sunburn as an adult (yes/no)	
USA	Up	3.8	1.4-10.4	NA	Number of blistering sunburns up to adult age, adjusted	Holly et al. (1987)
Denmark	Up	3.7	2.3-6.1	< 0.001	Sunburn causing pain for ≥ 2 days, < 15 years of age	Østerlind et al. (1988)
	Up	3.0	1.6-5.4	< 0.001	Sunburn causing pain for ≥ 2 days, during previous 10 years	
Italy	Up (men)	4.1	1.8-9.2	< 0.05	Sunburn in childhood (yes/no)	Zanetti et al. (1988)
	Up (women)	2.7	1.3-5.6	< 0.05		
Germany	No association	NR	NR	NR	Number of sunburns	Garbe et al. (1989)
Scotland	Up (men)	7.6	1.8-3.2	NR	Number of episodes of severe sunburn, any age, adjusted	MacKie et al. (1989)
	Up (women)	2.3	0.9-5.6	NR	Number of episodes of severe sunburn, any age, adjusted	

Table 23 (contd)

Place	Direction of association	OR[a]	95% CI	p value	Measurement of exposure	Reference
USA	Up	2.2	1.2–3.8	0.01	Number of blistering sunburns at ages 15–20	Weinstock et al. (1989)
Sweden	Up	1.7	1.0–2.9	NR	Erythema after sunbathing	Beitner et al. (1990)
UK	Up	3.6	1.4–11.2	< 0.05	Moderate sunburn at ages 8–12 (yes/no)	Elwood et al. (1990)
	No association	1.0	0.6–2.0	> 0.05	Moderate/maximum sunburn at ages 18–20 (yes/no)	
	Up	1.8	0.9–3.7	> 0.05	Moderate/maximum sunburn 18–20 yrs before diagnosis (yes/no)	
	Up	1.2	0.6–2.3	> 0.05	Moderate/maximum sunburn 5 years before diagnosis (yes/no)	

NR, not reported
[a]Odds ratio for maximal category
[b]Data calculated by Armstrong (1988)
[c]Exposure to fluorescent and other lighting sources

was specified as being of uveal origin. [The Working Group recognized that data collected from previously published case reports is not uniform and may not be typical of a true incidence or prevalence series. Furthermore, no information is available on the relationship between solar exposure and the occurrence of ocular melanoma in these patients.]

(b) *Descriptive studies*

As there is no separate ICD code for intra-ocular melanoma, descriptive data for cancer of the eye (ICD-9 190) as a whole have been used as a surrogate. Intra-ocular melanoma comprises some 80% of tumours of the orbit of the eye (Østerlind, 1987), and cancer of the eye has been used as a surrogate for adult ocular melanoma in previous studies (Swerdlow, 1983a,b).

(i) *Ethnic origin*

Examination of incidence figures from many parts of the world reveals higher rates of ocular tumours in whites than in blacks or Asians residing at the same latitude and under similar conditions (Waterhouse *et al.*, 1976; Muir *et al.*, 1987).

(ii) *Place of birth and residence*

When rates for whites are evaluated separately, no variation in incidence rates for ocular tumours is seen with decreasing latitude in the northern hemisphere (Table 24). Similarly, no incidence grading was seen among whites in the USA (Table 25). The more northerly states of Australia do not show higher incidence rates for ocular tumours than the southern states (Table 25).

Table 24. Trends in cancer of the eye for whites by latitude and by time period (rates per 100 000 age standardized to UICC 'world population')

Latitude	Area	~ 1968-72[a]		~ 1972-77[b]		~ 1977-82[c]	
		Men	Women	Men	Women	Men	Women
56°-61° N	Denmark	1.4	1.2	0.8	0.7	1.0	0.7
	Finland	0.9	1.0	0.9	0.7	1.0	0.7
	Sweden	1.3	1.2	0.9	0.8	0.9	0.6
47°-55° N	Canada						
	British Columbia	1.0	0.8	0.9	0.6	0.7	0.4
	Alberta	0.8	0.6	0.8	0.9	0.7	0.7
	Saskatchewan	1.3	0.8	1.1	1.0	1.0	0.7
	Manitoba	1.7	0.9	1.2	1.0	0.8	0.8
46° N	Geneva, Switzerland	0.4	0.2	0.8	1.1	0.6	1.1
38° N	San Francisco, CA, USA	0.9	0.9	0.9	0.5	0.9	0.8
35° N	New Mexico, USA	1.0	0.7	1.3	0.7	0.9	0.9
32°-38° S	Australia						
	New South Wales	NR	NR	0.8	0.8	0.9	0.5
	South Australia	NR	NR	0.9	1.0	0.7	0.6

Table 24 (contd)

Latitude	Area	~ 1968-72[a]		~ 1972-77[b]		~ 1977-82[c]	
		Men	Women	Men	Women	Men	Women
22 °S	Hawaii, USA	0.4	0.2	1.2	0.2	1.0	0.0
3 °S	Cali, Colombia	0.6	0.2	0.4	0.5	0.5	0.5

NR, not reported
[a]From Waterhouse et al. (1976)
[b]From Waterhouse et al. (1982)
[c]From Muir et al. (1987)

Table 25. Incidence of cancer of the eye (ICD-9 190) in US and Australian whites 1978-82 in various locations by latitude

Latitude	Location	Male rate/ 100 000	Female rate/ 100 000
USA			
47 °N	Seattle	0.9	0.8
42 °N	Detroit	0.7	0.6
42 °N	Iowa	1.0	0.7
41 °N	Connecticut	0.6	0.3
41 °N	New York City	0.5	0.4
41 °N	Utah	1.4	1.1
38 °N	San Francisco Bay Area	0.9	0.8
35 °N	New Mexico	0.9	0.9
34 °N	Los Angeles	0.7	0.6
33 °N	Atlanta	0.7	0.8
22 °N	Hawaii	1.0	0.0
Australia			
43 °S	Tasmania	1.2	0.8
38 °S	Victoria[a]	1.1	0.4
34 °S	South Australia	0.7	0.6
33 °S	New South Wales	0.9	0.5
32 °S	Western Australia	1.6	0.5
28 °S	Queensland[a]	0.6	0.7

From Muir et al. (1987); rates standardized to UICC 'world population'
[a]Data available only for 1982

Schwartz and Weiss (1988) compared the state of birth of 763 white (not of Spanish origin) US patients with uveal melanoma diagnosed between 1973 and 1984 and identified in nine cancer registries with those of the whites covered by the registries as recorded in the 1980 census. Patients with unknown or foreign birthplace or non-uveal ocular melanomas were excluded. Risk estimates were adjusted for age, sex and residence. The odds ratio for subjects born in the southern USA (south of 40 °N) was 1.1 (95% CI, 0.8–1.5). When states were classified according to average daily global solar radiation, a nonsignificant gradient was observed, only among women (odds ratio for > 15 500 kJ/m^2 *versus* ≤ 12 300 kJ/m^2, 1.6; 95% CI, 0.7–3.6).

Mack and Floderus (1991) examined birthplace and residence of patients diagnosed with intra-ocular melanoma among non-latino whites in 1972–82 in Los Angeles County. The proportional incidence ratio was not higher for cases born in California and Arizona than for those born in more northerly areas.

Doll (1991) observed a small rural excess in the incidence of cancer of the eye compared with urban residence, in a number of countries.

(iii) *Occupation*

Four studies of occupational mortality and one of incidence gave inconsistent results with regard to ocular cancer. Two investigations using proportional mortality ratios demonstrated more deaths from ocular cancer than expected among male farmers (Saftlas *et al.*, 1987; Gallagher, 1988), a group likely to have substantial exposure to solar UVR. These findings were not confirmed, however, in two other studies using similar methods (Milham, 1983; Office of Population Censuses and Surveys, 1986).

An investigation of ocular melanoma carried out on data from the cancer registry of England and Wales did not show an elevated incidence in farmers, but an increased risk was seen for professionals (relative risk, 124; 95% CI, 99–153), which was significant for teachers (177; 120–248) (Vågerö *et al.*, 1990).

(iv) *History of skin cancer*

Cancer registry-based studies (Østerlind *et al.*, 1985; Tucker *et al.*, 1985a; Holly *et al.*, 1991) found no or a nonsignificant (Lischko *et al.*, 1989) association between the occurrence of cancer of the eye and cutaneous melanoma or nonmelanocytic skin cancer. A single investigation of 400 sequential cases of uveal melanoma (Turner *et al.*, 1989) suggested that intra-ocular melanoma patients have an elevated frequency of prior cutaneous melanoma. Thus, although one study indicated a possible association, the overall evidence does not support an association between ocular melanoma and either melanoma or nonmelanocytic skin cancer.

(c) *Case–control studies*

Four case–control studies were evaluated. The first study (Gallagher *et al.*, 1985) evaluated all ocular melanomas, while the other three (Tucker *et al.*, 1985b; Holly *et al.*, 1990; Seddon *et al.*, 1990) studied uveal melanomas (excluding conjunctival melanomas).

Gallagher *et al.* (1985) conducted a study of ocular melanoma in patients diagnosed in the four western provinces in Canada between 1 April 1979 and 31 March 1981. Of the 90 ascertained cases, 87 were eligible by age for interview (20–79 years); of these, 65 cases (75%) were actually interviewed. For each case, a single control was randomly selected from the general population, matched by age (± 2 years), sex and province of residence. Response rates for controls were 59% for Alberta, Saskatchewan and Manitoba and 48% for British Columbia. Personal interviews were conducted in subjects' homes, and conditional logistic regression was used to control for matching variables and eye, hair and skin colour. No significant association was seen between ocular melanoma and either intermittent (occupational, recreational and holiday) or cumulative exposure to solar UVR. A strong association was detected between ocular melanoma and blue or grey iris colour (crude odds ratio, 3.0; $p = 0.04$) and blond or red hair colour (crude odds ratio, 7.7; $p = 0.03$). (In a multivariate analysis, these odds ratios became nonsignificant.) A nonsignificantly elevated risk (crude odds ratio, 2.8; $p = 0.08$) for ocular melanoma was also seen for subjects with light skin colour by comparison with subjects with darker skin.

A case–control study conducted by Tucker et al. (1985b) evaluated risk factors in 444 white patients with intra-ocular (uveal) melanoma treated at the Wills Eye Hospital in Philadelphia, USA, and 424 controls with detached retinas seen at the same centre. [The Working Group noted that use of a single disease category for the controls could introduce spurious associations with risk factors for that condition.] Response rates were 89% for cases and 85% for controls. Interviews were conducted by telephone; interviews were with next-of-kin for 17% of the cases and 14% of the controls. Logistic regression models were fitted which included sun-exposure variables, age, sex, eye colour and presence of cataracts, which was included to reduce bias in view of the association between cataracts and detached retina. Sunbathing appeared to increase the risk of intra-ocular melanoma, although no gradient of risk was noted with frequency of exposure (frequent versus never, odds ratio, 1.5; 95% CI, 0.9–2.3). A significantly elevated risk was detected for those who engaged in gardening (1.6; 1.0–2.4), but similar associations were not seen for other recreational outdoor activities, such as fishing, camping and hunting. Cases of intra-ocular melanoma also reported increased exposure to the sun during vacations in comparison with controls, with an odds ratio of 1.5 (95% CI, 0.97–2.3) for subjects 'frequently' experiencing increased exposure versus subjects never exposed (test for linear trend over four strata, $p = 0.01$). Cases reported less frequent use of eye protection (sunglasses, headgear, visors) when outdoors as compared with controls, but there was no dose–response relationship with frequency of use of these protective devices. A gradient of risk was seen with use of any eye shading when iris melanomas were examined separately, suggesting that eye shading may have been specifically important for lesions at the front of the eye (never versus occasional use of eye protection, odds ratio, 4.9; 95% CI, 1.4–13.7). [Numbers of iris melanomas were not given.] Subjects who were born in the southern USA (lower than 40 °N latitude) were found to have a significantly elevated risk of intra-ocular melanoma (2.7; 1.3–5.9) after adjustment for number of years spent in the south and for the presence of cataracts; with adjustment for all other sun-related variables, the odds ratio was 3.2 (95% CI, 1.8–5.7). The association persisted after excluding subjects not living close to Philadelphia. There was no relation between the number of years spent in the south and the risk of intraocular malignant melanoma, after adjustment for having been born in the south. Blue-eyed subjects had the highest risk of intra-ocular melanoma, with grey-green and hazel-eyed subjects at intermediate risk, and brown-eyed subjects at lowest risk (unadjusted odds ratio for brown- versus blue-eyed subjects, 0.6; 95% CI, 0.4–0.8). Cases were more likely than controls to have fair skin and blond or brown hair, although no odds ratios are given and the differences disappeared when eye colour was taken into account. Cases were also more likely to have 25 or more freckles (used as an indirect measure of sun exposure and sensitivity) than controls (odds ratio, 1.4; 95% CI, 1.0–2.0).

A case–control study by Holly et al. (1990) involved 407 white cases of uveal melanoma and 870 controls. The cases were diagnosed between January 1978 and February 1987 at the

Ocular Oncology Unit of the University of California, San Francisco, USA, were aged 20–74 at diagnosis and lived in 11 western states. Controls were selected by random digit dialling and were matched to cases on age and area of residence. Telephone interviews were conducted by interviewers unaware of the study hypotheses, most cases being interviewed within four years of their diagnosis. The response rate was 93% of cases and 77% of eligible controls. No clear association was seen between uveal melanoma and vacation time spent in sunny climates or high proportion of leisure time spent outdoors. Individuals who spent 50% of their leisure time indoors and 50% outdoors had a reduced risk for uveal melanoma (odds ratio, 0.6; 95% CI, 0.4–0.9) when compared to subjects who stayed mainly indoors. Significantly elevated risks were seen in subjects with grey, green, hazel or blue eyes, compared to those with brown eyes, with increasing frequency of large naevi (≥ 7 mm) ($p = 0.04$ for trend) and with a propensity to burn rather than tan in the sun.

Seddon *et al.* (1990) compared 197 white patients with uveal melanoma diagnosed in 1984–87, who were resident in the six New England states close to the Massachusetts Eye and Ear Infirmary, with 385 controls obtained through random digit dialling and matched to cases by age ($\pm$ 8 years), sex and area of residence. All subjects were interviewed by telephone using a standard questionnaire. The response rate was 92% among cases, and 85% of the eligible controls contacted agreed to participate in the study. Matched logistic regression techniques were employed to evaluate potential associations between exposure to UVR and risk of uveal melanoma, adjusting for age, sex, constitutional factors and socio-economic variables. An inverse association with southern birthplace (south of 40 °N latitude) was detected (odds ratio, 0.2; 95% CI, 0.0–0.7) after adjustment for constitutional and other factors. When cumulative lifetime residence in the south was examined, subjects who had lived for more than five years south of 40 °N had an odds ratio of 2.8 (95% CI, 1.1–6.9) after adjustment for birthplace. Several indices of sun exposure were computed for each subject. The first combined duration of residence in the north or south with self-reported severity of sun exposure (low, medium, high). Subjects in the highest exposure group appeared to have a higher risk of uveal melanoma by comparison with those in the lowest exposure category (1.7; 0.9–3.0) although no dose–response relationship was seen over the three categories of exposure. A further index was obtained by taking average values of solar radiation for each state in which the subject has resided and multiplying this value by the duration of residence within the state and the reported amount of time spent in the sun. No association was seen between this index and risk of uveal melanoma. Individuals who reported having spent a great deal of time working outdoors 15 years prior to diagnosis showed a somewhat lower risk of uveal melanoma than those who worked minimally outdoors or were retired (odds ratio, 0.6; 95% CI, 0.3–1.4) after control for age, skin, eye colour and southern residence. No association was seen with sunbathing, use of sunglasses or visors, or outdoor hobbies all conducted 15 years prior to diagnosis. Use of eye glasses was not related to uveal melanoma risk. Cases reported more cutaneous naevi and lighter skin colour than controls and were more likely to be of northern European or British ancestry than controls. An expanded analysis comparing 387 cases of uveal melanoma with 800 sibling controls was also conducted. There was a gradient of risk with cumulative years of intense sun exposure; the odds ratio for the highest exposure was 2.1 (1.4–3.2).

2.1.5 Other cancers

No adequate data were available to the Working Group.

2.2 Artificial sources of ultraviolet radiation

Epidemiological investigations that have attempted to assess exposure to artificial sources of UVR have neither measured actual UVR nor considered the emission spectra. It is presumed that in the studies described below, subjects were exposed to sources that varied in intensity and emission spectra.

2.2.1 Nonmelanocytic skin cancer

Three case–control studies, described in detail on p. 84, addressed this issue. In the study in Montréal, Canada, of Aubry and MacGibbon (1985), any use of a sunlamp gave an odds ratio of 13.4 [95% CI, 1.4–130.5] after adjustment for sun exposure and constitutional factors. O'Loughlin *et al.* (1985) in Ireland found that fewer cases than controls reported frequent exposure to 'artificial sunlight' (nonsignificant). In the study of Herity *et al.* (1989) in Ireland, a smaller proportion of cases than of controls reported ever having used sunlamps or sunbeds ($p = 0.2$).

2.2.2 Malignant melanoma of the skin[1]

The results of case–control studies of exposure to fluorescent light and melanoma are summarized in Table 26.

Beral *et al.* (1982) conducted a case–control study in Sydney, Australia, of 274 female cases aged 18–54 identified at a melanoma clinic between 1978 and 1980 and 549 hospital and population controls matched by age and, for population controls, residence. The response rate for cases was 71% [response rates for controls not given]. Each job lasting 12 months or longer was recorded, together with information about whether the work had been carried out predominantly indoors or outdoors, whether fluorescent lighting was present, and whether the fluorescent lights were switched on most of the time or less frequently. Among women who always worked indoors, the odds ratio increased with duration of working with fluorescent lights most of the time to a maximum of 2.6 (95% CI, 1.2–5.9) for 20 or more years' exposure. The effect was greater for office workers (odds ratio, 4.3) than for other indoor workers (2.0). Stratification by amount of time spent outdoors, main outdoor activity and amount of clothing worn, history of sunburn, place of birth, hair colour and skin colour did not diminish the association. Among cases exposed to fluorescent lights, there was a relative excess of melanomas on the trunk (a site likely to be covered at work); 24% in exposed cases *versus* 4% in unexposed cases. [The Working Group noted that crude estimates of sun exposure were used.]

Rigel *et al.* (1983) conducted a case–control study in New York, USA, described on p. 106. Cases had had shorter average daily exposure to fluorescent lights (4.9 h) than had

---

[1] After the meeting, the Secretariat became aware of a study by Walker *et al.* (1992) on the risk of cutaneous malignant melanoma associated with exposure to fluorescent light.

controls (5.4 h). Among office workers, average daily exposures were similar for cases and controls. The crude odds ratio for any exposure was 0.7 among all subjects and 0.6 among office workers.

English et al. (1985) conducted a study in 1980-81 of the exposure to fluorescent light of 337 cases and 349 age-matched controls who had already participated in a population-based case-control study in Western Australia (see Holman and Armstrong (1984a), p. 100). The response rate was 68% for cases and 91% for controls. Detailed information was obtained from telephone interviews about lifetime hours of residential and occupational exposure, the distance to the nearest light fixture and the presence of diffusers. Neither the duration of occupational exposure, the rate of total exposure (hours/year) nor cumulative total exposure was associated with risk for melanoma. Analyses by body site showed no consistent association with exposure to lights without diffusers. Adjustment for measures of total and intermittent exposure to the sun did not alter the results. Subjects were also asked about exposure to plan printers, laboratory equipment emitting UVR, insect tubes, black lights and photocopiers. No association was seen with any of these sources, although the number of exposed subjects was small. The odds ratio for any use of sunlamps was 1.1 (95% CI, 0.6-1.8), although few subjects had used sunlamps (Holman et al., 1986b).

Sorahan and Grimley (1985) examined fluorescent light exposure in 1980-82 in a case-control study in the United Kingdom, described in detail on p. 103. Information on exposure was confined to whether lights were 'mainly on' or 'sometimes on' at work. After adjustment for age and sex, no consistent association was seen for duration of exposure when cases were compared with electoral register controls.

Dubin et al. (1986) examined fluorescent light exposure in a subset of subjects in a case-control study in New York, USA, described on p. 108. Subjects were interviewed and/or sent postal questionnaires. In data obtained from interview, but not in data obtained from postal questionnaires, the odds ratios increased with average daily exposure in the five years before interview, after adjustment for age and sex (p value for linear trend, < 0.05). A similar pattern was seen for exposure 6-11 years and 11-20 years previously.

Elwood et al. (1986) examined fluorescent light exposure in their case-control study in the United Kingdom in 1981-84, described in detail on p. 103. Subjects were interviewed and later sent postal questionnaires to validate the responses. From the interview data, exposure to undiffused lights at work was associated with an odds ratio of 4.0 (95% CI, 0.8-19.2) for those maximally exposed (p value for trend $= 0.2$). Control for constitutional factors did not change the results. From the questionnaire data, the odds ratio for maximal exposure (undiffused lights) was 1.9 (95% CI, 0.4-8.4). No association was seen with exposure at home, and no association was seen for use of sunlamps. Subjects were also asked about exposure to particular or unusual light sources, such as vacuum or discharge lamps, insecticidal or germicidal lamps or welding equipment. The odds ratio for exposure to any such source was 2.2 (95% CI, 1.0-4.9). [The Working Group noted that the use of open-ended questions about lighting sources may have introduced recall bias.]

In the Western Canada case-control study in 1979-81 (see Elwood et al., 1984, 1985a,b, p. 107), no association was seen with use of sunlamps ($\chi^2 = 6.1$, 5 df) (Gallagher et al., 1986).

Østerlind et al. (1988b) examined exposure to fluorescent lighting at work and use of sunlamps and sunbeds in their case–control study in Denmark in 1982–85, described on pp. 103–104. The same proportions of cases and controls reported having been exposed to fluorescent lights at work, and no association was seen with age at first exposure, duration of exposure or type of work place. Past use of sunlamps was also not associated with melanoma, and a smaller proportion of cases than controls had ever used sunbeds (odds ratio, 0.7; 95% CI, 0.5–1.0).

In a case–control study in Scotland (Swerdlow et al., 1988), 180 cases aged 15–84 from three clinics during 1979–84 were compared with 197 age- and hospital-matched patients with various non-malignant diseases. Subjects were interviewed about exposure to fluorescent lights and UV lamps, use of sunbeds, sun exposure and constitutional factors. Controls with skin conditions were excluded from the analysis of UV lamps and sunbeds. No consistent association was seen with exposure to fluorescent lights at home or at work, with or without adjustment for constitutional factors and sun exposure. Significant, positive associations were seen for duration of use of UV lamps and sunbeds (p value for trend, < 0.05). The odds ratio for use for more than one year was 3.4 (95% CI, 0.6–20.3) after adjustment for constitutional factors and sun exposure. Amount of use within five years (1.9; 0.6–5.6) of the interview and more than five years (9.1; 2.0–40.6) before the interview were both positively associated with the risk for melanoma.

MacKie et al. (1989) examined use of sunbeds and sunlamps in their case–control study in Scotland described on p. 106. Use was associated with melanoma in men (odds ratio, 2.6; 95% CI, 0.9–7.3) but showed little association in women (1.5; 0.8–2.9). The effect on men largely disappeared after adjustment for sun exposure and constitutional factors.

In the study of Zanetti et al. (1988) from Turin, Italy, described in detail on p. 104, an odds ratio of 0.9 (0.4–2.0) was found for use of UVA lamps, although few subjects reported exposure.

A large population-based case–control study on occupational exposures was conducted during 1979–85 in Montréal, Canada (Siemiatycki, 1991). Overall, there were 3730 male cases of cancer aged 35–70, including 124 cutaneous melanoma cases; the participation rate was 82%. Each cancer site was compared with the other cancer sites. Exposure to 293 agents, including arc welding fumes and UVR, was assessed by a team of chemists and industrial hygienists on the basis of each individual's occupational history. Neither arc welding fumes nor exposures to UVR was associated with the risk for cutaneous melanoma (odds ratios, 0.5; 90% CI, 0.3–1.1 and 0.3; 0.1–1.5, respectively).

In a population-based study in southern Ontario, Canada (Walter et al., 1990), 583 cases identified from pathology laboratories and from the cancer registry between 1984 and 1986 were compared with 608 controls randomly sampled from property tax rolls. Participation rates were 90% for cases and 80% for controls. Odds ratios for any use of sunbeds or sunlamps were 1.9 (95% CI, 1.2–3.0) in men and 1.5 (0.99–2.1) in women. Adjustment for constitutional factors did not affect the results. The odds ratios increased with duration of use; for more than 12 months' use, the odds ratios were 2.1 (0.9–5.3) in men and 3.0 (1.1–9.6) in women.

Table 26. Case–control studies of melanoma of the skin and exposure to fluorescent lights

Country	Cases/controls	Odds ratio	95% CI	Definition of exposure	Reference
Australia	274/549	2.6[a,b] 4.3[a,b]	1.2–5.9 NR	Indoor workers, ≥ 20 years' occupational exposure Office workers, ≥ 20 years' occupational exposure	Beral et al. (1982)
USA	114/228	0.7 0.6	NS NS	Any exposure Any exposure, office workers	Rigel et al. (1983)
Australia	337/349	1.2[a,b] 1.2[a,b] 1.3[a,b] 1.2[a,b] 1.2[a,b]	0.8–1.9 0.7–1.9 0.8–1.9 0.8–1.9 0.6–2.6	≥ 35 000 h exposure ≥ 1600 h per year ≥ 22 500 h undiffused lights ≥ 1300 h per year undiffused lights ≥ 22 500 h head, neck, upper limbs, undiffused lights	English et al. (1985)
United Kingdom	58/333	0.6[a] 0.5[a]	NR NR	≥ 20 years, occupational exposure (mainly on) ≥ 20 years, indoor workers only (mainly on)	Sorahan & Grimley (1985)
USA	1103/585 508/222	2.3[a] 0.6[a]	1.0–5.8 0.3–1.3	≥ 9 h per day, 0–5 years previously (interview) ≥ 9 h per day, 0–5 years previously (postal questionnaire)	Dubin et al. (1986)
United Kingdom	83/83	1.4[a,b] 4.0[a,b]	0.4–5.1 0.8–19.2	≥ 50 000 h occupational exposure (total fluorescent light, interview) ≥ 50 000 h occupational exposure (undiffused lights, interview)	Elwood et al. (1986)
	67/66	1.2[a,b] 1.9[a,b]	0.3–5.7 0.4–8.4	≥ 50 000 h occupational exposure (total fluorescent light, postal questionnaire) ≥ 50 000 h occupational exposure (undiffused lights, postal questionnaire)	
Denmark	474/926	No association		Duration of exposure, age at first exposure, type of workplace	Østerlind et al. (1988b)
Scotland, United Kingdom	180/197	1.2[b] 0.8[b] 1.6[b] 1.4[b] 0.8[b]	0.7–1.9 0.4–1.4 0.9–2.6 0.9–2.3 0.4–1.4	Any occupational exposure < 5 years previously Any exposure at home < 5 years previously ≥ 5 h per day < 5 years previously at work and at home Any occupational exposure > 5 years previously Any residential exposure > 5 years previously	Swerdlow et al. (1988)

NR, not reported; NS, not significant
[a]Odds ratio for category with highest level of exposure
[b]Adjusted for sun exposure

2.2.3 *Malignant melanoma of the eye*

In the case–control study carried out in Philadelphia, USA, which is described in detail on p. 128, cases of uveal melanoma were more likely to report use of sunlamps than controls. After adjustment for age, eye colour and a history of cataracts, there was a trend to increasing risk with frequency of use (odds ratio for frequent *versus* never, 2.1; 95% CI, 0.3–17.9; test for linear trend over four levels: $p = 0.10$). The odds ratios for those who had ever worked as welders was 10.9 (2.1–56.5) (Tucker *et al.*, 1985b).

In the case–control study from San Francisco, USA, described on pp. 128–129, exposure to artificial UV light or 'black light' [details not given] conferred over three-fold risks for intra-ocular melanoma after adjustment for other significant factors (odds ratio, 3.7; 95% CI, 1.6–8.7). The odds ratios were 2.9 for 1–5 years of exposure and 3.8 for 6 or more years (Holly *et al.*, 1990).

In the case–control study from Boston, USA (Seddon *et al.*, 1990), described on p. 129, exposure to fluorescent lighting was associated with an elevated risk of uveal melanoma (odds ratio, 1.7; 95% CI, 1.1–2.5 for 40 h or more per week as compared to no exposure) in the larger data set, based on case–sibling comparison. In the population-based comparison, the corresponding odds ratio was 1.2 (95% CI, 0.6–2.1). A history of working with welding arcs was reported with similar frequency among cases and controls in both comparisons. Cases reported more frequent use of sunlamps in comparison with both sets of controls. After adjustment for constitutional factors and exposure to the sun, the odds ratios for frequent/occasional use *versus* never were 3.4 (1.1–10.3) in the population comparison and 2.3 (1.2–4.3) in the sibling comparison.

In the large Canadian study on occupational exposure, described on p. 132, 23 cases of ocular melanoma were included. Analysis only of French Canadians revealed four cases of eye melanoma with exposure to arc welding fumes (odds ratio, 8.3; 90% CI, 2.5–27.10) (Siemiatycki, 1991). No increase was found for substantial exposure; no increase in risk was reported for exposure to UVR.

2.3 Premalignant conditions

2.3.1 *Basal-cell naevus syndrome*

Basal-cell naevus syndrome is a hereditary condition (Gorlin, 1987) in which affected family members may show, among other major manifestations, an apparent excess of basal-cell carcinomas. These seem to occur more commonly in sun-exposed parts of the body or in unusual patterns. There is no other evidence that solar radiation plays a role in their development.

2.3.2 *Dysplastic naevus syndrome*

Dysplastic naevus syndrome is a hereditary condition in which affected family members have multiple dysplastic naevi and a greatly increased risk of malignant melanoma (Green *et al.*, 1985b). The distribution of tumours conforms to the usual distribution, and there is anecdotal evidence that solar radiation plays a role in their development (Kraemer & Greene, 1985).

2.4 Molecular genetics of human skin cancers

Analysis of mutations in DNA isolated from tumours and believed to be relevant to carcinogenesis can potentially help in making a causal link with exposures to carcinogens. Two important qualifications must, however, be borne in mind. Firstly, the changes detected may have arisen late in tumour development (whether or not the tumour is the result of exposure to UVR) and may not be involved in initiation or other early steps. Secondly, the spectrum of mutations that is seen may be constrained to those changes that can lead to a functional gene product. This qualification applies, for example, to mutations that activate *ras* genes but to only a lesser extent to tumour suppressor gene mutations in which inactivation of gene function is involved.

Experimental studies indicate that UV-induced mutations have a distinctive pattern of base-substitution mutations (see section 4.5):

- Virtually all mutations occur at dipyrimidine sites, especially 5'TC and 5'CC sequences.
- The majority of the base substitution mutations involve cytosine with the C→T transition predominating.
- Tandem 5'CC→5'TT mutations occur.

2.4.1 ras *Gene mutations*

Primary melanomas, metastases and cell lines derived from melanomas which developed at body sites characterized as exposed 'rarely', 'intermittently' or 'continuously' to the sun were analysed for the presence of N-*ras* mutations. Of 37 cutaneous melanomas, seven had N-*ras* mutations; all were from 'continuously' exposed sites. All mutations in the N-*ras* gene were at TT or CC sites, which are potential locations for mutagenic UV photoproducts, suggesting a role of sun exposure in N-*ras* mutation (van't Veer *et al.*, 1989).

In several investigations, base-substitution mutations were found in Ha-, Ki- and N-*ras* genes in human skin melanomas (Table 27) and in squamous-cell and basal-cell carcinomas (Table 28) from xeroderma pigmentosum and normal patients. In single studies, Ha- and N-*ras* gene amplification was found in squamous-cell carcinomas of the skin (Ananthaswamy & Pierceall, 1990), and loss of the Ha-*ras* allele was seen in basal-cell and squamous-cell carcinomas (Ananthaswamy *et al.*, 1988). Whether exposure to the sun was involved in tumour induction in these studies is, however, less clear.

2.4.2 p53 *Gene mutations*

Brash *et al.* (1991) found p53 mutations at various codons in 14 out of 24 (58%) invasive squamous-cell carcinomas from sun-exposed skin (Table 29). The mutations found were predominantly C→T (5 of 14 total mutants, 36%) and CC→TT (3 of 14, 21%) transitions, exclusively at tandem pyrimidine stretches. This finding is consistent with the hypothesis that these mutations are induced by UV irradiation. CC→TT double-base changes in the p53 gene have not yet been found in tumours in any internal organ. These results strongly suggest that solar radiation plays a role in the induction of p53 gene mutations.

Pierceall *et al.* (1991) found p53 mutations in exon 7 in 2 out of 10 squamous-cell carcinomas from sun-exposed body sites; one was a C→T transition and the other a C→A transversion.

Table 27. *ras* Gene mutations detected in human naevi and primary and secondary melanomas that developed at sites subject to sun exposure

Oncogene codon	Base change	Base-substitution mutation	Site of original tumour	Reference
N-*ras*-61	G*GA CAA GAA*			
	AAA	C to A	Neck	van't Veer et al. (1989)
	AAA	C to A	Lower leg	van't Veer et al. (1989)
	AAA	C to A	Nose	van't Veer et al. (1989)
	AAA	C to A	Cheek	van't Veer et al. (1989)
	CGA	[T to C]	Lower leg	van't Veer et al. (1989)
	CAT	[T to A/G]	Xeroderma pigmentosum patient[a]	Keijzer et al. (1989)
	CAT	[T to A]	Site unspecified, probably metastasis	Sekiya et al. (1984)
N-*ras*-13	G*GT GGT GTT*			
	GAT	[C to T]	Finger	van't Veer et al. (1989)
	GTT	[C to A]	Finger	van't Veer et al. (1989)
	GTT	[C to A]	Lower leg	van't Veer et al. (1989)
N-*ras*-12	GAT	[C to T]	Leg	van't Veer et al. (1989)
N-*ras*-61	CAT/C	[T to A/G]	Back	Shukla et al. (1989)
Ki-*ras*-61	G*GA CAA GAA*			
	AAA	C to A	Lower leg	Shukla et al. (1989)
Ki-*ras*-12	G*CT GGT GGC*			
	TGT	[C to A]	Abdomen	Shukla et al. (1989)
	TGT	[C to A]	Knee	Shukla et al. (1989)
	TGT	[C to A]	Site unspecified, probably metastasis	Shukla et al. (1989)
	TGT	[C to A]	Site unspecified, probably metastasis	Shukla et al. (1989)
	TGT	[C to A]	Site unspecified, probably metastasis	Shukla et al. (1989)
	TGT	[C to T]	Buttock	Shukla et al. (1989)
		[C to T]	Site unspecified, probably metastasis	Shukla et al. (1989)
		[C to T]	Forearm (naevus)	Shukla et al. (1989)
		[C to T]	Abdomen (naevus)	Shukla et al. (1989)
Ha-*ras*-12	GCC *GGC GGT*			
	TGC	[C to A]	Abdomen	Shukla et al. (1989)

Italics indicate potential pyrimidine dimer site including neighbouring codon; [ ], base changes occurring in anti-sense strand

[a]Malignant melanoma probably resulting from metastasis of a primary skin tumour

Table 28. *ras* Gene mutations detected in human keratoacanthomas (KA), basal-cell carcinomas (BCC) and squamous-cell carcinomas (SCC) that developed at sites subject to sun exposure

Oncogene codon	Base change	Base-substitution mutation	Tumour	Site	Reference
Ki-*ras* 12	GCT *GGT* GGC 　　TGT	[C to A]	SCC	Lip	van der Schroeff *et al.* (1990)
			BCC	Shoulder	van der Schroeff *et al.* (1990)
			BCC	Neck	van der Schroeff *et al.* (1990)
	GAT	[C to T]	BCC	Face	van der Schroeff *et al.* (1990)
Ha-*ras* 61	*GGC CAG GAG* 　　CTG	[T to A]	SCC	Not specified	Corominas *et al.* (1989)
	CTG	[T to A]	KA	Not specified	Corominas *et al.* (1989)
	CAT	[C to A]	BCC	Face	van der Schroeff *et al.* (1990)
	AAG	C to A	KA	Not specified	Corominas *et al.* (1989)
Ha-*ras* 12	GCC *GGC GGT* 　　AGC	[C to T]	SCC	Not specified	Corominas *et al.* (1989)
	AGC	[C to T]	KA	Not specified	Corominas *et al.* (1989)
	AGC	[C to T]	KA	Not specified	Corominas *et al.* (1989)
	TGC	[C to A]	SCC	Not specified	Corominas *et al.* (1989)
	TGC	[C to A]	SCC	Not specified	Corominas *et al.* (1989)

Italics indicate potential pyrimidine dimer site including neighbouring codon; [ ], base changes occurring in anti-sense strand

Table 29. p53 Tumour suppressor gene mutations in human squamous-cell carcinomas that developed at sites subject to sun exposure

Codon	Nucleotide sequence	Base-substitution mutation	Incidence[a]	Site of tumour origin	Reference
7	TCT	TGT; C→G	1/14/24	Preauricular	Brash *et al.* (1991)
56	*T* TCA	TAA; C→A	1/14/24	Chest	Brash *et al.* (1991)
104/105	C*G* CCT	deletion of a C	2/14/24	Preauricular/temple	Brash *et al.* (1991)
151	*CCC* CC	CAC; C→A	1/14/24	Scalp	Brash *et al.* (1991)
152	CC *CCC*	CAC; C→T	1/14/24	Hand	Brash *et al.* (1991)
179	A *CCA*	CAA; C→A	1/14/24	Scalp	Brash *et al.* (1991)
244	*CCG* G	TCG; C→T	1/2/10	Face	Pierceall *et al.* (1991)
245	G *CCG*	CAG; C→A	1/14/24	Cheek	Brash *et al.* (1991)
245	G *CCG*	T T; CC→TT	1/14/24	Chest	Brash *et al.* (1991)
247/248	AC *CG*	T T; CC→TT	1/14/24	Nose	Brash *et al.* (1991)
248	G*CC*	GAC; C→A	1/2/10	Face	Pierceall *et al.* (1991)
258	*T* TCC	TTC; C→T	1/14/24	Face	Brash *et al.* (1991)
278	*T* CCT	TCT; C→T	1/14/24	Cheek	Brash *et al.* (1991)
285/286	T*C CT*	T T; CC→TT	1/14/24	Face	Brash *et al.* (1991)
286	TC *CT*	CTT; C→T	1/14/24	Forehead	Brash *et al.* (1991)
317	CC *CCA*	TCA; C→T	1/14/24	Postauricular	Brash *et al.* (1991)

Italics indicate potential pyrimidine dimer site

[a]No. of specific mutations/no. of total mutations found/Total number of samples tested only from sites continuously exposed to the sun

3. Studies of Cancer in Animals

3.1 Experimental conventions

3.1.1 *Species studied*

The experimental induction of skin cancers in mice following exposure to a mercury-arc lamp was first reported by Findlay (1928). Initially, haired albino mice were used, but hairless and nude mice are now preferred.

An important development was the use of the hairless mouse as a model (Winkelmann *et al.*, 1960, 1963). In haired animals, the fur provides effective protection of the skin against UVR. This limits investigations to sparsely haired skin regions, mainly the ears, as, in long-term experiments with frequent exposures, the mechanical trauma caused by shaving might influence the process of tumorigenesis. The skin of hairless mice differs, however, from human skin in many respects. It is, for instance, much thinner and has abnormal hair follicles. The hairless mouse does, however, have a thymus and a functioning immune system, in contrast to the nude mouse (Eaton *et al.*, 1978; Hoover *et al.*, 1987). Many recent studies on carcinogenesis induced by UVR used the hairless mouse model (Forbes *et al.*, 1981; de Gruijl *et al.*, 1983; Gallagher *et al.*, 1984b). The changing designations of 'Skh' mice are listed in Table 30. Skin tumorigenicity has been evaluated experimentally in only a relatively small number of species other than the mouse.

Table 30. Alternative designations used for 'Skh' outbred stocks of hairless mice

Phenotype	1970–86	After 1986	Synonyms used in the literature	Inbred strains derived from Skh:hr stock[a]
Albino[b]	Skh: hairless-1	Skh:hr I	Sk-1; Skh-1; Skh/Hr-1; Skh:HR; HRA/Skh-1; Skh-hr1	HRA/Skh (Temple University, Philadelphia, PA, USA)
Pigmented[c] (any colour)	Skh: hairless-2	Skh:hr II	Sk-2; Skh-2; Skh/Hr-2	HRA/Skh-1 (University of Sydney, Sydney, Australia)

[a]From Forbes *et al.* (1990)
[b]Forbes *et al.* (1981); de Gruijl *et al.* (1983)
[c]Davies & Forbes (1988)

3.1.2 *Wavelength ranges*

As noted in section 1.1, for the purposes of this monograph, the UV wavelength range is subdivided according to the convention of the Commission Internationale de l'Eclairage (1987) into: UVA (315–400 nm), UVB (280–315 nm) and UVC (100–280 nm). The UVB range is generally found to be most effective in inducing skin cancer, i.e., tumorigenesis may be achieved with smaller doses of radiant exposure than with UVA and UVC. A complete discussion of wavelength ranges is given in section 1.1.

3.1.3 *Measured doses*

Many investigators of the carcinogenicity of UVR have reported the type of lamps they used, which are frequently broad-spectrum lamps, sometimes in combination with filters. When estimates of the doses of UVR administered are given, the measuring instrument is usually mentioned and the result is given in terms of irradiance or dose, with no further detail. Such information is of some value, especially for comparing the results of experiments in which the same type of lamps were used.

The action spectrum (see section 1) given in Figure 10 shows that the carcinogenic effectiveness of UVR in hairless mice changes steeply, even by orders of magnitude, over a wavelength range of 10 or 20 nm. This pattern indicates that irradiance must be spectrally specified in order to be meaningful, and not integrated into one value over a broad spectrum. One approach is to give irradiance weighted according to the action spectrum for UV carcinogenesis, but this is available only in provisional form (see Fig. 10 and discussion on pp. 46–47). Another approach is to provide data on erythemally weighted irradiance, since the action spectrum for erythema corresponds approximately to that for carcinogenesis (Forbes *et al.*, 1978). A simple, direct way of calculating this is to relate the doses administered to the minimal erythema dose or to the minimal oedemic dose for the animal being investigated. When investigators supplied such measures of effect, they are mentioned in the summaries below.

In experimental situations, there is never a perfectly sharp cut-off of wavelengths. The expression 'mainly UVA' is of questionable value, because even if UVB represents only 0.1% of the emission spectrum, it may still dominate the effect (see pp. 144–147, 151 and Fig. 10). Terms such as 'mainly UVB' are used below only when there are good reasons to assume that the effects considered are due mainly to UVB radiation.

3.1.4 *Protocols*

Experimental investigations on the carcinogenicity of UVR, conducted mostly on mice, have been reviewed (Blum, 1959; Urbach *et al.*, 1974; Kripke & Sass, 1978; WHO, 1979; van der Leun, 1984; Epstein, 1985).

Hundreds of studies have been reported. Most were not designed to test whether or not the radiation used was carcinogenic *per se* but to investigate the process of UV carcinogenesis. The methods used in these studies differ in many respects from those in standard lifetime studies to evaluate the carcinogenicity of chemicals. For example, many studies do not give complete details of the UVR emission spectrum used or exposure dose, do not enumerate all tumours, do not provide data on survival or do not provide histological details of tumours. Control groups are not always included; however, spontaneous skin tumours are rare in mice and rats. In many of the studies presented in detail below, appropriate statistical analyses have been done demonstrating clear dose-related trends in numbers of tumour-bearing animals, number of tumours per animal and/or median time to first tumour.

Fig. 10. Sterenborg–Slaper action spectrum for ultraviolet-induced skin carcinogenesis (1.0-mm tumours) in albino hairless mice. Effectiveness is defined as the reciprocal of the daily dose at each wavelength that leads to tumours of 1-mm diameter in 50% of animals in 265 days, relative to the corresponding value at the wavelength of maximal effectiveness. The effectiveness between 340 and 400 nm represents an average value for that wavelength range.

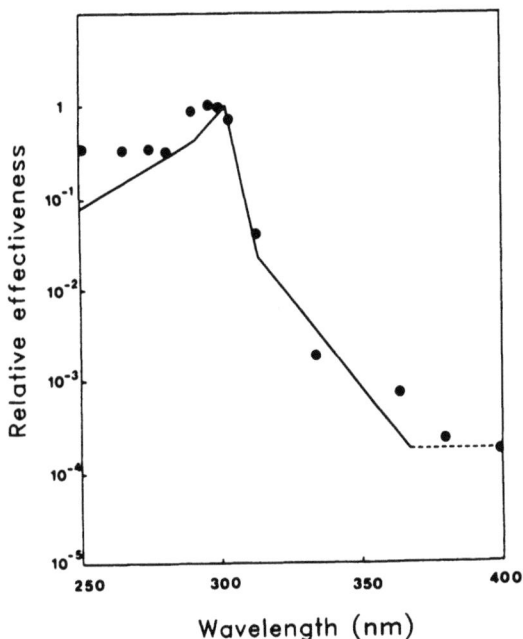

From van der Leun (1987a)

3.2 Broad-spectrum radiation

3.2.1 *Sunlight*

In one study by Roffo (1934), 600 rats [sex and strain unspecified] were exposed to solar radiation (sunlight) at a latitude of 35 °S in Buenos Aires, Argentina. The average exposure was for 5 h per day, with avoidance of the hours around solar noon in the summer. In the first days, 365 rats died from sunstroke. Of the 235 remaining animals, 165 (70%) developed tumours. There were 140 tumours of the ear (58% squamous-cell carcinomas; 36% spindle-cell sarcomas; 6% carcinosarcomas); 58 eye tumours (tumours of the conjunctiva, 100% spindle-cell sarcomas; tumours of the eyelid, 50% squamous-cell carcinomas and 50% spindle-cell sarcomas); and 15 other tumours, mainly squamous-cell carcinomas, at sites including the nose, tail, paw and neck. In complementary experiments reported in the same paper, groups of animals were exposed either to sunlight filtered through various colours of glass, to radiation from various types of lamp (quartz mercury, glass mercury, neon gas and filament lamps) or to short Hertzian wavelengths. Tumours [types and sites unspecified] were observed in all 150 animals exposed to quartz mercury lamps; no tumour was induced in any other experimental group. On the basis of this evidence, the author concluded that the carcinogenicity of sunlight could be attributed to UVR.

In another report by Roffo (1939), 2000 white rats and mice [exact numbers unspecified] were exposed to sunlight for an average of 5 h per day. After three to six months, benign neoplasms and, after seven to nine months, malignant neoplasms of the skin of the ear (88% of all malignant tumours), the forepaw (7.25%), the tail (2%) and nose (one tumour) developed in 600 animals; 25% of the tumours were seen on the eyes. The ear tumours were diagnosed as squamous-cell carcinomas (58%), spindle-cell sarcomas (36%) and carcinosarcomas (6%) by detailed histological examination. Similarly, the paw tumours were diagnosed as squamous-cell carcinomas (42%) and spindle-cell sarcomas (58%); the tumours of the tail were all squamous-cell carcinomas. The distribution of tumours of the eye was similar to that in the study of Roffo (1934). [The Working Group considered that these are exceptional studies which fully document the carcinogenicity of solar radiation in rats and mice, even though quantitative detail is lacking. The resulting neoplasms are described and photographically illustrated in exact detail. The Working Group accepted the weight of evidence contained in these studies as to the carcinogenicity of solar radiation to rats and mice.]

Domestic and other animals of many species (cows, goats, sheep (reviewed by Emmett, 1973), cats (Dorn *et al.*, 1971) and dogs (Madewell *et al.*, 1981; Nikula *et al.*, 1992)) develop skin tumours, and there are good indications that sunlight is involved. The tumours described generally developed in sparsely haired, light-coloured skin. Cancers of the eye occur in many species, including dogs, horses, cats, sheep and swine, but are particularly frequent in cattle (Russell *et al.*, 1956).

3.2.2 *Solar-simulated radiation*

In several investigations on carcinogenesis by UVR, 'solar-simulated radiation' was used (Forbes *et al.*, 1982; Staberg *et al.*, 1983a; Young *et al.*, 1990; Menzies *et al.*, 1991). In one large, particularly informative experiment (Forbes *et al.*, 1982), more than 1000 hairless albino Skh-hr1 mice were exposed to solar-simulated radiation from a xenon arc lamp, with various filters to make the spectral distribution in the UV region similar to that of sunlight under various thicknesses of the ozone layer. The exposures lasted for up to 80 weeks. More than 90% of the mice developed skin tumours, predominantly squamous-cell carcinomas. The time to development of 50% of first tumours was shorter after exposure to the spectra that included higher irradiance in the wavelength range 290–300 nm. The other experiments mentioned were more limited and dealt with more specialized aspects of UV carcinogenesis.

3.2.3 *Sources emitting UVC, UVB and UVA radiation*

Sources emitting radiation in the entire UV wavelength range were used in experiments on UV carcinogenesis mainly between 1930 and 1960.

(a) Mouse

Grady *et al.* (1943) exposed 605 strain A mice to broad-spectrum UVR at a wide range of doses and irradiances (weekly doses, 3.6–43 × 10^7 ergs/cm^2 [40–430 kJ/m^2]; Blum &

Lippincott, 1942). The investigation dealt primarily with skin tumours (mainly spindle-cell sarcomas). About 5% of the mice developed tumours of the eye. Histological examination by Lippincott and Blum (1943) showed that the eye tumours arose mostly in the cornea and were spindle-cell sarcomas or fibrosarcomas; haemangioendotheliomas were also found.

A particularly large, informative series of investigations was carried out with unfiltered medium-pressure mercury arc lamps which emitted UVC, UVB and UVA (Blum, 1959). More than 600 strain A mice were irradiated (daily dose, $0.32–8.6 \times 10^7$ ergs/cm^2 [3–86 kJ/m^2]) in a series of investigations dealing with various aspects of UV carcinogenesis; the dose–effect relationship was addressed particularly. In most of the experiments, more than 90% of mice developed skin tumours, mainly of the ears, the only site for which quantitative data were given.

(b) Rat

Findlay (1930) exposed six epilated albino rats to broad-spectrum UVR from a mercury-vapour lamp at a distance of 18 in [46 cm] for 1 min three times a week. Rapidly growing papillomas were reported in one rat. The time required was, however, much longer than in mice exposed similarly, namely, 21 months as compared to eight months for mice.

Putschar and Holtz (1930) exposed 35 rats [strain unspecified] with very low spontaneous tumour incidence to almost continuous irradiation with broad-spectrum UVR from a quartz mercury lamp for 11 months. They reported regular occurrence of skin tumours, including papillomas, squamous-cell carcinomas and, occasionally, basal-cell carcinomas. The tumours were first seen after 27 weeks of exposure.

Huldschinsky (1933) exposed seven white rats to UVR from a solar lamp for 2 h per day, six days per week for one year or more. Another group of five rats was exposed to a quartz lamp emitting a predominantly UVC waveband (< 270 nm). The doses given per session were about 10 times higher than those used in phototherapy. Spindle-cell sarcomas of the eye were found in 2/7 and 5/5 rats in each group, respectively.

Hueper (1942) reported squamous-cell carcinomas and, rarely, spindle-cell carcinomas and sarcomas, round-cell carcinomas and basal-cell carcinomas of the skin in 20 rats [strain unspecified] exposed for up to 10 months to broad-spectrum UVR from a mercury vapour burner (a Hanovia Super S Alpine lamp) at a distance of 75 cm.

In a study by Freeman and Knox (1964), a group of 78 rats (66 pigmented and 12 unpigmented) was exposed to broad-spectrum UVR from mercury lamps at 50 cm from the skin on five days a week for one year; the doses per session corresponded to approximately 1 MED for rat skin. A total of 98 eye tumours developed, with more tumours in pigmented rats. The tumours arose in the corneal stroma; two-thirds were diagnosed as fibrosarcomas and one-third as haemangioendotheliomas.

(c) Hamster

Hamsters exposed to an irradiation regimen similar to that described above also developed eye tumours (Freeman & Knox, 1964). In 19 animals (9 pigmented, 10 unpigmented) exposed for one year, haemangioendotheliomas and fibrosarcomas developed in 14 eyes.

(d) Guinea-pig

Guinea-pigs were exposed to the same regimen as described above. None of 17 animals developed a tumour of the eye (Freeman & Knox, 1964).

3.3 Sources emitting mainly UVB radiation

Many experiments have been carried out with sources emitting mainly UVB radiation, in which increases in the number of tumour-bearing animals and/or in the number of tumours per animal were seen (Blum, 1959; Winkelmann et al., 1963; Freeman, 1975; Stenbäck, 1975a; Daynes et al., 1977; Kripke, 1977; Spikes et al., 1977; Forbes et al., 1981; de Gruijl et al., 1983; Gallagher et al., 1984b). The most informative studies are described below.

3.3.1 Mouse

Freeman (1975) studied carcinogenesis induced by chronic exposure to narrow-band UVB produced by a high-intensity diffraction grating monochromator with a half-power band-width of 5 nm. Exposure was three times per week to one ear of each haired albino mouse. Four wavelengths were used, and the doses were determined as the MED. Of a group of 30 mice exposed to 300 nm (weekly dose, 60 mJ/cm^2), 16 developed squamous-cell carcinomas of the ear. Of a group of 30 mice exposed to 310 nm (weekly dose, 750 mJ/cm^2), 16 survived to 450 days and eight developed five squamous-cell carcinomas, two fibrosarcomas and one angiosarcoma of the ear. No skin tumour was observed among 30 mice irradiated with UVR at 290 nm (weekly dose, 42 mJ/cm^2); of five mice irradiated with 320 nm (weekly dose, 4950 mJ/cm^2), two developed squamous-cell carcinomas of the ear.

Two fibrosarcomas and one unspecified tumour of the eye were reported in 24 C3H/HeN mice bearing 25 skin tumours (mostly fibrosarcomas) after exposure to UVR (168 J/m^2 three times a week) from Westinghouse FS40T12 sunlamps (280–340 nm) (Kripke, 1977).

In the experiment of Forbes et al. (1981), groups of 24 male and female hairless albino Skh:HR mice (the changing designations of sources of 'Skh' mice are listed in Table 30), six to eight weeks old, were irradiated on five days per week with Westinghouse FS40T12 sunlamps (see Fig. 9c, p. 64), emitting mainly UVB (with < 1% below 280 nm; two-thirds at 280–320 nm and one-third at > 320 nm). All animals had developed tumours by the end of the experiment (up to 45 weeks), and a dose–response effect was demonstrated, as assessed by time to tumours in 50% of animals (Table 31). Histological examination showed tumours of 4 mm or more in diameter to be squamous-cell carcinomas; those of about 1–4 mm formed a continuum from carcinoma *in situ* to squamous-cell carcinoma, and those less than 1 mm comprised epidermal hyperplasia and squamous metaplasia tending toward carcinoma *in situ*. Less than 1% of tumours were fibrosarcomas.

Six groups of 22–44 male and female Skh-hr 1 hairless albino mice (total, 199), six to eight weeks of age, were exposed to daily doses ranging from 57 to 1900 J/m^2 of mainly UVB radiation from Westinghouse FS40TL12 sunlamps; this dose range encompassed a factor of 33. Most of the animals developed skin tumours, although even the highest daily dose was sub-erythemic. A clear-cut relationship was shown between daily dose and time required for 50% of animals to develop skin tumours, which were predominantly squamous-cell carcinomas (Fig. 11). Squamous-cell carcinomas developed in 71% of the mice in the lowest

Table 31. Dose–response to ultraviolet radiation of hairless Skh:HR mice

Daily dose (J/m^2)	Time to 50% tumour incidence (weeks)	Terminated at week
420	38.6	45
587	33.3	45
822	29.2	45
1152	20.0	36
1613	17.6	36
2259	12.9	25

From Forbes et al. (1981)

Fig. 11. Dose–effect relationship for the induction of < 1-mm skin tumours in hairless mice by exposure to UVB radiation over a wide range of daily doses; t_m, median induction time

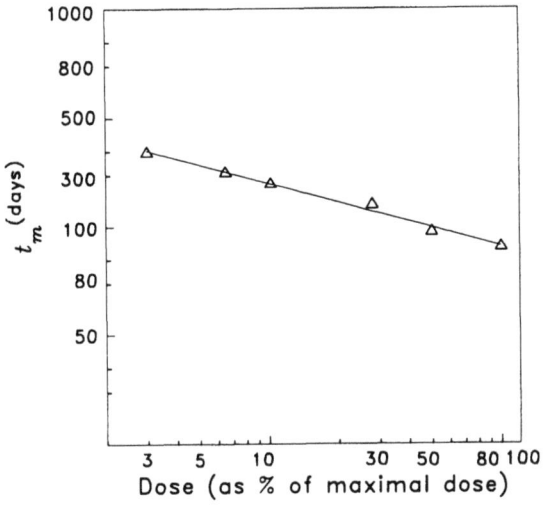

From de Gruijl et al. (1983)

dose group, and two skin tumours were reported in a total of 24 nonirradiated control mice (de Gruijl et al., 1983).

In albino hairless Skh:Hr-1 mice irradiated with UVB or UVB plus UVA radiation three times a week for 16 weeks, with a 17-week recovery period, the spectrum for UV tumorigenesis was sharp and had a maximum near 300 nm (Bissett et al., 1989).

3.3.2 Rat

Skin tumour induction was studied in a group of 40 shaven female NMR rats, 8–10 weeks old at the start of the experiment. The animals were irradiated chronically at a distance of 37.5 cm for 60 weeks with Westinghouse FS40T12 sunlamps (Fig. 9c), emitting mainly UVB (weekly dose, $5.4–10.8 \times 10^4$ J/m^2). A total of 25 skin tumours, most of which were papillomas of the ears, developed in 16/40 animals (Stenbäck, 1975a).

3.3.3 Hamster

Stenbäck (1975a) irradiated 40 shaven female Syrian golden hamsters, 8–10 weeks of age, using the same protocol as described above. A total of 30 skin tumours developed in 14/40 animals; 22 were papillomas (14 animals), four were keratoacanthomas (three animals), one was a squamous-cell carcinoma of the skin and three were papillomas of the ear (one animal).

3.3.4 Guinea-pig

Stenbäck (1975a) exposed guinea-pigs using the same protocol as above and found skin tumours in 2/25 animals (a fibroma and a trichofolliculoma).

3.3.5 Fish

Two hybrid fish strains susceptible to melanocytic neoplasms by UVR were developed by Setlow et al. (1989) by crossing platyfish and swordtails. A group of 460 fish were exposed to mainly UVB radiation from Westinghouse FS40 sunlamps, filtered with acetate sheets transmitting > 290 nm or > 304 nm at various doses (150 and 300 J/m^2 per day for > 290 nm; 850 and 1700 J/m^2 per day for > 304 nm) for 1–20 consecutive days. There were 103 controls. Depending on the wavelength, the level, the number of days of exposure and the strain, 19–40% of the irradiated fish developed melanocytic tumours; 13 and 2% of the controls in the two strains, respectively, developed such tumours.

3.3.6 Opossum

Monodelphis domestica, a South American opossum, is unusual in showing the phenomenon of photoreactivation (see Glossary) of pyrimidine dimers and erythema (Ley, 1985); it also developed actinic keratoses and skin tumours (mainly fibrosarcomas and squamous-cell carcinomas) on exposure to UVR from an FS-40 sunlamp (280–400 nm) (Ley et al., 1987). Animals were shaved regularly and exposed to mainly UVB radiation from Westinghouse FS40 sunlamps, with relative emissions of 0.04, 0.27, 0.69, 1.0 and 0.09 at a dose of 250 J/m^2 (which is approximately half of an average MED; see Fig. 9c) at 280, 290, 300, 313 and 360 nm, respectively. Eight of 13 animals developed localized melanocytic hyperplasia; 100 weeks after the start of the experiment, melanomas were found in 5/13 surviving animals. *M. domestica* do not develop spontaneous melanomas, as was apparent in a much larger colony not exposed to UVR. Exposure of another group to photoreactivating light after UV irradiation reduced the incidence of melanocytic hyperplasia (3/17); this was considered to be a precursor lesion of the melanomas, although photoreactivation could not be demonstrated in the melanoma (Ley et al., 1989).

[The Working Group noted that the melanocytic lesions induced in fish and the South American opossum differ histologically from human melanoma: they grow to a larger size and do not metastasize readily.]

Ley et al. (1991) exposed groups of M. domestica to UVR from fluorescent sunlamps (Westinghouse FS40; 280–400 nm with a peak at 313 nm) three times a week for 70 weeks at a dose of 250 J/m^2. Besides skin tumours, tumours of the anterior eye were observed beginning 30 weeks after the start of exposure. At 69 weeks, 50% of the animals had eye tumours, which were classified as fibrosarcomas of the corneal stroma. In animals exposed to UVR followed immediately by photoreactivating light, tumours appeared later and in reduced numbers.

'Cancer eye' in cattle, which includes squamous-cell carcinoma of the eye and the circumocular skin, is thought to be caused by solar UVR. In an attempt to confirm this relationship experimentally (Kopecky et al., 1979), four Hereford cattle (which lack pigment around the eyes) were exposed to UVB radiation from Westinghouse FS40 lamps. Three cows developed grossly observable tumours of the eye, one of which was histopathologically confirmed as a preneoplastic growth.

3.4 Sources emitting mainly UVC radiation

3.4.1 Mouse

Carcinogenicity studies have been performed mainly in mice, but no study is available in which animals were exposed solely to UVC radiation. Several studies have been reported in which the source of UVC radiation was low-pressure mercury discharge germicidal lamps, which emit 90–95% of their radiation at wavelength 254 nm and weaker spectral lines in the UVB, UVA and visible light regions (Rusch et al., 1941; Blum & Lippincott, 1942; Forbes & Urbach, 1975; Lill, 1983; Joshi et al., 1984; Sterenborg et al., 1988). In all of these investigations, the exposures induced tumours. Two of the most informative studies are described in more detail below.

A group of 40 female C3H/HeNCr1Br mice were irradiated with these lamps at a weekly dose of 3×10^4 J/m^2. Three animals died without tumours after 9, 43 and 63 weeks of irradiation; all of the other animals had tumours. By 52 weeks, 97% of the animals had developed skin tumours, with a median time to appearance of 43 weeks. The mean number of tumours per tumour-bearing mouse was 2.9. Tumour histology was carried out in 29/37 mice. Of a total of 83 suspected tumours, 66 were squamous-cell carcinomas, 10 were proliferative squamous lesions and 6 were invasive fibrosarcomas; one had the appearance of a cystic dilatation (Lill, 1983). [The Working Group that resulted in *IARC Monographs* volume 40 (IARC, 1986a) noted that the 4% UVB content of the source, representing a weekly dose of 1170 J/m^2, could not be excluded as contributing to the induction of skin tumours.]

Sterenborg et al. (1988) presented evidence that the tumours they induced in albino hairless mice were indeed due to UVC radiation. Groups of 24 male and female hairless albino mice (Skh-hr1), 6–10 weeks of age, were exposed to UVC radiation from Philips germicidal TUV 40W low-pressure mercury discharge lamps (mainly 254 nm) on seven days a week for 75 min per day at 230, 1460 or 7000 J/m^2 (30 times the MED); this dose was 60% less during the first seven days of the experiment. A total of 65 squamous-cell carcinomas of the skin were found [number of animals with tumours not specified]. Both the percentage of tumour-bearing animals and the number of tumours per mouse were strongly dose-related. By comparing their results with those of experiments with UVB, the investigators concluded that (i) the UVB emitted by the low-pressure mercury discharge lamps was insufficient to account for the induction of tumours at the rate found, as at least 850 days of exposure to the UVB radiation present would be required to induce skin tumours at the rate observed, as compared to 161 days with the low-pressure mercury discharge lamp used; (ii) there is a qualitative difference between the effects of low-pressure mercury discharge and UVB lamps, in that the tumours induced by the mercury discharge lamps were scattered more widely over the skin of the mice than in the experiments with UVB; and (iii) the dose–effect relationship for tumorigenesis was less steep with the mercury discharge lamps than with UVB sources. [The Working Group noted that the evidence given to exclude UVB as contributing to the induction of skin tumours does not obviate the possibility that some interaction between UVC and UVB radiation led to tumour induction.]

3.4.2 *Rat*

Nine groups of 6 or 12 male CD-1 rats, 28 days of age, were shaved and exposed to varying doses of UVC from Westinghouse G36T6L sterilamps emitting predominantly 254 nm (dose range, 0.08–26.0×10^4 J/m^2). Survival ranged from 75 to 92% for the nine experimental groups. Keratoacanthoma-like skin tumours developed at a yield that was approximately proportional to dose throughout the dose range 0.65–26.0×10^4 J/m^2, although no tumour was observed at 0.32×10^4 J/m^2 or below (Strickland *et al.*, 1979).

3.5 Sources emitting mainly UVA radiation

The carcinogenic properties of UVA radiation received little attention before the introduction of UVA equipment for tanning, which led to the development of powerful sources of UVA. Many experiments have now been performed, using mainly hairless mice, to examine the possible carcinogenicity of UVA radiation (Zigman *et al.*, 1976; Forbes *et al.*, 1982; Berger & Kaase, 1983; Staberg *et al.*, 1983a,b; Kaase *et al.*, 1984; Santamaria *et al.*, 1985; Strickland, 1986; van Weelden *et al.*, 1986; Slaper, 1987; Kligman, 1988 [abstract]; van Weelden *et al.*, 1988; Kligman *et al.*, 1990 [abstract]; Sterenborg & van der Leun, 1990; van Weelden *et al.*, 1990a; Kelfkens *et al.*, 1991a; Kligman *et al.*, 1992). Some have shown no induction of tumours (Staberg *et al.*, 1983a,b; Kaase *et al.*, 1984; Kligman, 1988 [abstract]). [The Working Group noted that the doses may have been too small (daily doses in the range of 160 kJ/m^2) (Staberg *et al.*, 1983b) or the exposure period too short (Berger & Kaase, 1983; Kaase *et al.*, 1984; Kligman, 1988 [abstract]), as noted by the authors in a subsequent report (Kligman *et al.*, 1992).] In the other experiments, tumours were induced. [The Working Group noted that in some of the latter experiments either it is unclear whether UVB radiation was sufficiently excluded from the spectrum (Zigman *et al.*, 1976; Berger & Kaase, 1983; Staberg *et al.*, 1983a; Santamaria *et al.*, 1985) or the exclusion of UVB radiation was not fully convincing (Strickland, 1986).]

Studies in which the exclusion of UVB radiation was documented to be sufficient and which led to the induction of tumours by UVA in hairless mice were reported by van Weelden *et al.* (1986, 1988, 1990a), Slaper (1987), Kligman *et al.* (1990 [abstract], 1992), Sterenborg and van der Leun (1990) and Kelfkens *et al.* (1991a). A few of the most informative studies are described below.

Groups of 24 male and female albino hairless Skh-hr 1 mice were exposed to UVA radiation from a bank of Philips TL40W/09 fluorescent tubes, filtered through a 10-mm glass plate selected for strong absorption of UVB radiation, for 12 h a day on seven days a week for about one year, at which time the experiment was terminated. The daily dose was 220 kJ/m^2. Most animals developed scratching lesions before they contracted skin tumours, which occurred in all animals; the median time to tumour appearance was 265 days. At the end of the experiment, the larger lesions were examined histologically: 60% were classified as squamous-cell carcinomas, 20% as benign tumours, including papillomas and keratoacanthoma-like lesions, and 20% as mild cellular and nuclear atypia. The histological findings were similar to those observed in a parallel experiment with UVB, but the tumours in the UVA-exposed group appeared over a longer time span. Residual UVB radiation was excluded as the cause of tumours in UVA-exposed mice on quantitative considerations: the authors concluded that more than 100 000 times the UVB present would have been required in order to induce tumorigenesis at the rate observed (van Weelden *et al.*, 1986, 1988).

Groups of 48 male and female hairless albino Skh-hr 1 mice were exposed to 220 kJ/m^2 UVA radiation (> 340 nm) from four high-pressure mercury metal-iodine lamps (Philips HPA 400 W), passed through liquid filters, for 2 h per day on seven days per week for up to 400 days. The spectrum matched that of a lamp used for tanning (the UVASUN 5000); UVB was effectively excluded by the filters. Skin tumours developed in most of the animals, and 31 developed tumours before any scratching was observed. The largest tumours were examined histologically at the end of the experiment: 15/20 tumours examined were squamous-cell carcinomas (Sterenborg & van der Leun, 1990).

The desire to tan safely has raised interest in the possible carcinogenicity of long-wavelength UVA (340-400 nm). In some experiments, UVB was excluded so rigorously that there was also very little UVA in the range 315-340 nm; exposure was therefore mainly to wavelengths in the region of 340-400 nm (van Weelden *et al.*, 1988; Sterenborg & van der Leun, 1990; van Weelden *et al.*, 1990a). These experiments yielded squamous-cell carcinomas in most animals. [The Working Group noted that if these were to be ascribed to the small proportion of shorter-wavelength UVA present in the spectra, a sharp peak in the action spectrum for UV carcinogenesis would have to occur between 330 and 340 nm, which does not appear likely.] In experiments by Kligman *et al.* (1990 [abstract], 1992), wavelengths shorter than 340 nm were filtered out rigorously. Female hairless albino Skh-hr 1 mice were exposed several times per week for 60 weeks to UVA at wavelengths of 340-400 nm at daily doses of 360 and 600 kJ/m^2, as used in artificial suntanning. Eighteen weeks later, 44 surviving mice had 19 skin tumours, mostly papillomas. At week 100, 22 surviving mice had 40 tumours, many of which were considered clinically to be squamous-cell carcinomas.

The carcinogenicity of short-wavelength UVA (315-340 nm) was investigated in one experiment. Groups of 24 male and female albino hairless Skh:hr 1 mice were exposed to average daily doses of 20 or 56 kJ/m^2 radiation from specially developed fluorescent tubes with peak emission near 330 nm (UVB radiation was filtered out efficiently using a glass filter) on seven days a week for 650 days. All mice in the high-dose group developed multiple tumours, first mainly papillomas and later predominantly squamous-cell carcinomas. In the lower-dose group, three mice developed skin tumours, all of which were papillomas. The lamps also emitted long-wavelength UVA (340–400 nm), but in a proportion considered by the authors to be too small to account for the rate of tumorigenesis observed (Kelfkens *et al.*, 1991a). The investigators estimated the carcinogenic effectiveness of short-wavelength UVA (315–340 nm) to be approximately five times greater than that of long-wavelength UVA (340–400 nm).

3.6 Interaction of wavelengths

In daily life, the skin is exposed frequently to several wavelength ranges (UVA, UVB, UVC) simultaneously, or to different combinations at different times. The simplest explanation of an effect of such combined exposures is 'photoaddition', i.e., each exposure contributes to the effective dose in an additive way. The validity of this hypothesis is one of the assumptions underlying widely used concepts such as 'erythemal effective energy' and the derivation of the action spectrum shown in Figure 10 (p. 141). It implies that any additional exposure to an effective dose, in any wavelength region, increases the carcinogenic effect.

Several studies provide indications, however, that the situation is more complicated. Interactions are seen between the effects of different wavebands that result in deviations from photoaddition (for reviews, see van der Leun, 1987b, 1992). The literature on this topic is controversial and cannot be summarized in detail here. The following two sections form an attempt to give an overview and interpretation.

3.6.1 *Interaction of exposures given on the same day*

Several types of interactions have been reported between different wavelength ranges administered simultaneously or in close temporal proximity. These have led to concepts of processes such as:
- photorecovery: the effect of UVB or UVC is reduced by simultaneous or immediately subsequent exposure to UVA or visible light [The Working Group noted that photoreactivation is a special case of photorecovery but applies only to species that have the 'photoreactivating enzyme', photolyase (see Glossary).];
- photoprotection: the effect of UVB or UVC is reduced by prior administration of UVA or visible light;
- photoaugmentation: the effect of UVB or UVC is enhanced by prior, simultaneous or subsequent administration of UVA or visible light.

Photoaugmentation of UVB carcinogenesis by UVA was suggested by several investigators (Urbach *et al.*, 1974; Willis *et al.*, 1981, 1986; Kligman, 1988 [abstract]; Talve *et al.*, 1990) but could not be confirmed by others (Forbes *et al.*, 1978; van Weelden & van der Leun, 1986). The latter investigators found evidence of photorecovery: the effect of UVB plus UVA was smaller than that of the same UVB exposure given alone. The reduction was small; however, UVA reduced the carcinogenic effective dose of UVB by 16%.

Interactions of different wavelength ranges when given simultaneously, prior to or immediately after each other appear to be either nonexistent or unproven, as in the case of photoaugmentation, or small, as in the case of photorecovery. Such interactions currently play a small role in the evaluation of risks (see, for example, Health Council of the

Netherlands, 1986). Other uncertainties in the estimates, such as the dose received, are likely to have a greater influence than interactions. Photoreactivation, is, however, a well-defined process in those species which possess photolyase and may result in reduction of effects.

3.6.2 *Long-term interactions*

A different type of interaction occurs when exposures to one wavelength band are separated temporally from exposures to another. For example, a prolonged course of UVB exposures, by itself sufficient to induce tumours, is compared with an identical UVB course that is preceded or followed by a course of UVA exposures, usually over several weeks.

Forbes *et al.* (1978) exposed hairless mice to tumorigenic UVB or to UVB followed by UVA and visible light for 30 weeks. The longer-wavelength exposures reduced the tumorigenic effect of the UVB. Staberg *et al.* (1983b) gave mice a tumorigenic combination of UVB and UVA and found that subsequent exposures to UVA increased the tumorigenic effect. The UVA was derived from Philips TL40W/09 lamps filtered through 2-mm plain glass to remove the UVB. [The Working Group noted that since the glass transmitted some UVB the increased carcinogenic effect may have been due to added UVB radiation.] Bech-Thomsen *et al.* (1988a) pretreated lightly pigmented hairless female hr/hr C3H/Tif mice with UVA for four weeks before exposure to broad-spectrum UVR. The UVA reduced the carcinogenic effect of the broad-spectrum UVR. This result was not corroborated in a subsequent, similar experiment by the same investigators (Bech-Thomsen *et al.*, 1988b), in which mice were pretreated with radiation from various UVA sources. The purest UVA radiation neither increased nor decreased the carcinogenic effect of UVB.

Slaper (1987) exposed one group of mice daily to UVB and a second group daily to UVA at doses matched for approximately equal carcinogenic effect. In a third group of mice that received the two regimens alternately every week, the carcinogenic effect was less than that in the UVA- or the UVB-exposed group. The effective dose in the alternating regimen was estimated to be 80% that in the UVB regimen. The investigator concluded that both UVA and UVB contributed to the carcinogenic effect of the alternating regimen.

[The Working Group noted that the effect of long-term interactions appears to be similar to that of interactions of exposures given on the same day. Photoaddition gives a reasonable prediction, but the combined effects tend to be slightly less than would be predicted.]

3.7 Additional experimental observations

3.7.1 *Tumour types*

Skin tumours in UV-exposed animals are commonly epidermal, benign papillomas and malignant squamous-cell carcinomas; adnexal neoplasms, mainly basal-cell carcinomas, are less common. Attempts have been made to induce naevi and malignant melanomas. Many tumours are found, since the animals are followed for long periods of time; however, tumours coalesce and regress, and all tumours are not examined histologically.

Squamous-cell carcinoma is the commonest type of tumour found after exposure to UVR. These tumours have been reported in mice exposed to predominantly UVB radiation (Winkelmann *et al.*, 1960, 1963; Epstein & Epstein, 1963; Freeman, 1975; Forbes *et al.*, 1981;

de Gruijl *et al.*, 1983), to predominantly UVA radiation (van Weelden *et al.*, 1988; Sterenborg & van der Leun, 1990) and to predominantly UVC radiation (Lill, 1983; Sterenborg *et al.*, 1988). They have also been found in rats (Putschar & Holtz, 1930; Roffo, 1934, 1939; Hueper, 1942), hamsters (Stenbäck, 1975a) and opossums (Ley *et al.*, 1989) following exposure to broad-spectrum UVR.

Papillomas were reported to be the commonest tumour after exposure of hairless mice to UVR consisting of UVB and UVA (Gallagher *et al.*, 1984b). Papillomas were also reported to precede or accompany squamous-cell carcinomas induced in hairless mice by UVA (van Weelden *et al.*, 1988), UVB (Stenbäck, 1978) or UVC radiation (Sterenborg *et al.*, 1988). Papillomas were also common in rats (Findlay, 1930; Putschar & Holtz, 1930; Stenbäck, 1975a) and hamsters (Stenbäck, 1975a) exposed to broad-spectrum UVR.

The main type of tumour diagnosed after exposure of haired mice to broad-spectrum UVR was *fibrosarcomas* (Grady *et al.*, 1941, 1943). Squamous-cell carcinomas were less common, but the ratio of carcinomas to sarcomas increased with the number of exposures per week (Grady *et al.*, 1943). Spikes *et al.* (1977) reported many squamous-cell carcinomas in clipped C3Hf mice irradiated with UVB, especially at low doses; the high-dose group had a much higher proportion of fibrosarcomas. The investigators suggested that the type of tumour induced might be dose-dependent. Norbury and Kripke (1978) found that the type of tumour might depend on immunological factors. They compared UVB tumorigenesis in normal C3H/HeN (MTV⁻) mice, in T cell-depleted mice and in T cell-depleted mice reconstituted with thymus grafts. In the normal mice, fibrosarcomas predominated; in the T-cell depleted, reconstituted mice, squamous-cell carcinomas predominated. Spindle-cell sarcomas were reported in rats irradiated with sunlight (Roffo, 1934), and fibrosarcomas were seen in opossums irradiated with UVB (Ley *et al.*, 1989).

The diagnosis of fibrosarcoma was questioned by Morison *et al.* (1986). After C3H/HeNCr (mammary tumour virus-free) haired pigmented mice were exposed to mainly UVB radiation, the tumours induced were almost all squamous-cell carcinomas. The investigators noted that the same type of tumour had been diagnosed in many previous reports as fibrosarcoma; they diagnosed squamous-cell carcinomas by studying specific markers for cell differentiation in the tumours. In a study by Phelps *et al.* (1989) in which hairless albino Skh/hr-1 mice were exposed to UVA and UVB at 0.3 J/cm^2 [30 kJ/m^2], all mice developed epidermal neoplasia and 25% of animals developed spindle-cell tumours that resembled human atypical fibroxanthoma. [The Working Group noted that earlier studies did not use presently available cellular markers.]

Keratoacanthomas and similar benign epidermal neoplasms have been reported in mice exposed to UVB (Stenbäck, 1978), rats exposed to UVB and UVC (Strickland *et al.*, 1979) and hamsters exposed to UVB (Stenbäck, 1975a).

Actinic keratosis, or solar keratosis, a precursor lesion of squamous-cell carcinomas, has been reported in hairless mice exposed to UVA and UVB (Kligman & Kligman, 1981) and in haired mice exposed to UVB (Stenbäck, 1978).

Basal-cell carcinomas have not been reported in studies in mice. A few studies on UV carcinogenesis in nude mice, which have a deficient immune system, have been reported (Eaton *et al.*, 1978; Anderson & Rice, 1987; Hoover *et al.*, 1987). The skin tumours induced by mainly UVB radiation in these studies were mostly squamous-cell carcinomas, but in the experiments reported by Anderson and Rice (1987) in nude mice of BALB/c background there were several basal-cell carcinomas. Basal-cell carcinomas were found occasionally in rats exposed to broad-spectrum UVR (Putschar & Holtz, 1930; Hueper, 1942). [The Working Group noted that the classification of these neoplasms and their relation to the corresponding neoplasms in humans is not clear.]

There is no report in which cutaneous *malignant melanoma* was induced in mice by UVR alone (Epstein, 1990; van Weelden *et al.*, 1990b; Husain *et al.*, 1991), in spite of concerted attempts to achieve this.

No study was found in which the primary objective was to examine the susceptibility of the eye to UVR; rather, eye tumours were found incidentally in studies designed to investigate skin carcinogenesis. All of the tumours of the eye identified in these reports involved superficial parts of the eye (cornea and conjunctiva); no tumour of the interior eye was reported.

Studies of the effect of UVR on tumour induction in other organs (lymphoma in mice) are few and were not designed to determine this effect (Ebbesen, 1981; Joshi *et al.*, 1986). [The Working Group considered that the data were inadequate for evaluation and that data on survival among treated and control groups, sample selection and analysis of data were limited.]

3.7.2 *Dose and effect*

Quantitative information is available mainly on the induction of squamous-cell carcinoma in mice. In most of the experiments, exposure was regular, several times per week or every day, until tumours developed. The daily doses of UVR required for skin tumorigenesis are usually well below those present outdoors in the environment, and most experiments have been conducted with UVB doses lower than those required to elicit acute reactions in mouse skin (erythema or oedema). In one experiment in hairless mice, with a UVB dose 33 times lower than that required for acute reactions, 71% of the skin tumours were squamous-cell carcinomas (de Gruijl *et al.*, 1983). The effectiveness of UVB radiation is increased at lower dose rates (Kelfkens *et al.*, 1991b).

The higher the dose given, the less time it takes for tumours to appear. In most experiments, the time required for 50% of mice to develop tumours ranged between a few months and one year. By maximizing the exposure regimen in hairless mice (escalating doses of UVB radiation), the time could be reduced to 18 days (Willis *et al.*, 1981). In a few experiments, in both mice and rats, skin tumours resulted from a single exposure to UVB radiation (Hsu *et al.*, 1975; Strickland *et al.*, 1979); in mice, this required a dose that first caused skin ulceration: hairless mice, 60 kJ/m^2 (Hsu *et al.*, 1975); Sencar mice, 29 kJ/m^2 (Strickland, 1982).

Quantitative dose–effect relationships have been derived for mice exposed regularly (usually daily) to UVR. The median time to first tumour, t_m, has been used as a measure of the effect and is related to dose level. Dose–effect relationships of the form

$$t_m = c\, D^{-r},$$

where c is a constant incorporating the susceptibility of the strain of mice as well as the effectiveness of the radiation spectrum, D is the daily dose of radiation and r is a numerical exponent giving the steepness of the relationship, have been proposed by several authors. Estimates of r vary from 0.2 (Sterenborg et al., 1988) for small tumours of the skin induced by UVC radiation in hairless mice, to 0.5 (Blum et al., 1959) for large tumours on the ears of haired mice induced by broad-spectrum UVR and to 0.6 (de Gruijl et al., 1983) for small tumours induced by broad-band UVB in hairless mice. Figure 11 (p. 145) illustrates the shape of this dose–response relationship for $r = 0.6$; other forms of the relationship have been proposed (Forbes et al., 1982). All of them provide adequate descriptions of the dose–response within the range of the available data, although extrapolations outside this range differ substantially.

3.7.3 *Dose delivery*

The tumorigenic effect of UVR depends not only on the dose but also on the temporal pattern of exposure. In general, the effectiveness of treatment increases with the number of fractions of the dose per week (Forbes et al., 1981), for both daily and accumulated doses. A daily dose administered over 12 h is more effective than the same daily dose administered in 1 h (Kelfkens et al., 1991b). The same weekly dose is more effective when given over three to five days than if given in one day (Forbes et al., 1981).

3.7.4 *Action spectra*

Ideally, the carcinogenic effectiveness of UVR can be expressed as a continuous function of wavelength. That function, called the action spectrum for UV carcinogenesis, is not yet completely delineated. Freeman (1978) made an early attempt to determine this spectrum and found that it was limited to a few narrow bands around the wavelengths 290, 300, 310 and 320 nm. Narrow-band monochromatic sources are difficult to achieve.

Since that time, various action spectra have been proposed to weight the spectral irradiance of a source. Forbes et al. (1982) and Cole et al. (1986) determined dose–effect relationships similar to that shown in Figure 11 for many different UV spectra. By weighting these lamp spectra with various existing action spectra for photobiological effects, effective doses were computed for each experiment. In this way, the investigators tried to align the results from the experiments with different UV spectra into one dose–effect relationship. One of the action spectra (MEE48), originally determined for the induction of oedema in mice 48 h after exposure to UVR and which is similar to the human erythema action spectrum, fitted well. The authors concluded that the mouse oedema spectrum was also appropriate for describing skin cancer induction (Cole et al., 1986).

Sterenborg and van der Leun (1987) attempted to determine an action spectrum directly from observations on UV carcinogenesis. They exposed hairless albino mice to seven different lamp spectra under otherwise identical circumstances. The lamp spectra overlapped to some extent, and the action spectrum was derived by mathematical fitting. The analysis yielded an action spectrum for the wavelength range 250–360 nm. Slaper (1987) added observations in the UVA region and extended the action spectrum throughout the UVA range (see Fig. 10, p. 141).

The action spectrum shown in Figure 10 is for albino hairless Skh-hr 1 mice with an end-point of 1.0-mm tumours. Although different end-points may yield different action spectra, this curve shows good agreement in the UVB range with the MEE48 spectrum and also with the observations of Freeman (1978) for wavelengths 300, 310 and 320 nm. [The Working Group noted that the action spectrum for UV carcinogenesis in the wavelength range 300–320 nm may be considered a good approximation.] The different shapes of Figure 10 and MEE48 in the UVC reflect a scarcity of data in this wavelength range. [The Working Group noted that the action spectrum for carcinogenesis by UVC is still highly uncertain.] The MEE48 left widely different options open for the action spectrum of long-wavelength UVA: the effectiveness in the wavelength range 330–400 nm could be either zero or as high as 0.0002 (Cole *et al.*, 1986). More recent data on the carcinogenesis of UVA, used to construct the curve in Figure 10, indicate a mean effectiveness of 0.00015 in this range (Slaper, 1987). [The Working Group noted that this value for the carcinogenic effectiveness for UVA may be regarded as an estimate of the order of magnitude.]

3.7.5 *Pigmentation*

Pigment was reported to be protective against tumours arising from the conjunctiva in cattle (Anderson, 1963).

Freeman and Knox (1964) also examined the influence of pigmentation in a group of 78 rats composed of 66 pigmented rats of various strains (black, black and white, grey-brown, grey and white) and 12 albinos. Under the same irradiation regimen, the pigmented rats developed tumours in 73% of eyes and the albinos in only 8%. The tumour yield was consistently higher in the pigmented strains than in the albinos. In nine pigmented and 10 albino hamsters exposed for one year, 50% of pigmented animals and 25% of non-pigmented animals developed eye tumours.

Davies and Forbes (1988) exposed closely related albino hairless Skh-hr 1 mice and pigmented hairless Skh-hr 2 mice to broadband UVR from a filtered xenon arc lamp. Especially at high doses, the latent period until 50% of animals had first tumours was longer in Skh-hr 2 mice.

van Weelden *et al.* (1990a) derived mice of different degrees of pigmentation—'browns' and 'blacks'—by selective breeding from Skh-hr 2 stock and exposed 24 albinos (Skh-hr 1) (van Weelden *et al.*, 1988), 16 'browns' and eight 'blacks' to UVA radiation. The brown mice were less susceptible to skin tumours than the albinos, but the more heavily pigmented blacks were as susceptible as the albinos: the median times for tumour induction were 265 days for albinos, 267 days for blacks and 375 days for browns (van Weelden *et al.*, 1990a).

3.8 Administration with known chemical carcinogens

Since UVR alone produces tumours, it is a 'complete' carcinogen and may thus be involved in cocarcinogenicity. Several investigators have attempted to determine whether UVR has tumour 'initiating' and/or tumour 'promoting' activity when tested in a traditional two-stage protocol. For the purposes of this monograph, a 'tumour initiator' is defined as an agent that, at a stated amount and upon administration once, is incapable of causing tumours in the population of animals unless the skin is subsequently treated with a 'tumour promoter'. A 'tumour promoter' is defined as an agent that, under stated conditions is incapable of causing tumours unless the skin was previously treated with a 'tumour initiator'. The test systems used embody a number of variables, not all of which were necessarily considered by the authors. For example, UVR has also been shown to influence the immune system, and polycyclic aromatic hydrocarbons are photochemically active.

3.8.1 *Administration with polycyclic aromatic hydrocarbons*

Most of the studies summarized below demonstrate that UVR has a cocarcinogenic action with other carcinogens. Other reports provide additional information on cocarcinogenesis, on photolysis of polycyclic aromatic hydrocarbons and on other interference with chemical carcinogenesis (Clark, 1964; Ito, 1966; Santamaria *et al.*, 1966; Davies *et al.*, 1972 [abstract]; Shabad & Litvinova, 1972; Stenbäck & Shubik, 1973; Stenbäck, 1975b; Roberts & Daynes, 1980; Gensler & Welch, 1992).

(a) 3,4-Benzo[a]pyrene

Groups of 18 female SPF (specific pathogen-free) BALB/c mice, six weeks of age, received 30-min exposures on the shaved dorsal skin to UVB from a Westinghouse FS40 sunlamp (280–320 nm) five times a week for 13 weeks (total dose, 7.0×10^5 J/m^2) or no UVB exposure followed one week later by twice weekly applications of 0, 0.1 or 1.0 mg 3,4-benzo-[*a*]pyrene in acetone on the shaved ventral skin for 20 (acetone only), 20 or 10 weeks, respectively. Pre-exposure to UVB enhanced tumour growth in the high-dose group: 29 tumours (of 20 examined histologically, 90% were squamous-cell carcinomas and 10% undifferentiated sarcomas) in the UVB-pretreated group compared to two (squamous-cell carcinomas) in the non-irradiated 3,4-benzo[*a*]pyrene-treated animals 18 weeks after the first treatment with 3,4-benzo[*a*]pyrene. No such effect was seen in the low-dose group (Gensler & Bowden, 1987; Gensler, 1988a).

(b) 7,12-Dimethylbenz[a]anthracene

In an attempt to assess the promoting effects of UVR, groups of 15–31 male and 16–22 female Swiss albino mice, 11–18 weeks of age, received a single application of two drops (0.1 ml) of 0 or 0.5% 7,12-dimethylbenz[*a*]anthracene (DMBA) in acetone on the posterior half of the dorsal skin, followed 14 days later by exposures to UVB (280–320 nm; high-pressure Hanovia hot quartz contact lamp) twice a week for 67 weeks (total dose, 13.33×10^7 ergs/cm^2 [133 kJ/m^2]) or no exposure. At the end of the UVB treatment, 16/31 mice treated with DMBA and UVB had developed 19 skin tumours, compared to 4/41 and 0/47, respectively, among mice treated with DMBA alone and UVB alone. Exposure to UVB also enhanced the multiplicity and degree of malignancy of DMBA-induced tumours (Epstein & Epstein, 1962).

Groups of 26–42 male and female outbred hairless mice, 7–12 weeks old, received a single application of two drops (0.1 ml) of 0 or 0.5% DMBA in acetone, followed six weeks later by exposures to UVB (280–320 nm; high-pressure Hanovia hot quartz contact lamp) three times a week for 29 weeks (total dose, 15.34×10^7 ergs/cm^2 [153 kJ/m^2]) or no exposure. All animals were observed for 63 weeks. UVB exposure produced skin tumours in 22/26 animals, and DMBA treatment alone in 3/41; acetone alone produce no skin tumour. Exposure to UVB following DMBA treatment enhanced carcinogenicity with regard to appearance time (first tumour observed at 14 weeks compared to 30 in the group treated with DMBA alone and 20 in that given UVB alone), multiplicity at 58 weeks after DMBA treatment (40 in 24 animals compared to 22 in 26 animals treated with UVB alone and 3 in 41 animals treated with DMBA alone) and degree of malignancy. Two 'melanomas' appeared in the group receiving the combined treatment (Epstein, 1965).

Groups of 18–46 outbred hairless pigmented mice [sex unspecified], 8–11 weeks old, received a single application of 0.05 ml of 0.4% DMBA (0.2 mg) in acetone or no DMBA. After 13 months, mice treated with DMBA had developed pigmented lesions ('blue naevi') in the treated areas. For the following seven months, mice received UVB (280–320 nm; high-pressure Hanovia hot quartz contact lamp) three times a week or no UVB treatment. Exposure to UVB following DMBA treatment enhanced the growth of naevi into malignant-appearing pigmented tumours ('melanomas'): 5/18 *versus* 0/41 in the group treated with DMBA alone and 0/39 in the group treated with UVB alone (Epstein *et al.*, 1967). [The Working Group noted the limited reporting on metastases.]

A group of 56 $B6D2F_1$/J mice [sex unspecified], six weeks of age, was irradiated with UVB (280–340 nm; Westinghouse FS40 sunlamp) dorsally for 30 min per day on five days per week (Roberts & Daynes, 1980) for 11.5 weeks (total dose, 6.2×10^5 J/m^2). A control group of 41 mice received no irradiation. Both groups subsequently received a single application of 100 μg DMBA in 0.1 ml acetone on the shaved ventral skin, followed four days later by applications of 5 μg 12-*O*-tetradecanoylphorbol 13-acetate (TPA) three times a week for 32 weeks. Tumour yield was significantly decreased at 32 weeks (2.2 *versus* 4.8 tumours/mouse) in the pre-irradiated mice (Gensler, 1988b).

Groups of 20–24 female hairless Skh-hr 2 mice, six to eight weeks old, received a single application of 0 or 0.5% DMBA in acetone on the dorsal skin. Two weeks later, the animals were irradiated with UVB (290–320 nm; Westinghouse FS40-T12 sunlamp), UVA (320–400 nm; GTE-Sylvania fluorescent black light tubes) or a combination of UVA plus UVB three times a week for 30 weeks or were not irradiated, and were observed for 12 months. All mice receiving DMBA treatment developed multiple 'blue naevi'; virtually none of the untreated mice or mice that received UVR treatment only showed this effect. Irradiation of DMBA-treated animals induced a higher incidence of papillomas (70–100%), squamous-cell carcinomas (30–80%), melanomas (25–33%) and lymphomas (21–50%), than exposure to UVA alone (0–32% papillomas, 0–47% squamous-cell carcinomas, no melanoma and no lymphoma) or to DMBA alone (90, 25, 0 and 5% of these tumours, respectively). The authors also examined selected lesions induced by DMBA alone or by DMBA with UVR for the presence of H- or N-*ras* mutations. Mutations at codon 61 in N-*ras* were present in three (two induced by DMBA plus UVR, one by DMBA alone) out of eight of the early pigmented lesions examined and in one out of three of the malignant melanomas examined (induced by DMBA plus UVR); no H-*ras* mutation was observed (Husain *et al.*, 1991). [The Working Group noted that lesions were not induced by UVR alone.]

3.8.2 *Administration with other agents with promoting activity*

These studies were designed to evaluate the action of UVR as a tumour initiator.

(*a*) *Croton oil*

Groups of 15–53 male and 9–30 female random-bred hairless mice, 9–12 weeks old, received a single exposure to UVB (280–320 nm; high-pressure Hanovia hot quartz contact lamp) for 30 s (1.3×10^7 ergs/cm^3 [13 kJ/m^2]) or no exposure, followed two weeks later by applications to the dorsal skin of 0 or 0.1 ml croton oil in acetone twice a week for 18 months. Neither UVB exposure nor croton oil alone produced any skin tumour over the course of the study. The group of 79 mice that received both UVB exposure and croton oil had eight persistent skin tumours (one per mouse) (Epstein & Roth, 1968).

Groups of 30 female Swiss mice, eight weeks old, received UVB once (5.5×10^7 ergs/cm^2 [55 kJ/m^2]) from Westinghouse FS40T12 lamps or croton oil (0.02 ml of a 2.5% solution, twice a week for 30 weeks); a group of 60 mice received UVB followed after 10 days by croton oil for life. UVB alone produced no tumour; croton oil alone produced regressing tumours, and the combination produced 11 tumours (four papillomas, four fibromas and three regressing tumours) in seven mice (Stenbäck, 1975c).

Groups of 40 male haired mice (random-bred 'Hall' strain), 18 weeks of age, were clipped and exposed once to UVC (medium-pressure mercury discharge lamp). One group received no further treatment; the other received one application of croton oil one day before irradiation and, beginning two weeks later, received applications of 0.25 ml croton oil (0.5% solution) once a week for 30 weeks. By 35 weeks, the groups had 20 and 23 survivors, with 0 and 12 skin tumours, respectively (Pound, 1970).

(b) 12-O-Tetradecanoylphorbol 13-acetate

Six groups of 25 eight-week-old female C3H/HeNCr(MTV$^-$) mice were irradiated with UVB (Westinghouse FS40 sunlamps) on the shaved dorsum for 30 min, five times a week for two weeks (total dose, 1.44×10^5 J/m^2), followed two weeks later by 'promotion' with applications of 0 or 5 µg TPA in acetone twice a week. Ventral irradiation for 30 min, three times a week for 12 weeks (total dose, 4.54×10^5 J/m^2) (to produce a 'systemic' effect) was begun two weeks after completion of dorsal initiation. At 70 weeks, UVB exposure of the dorsum alone had produced no tumour, and dorsal applications of TPA alone had produced a 5% incidence of tumours. The combination of these treatments produced a 41% tumour incidence. Ventral irradiation of animals that had received TPA only produced a 33% incidence, and ventral irradiation of mice that had received both UVB and TPA produced a 100% incidence. The authors suggested that these findings reflect a systemic effect—possibly suppression of immune surveillance or a biochemical influence on the epidermal growth regulatory system (Strickland *et al.*, 1985).

(c) Benzoyl peroxide

Benzoyl peroxide is considered to be a prototype promoter of two-stage chemical carcinogenesis in the skin (Slaga *et al.*, 1981). The studies summarized below were motivated, however, by concerns about the safety of using this compound for treating acne vulgaris.

Groups of Uscd (Hr) stock hairless albino mice (total, 148) [sex unspecified], three to four months old, were exposed on the posterior half of the back to UVR (Hanovia hot quartz contact lamp emitting primarily UVB; 270 mJ/cm^2 [2.7 kJ/m^2]) three times a week for eight weeks. Four weeks later, the mice were divided into four groups. The final skin tumour incidences at the irradiated sites were: 38% in the group that received applications of 0.1 ml of a 0.1% solution of croton oil in acetone on the back skin five times a week for the duration of the experiment (62 weeks); 5% in the group that received applications of acetone alone;

8% in mice that received applications of the benzoyl peroxide base; and 8% in those that received applications of a 5% lotion of benzoyl peroxide in water five times a week for the duration of the study (Epstein, 1988).

Five groups of Oslo hairless mice (16 males and 16 females) were irradiated under Philips HP3114 sunlamps (mostly UVB) twice a week for 52 weeks (total dose, 26.5 J/cm^2 [265 kJ/m^2]). The mice were treated before or after each exposure with 5% benzoyl peroxide in gel, with the gel alone or with no chemical. Throughout the study, the groups were indistinguishable in terms of the proportion with one of more tumours (median latent period, approximately 40 weeks) and of the total number of tumours per survivor (approximately 1.5 at 40 weeks and approximately 4 at 48 weeks). Thus, benzoyl peroxide did not enhance photocarcinogenesis. The study also included several groups of SENCAR mice treated topically with DMBA once (51.2 µg) or with vehicle followed by benzoyl peroxide twice a week. Benzoyl peroxide reduced the number of DMBA-induced tumours (Iversen, 1988). Two unresolved concerns were raised by the author: Firstly, the fact that benzoyl peroxide reduced the tumorigenicity of DMBA was contrary to the author's previous experience (Iversen, 1986) and to that of several others; secondly, the UVR dose used in this study was lower (total dose, 265 kJ/m^2) than that used in the 1986 study (total dose, 480 kJ/m^2), but the tumour response was significantly greater.

(d) Methyl ethyl ketone peroxide

A postulated mechanism for tumour promotion involves the generation of free radicals, possibly with reactive oxygen species, leading to enhanced lipid peroxidation and DNA damage and/or cell phenotype. A study was therefore designed to test whether methyl ethyl ketone peroxide (MEKP), which is known to produce lipid-peroxidizing activity *in vivo*, acts as a tumour promotor in skin 'initiated' by UVR. Furthermore, since glutathione has been shown to be a major endogenous reducing agent which protects against lipid peroxidation, the study also tested diethyl maleate (DEM), which is known to deplete the intracellular level of glutathione in mouse skin.

Groups of 24 male and female hairless albino mice (14–16 weeks old) were irradiated with UVB (280–320 nm; Westinghouse FS40 fluorescent sunlamps; 2054 J/m^2 daily) for 18 weeks. Three weeks later, topical application of MEKP (20 µl containing 0 or 10 µg MEKP) was begun and continued twice a week for 25 weeks. Other groups received DEM (0 or 1 µg in dibutyl phthalate) 1 h before each MEKP application. Otherwise identical control groups received either the chemical treatments or UVB alone. At 46 weeks, the groups that did not receive UVB irradiation had at most two tumours on two mice (among 21 survivors in mice exposed to MEKP plus DEM). Exposure to UVB produced five tumours in four mice exposed to the solvent, out of 19 survivors; 11 tumours in eight mice exposed to MEKP, out of 21 survivors; and 18 tumours in nine mice exposed to MEKP plus DEM, out of 16 survivors. Using tumour onset rate analysis (Peto *et al.*, 1980), the overall effect of MEKP was statistically significant. Tumour enhancement by MEKP was greater in the presence of DEM (Logani *et al.*, 1984).

3.9 Interaction with immunosuppressive agents

Investigations have been reported on agents known to influence immunological responses in humans and on agents chosen to test some aspect of immunological response in mice. [The Working Group noted that in most cases the effect on the immune system of the animals was not evaluated directly; these agents have effects other than immunosuppression, which may explain their interaction with photocarcinogenesis.]

Three groups of 12 male Skh-Hr1 hairless mice, eight weeks of age, were irradiated with 280–320 nm UVB (Westinghouse FS40T12 sunlamps) on five days per week for 30 weeks at daily doses of 470 J/m^2. Two weeks after the first UVB exposure, one group received subcutaneous injections of 0.1 ml *anti-mouse lymphocytic serum* twice a week for 20 weeks; a second received intraperitoneal injections of 12 mg/kg bw *6-mercaptopurine* (Purinethol) five times a week for 20 weeks; and a third received intraperitoneal injections of 0.1 ml isotonic saline five times a week for 20 weeks. Treatment with anti-mouse lymphocytic serum resulted in an earlier appearance and a greater numbers of tumours than did treatment with saline; in contrast, 6-mercaptopurine appeared to delay the appearance of tumours (Nathanson *et al.*, 1976).

Groups of 24–28 female albino HRA/Skh-1 hairless mice, 21–35 weeks of age, were irradiated with UVR (UVB from an Oliphant FL40SE tube and UVA from six Sylvania 40BL tubes) to simulate the UVR portion of terrestrial sunlight on five days per week for 10 weeks to achieve a MED. At the same time, the animals received intraperitoneal injections of 15 mg/kg bw *azathioprine* in 0.1 ml glycine buffer, 10.6 mg/kg bw *cyclophosphamide* in 0.1 ml glycine buffer or 0.1 ml vehicle alone. At day 200, mice receiving UV irradiation alone had a tumour incidence of 77%; those also receiving azathioprine had an incidence of 96% (marginally significant enhancement of tumour growth); and those receiving cyclophosphamide had an incidence of 85% (nonsignificant increase) (Reeve *et al.*, 1985).

Groups of 15 female albino HRS/J hairless hr/hr mice, eight weeks old, were irradiated with UVB (280–320 nm; Westinghouse FS40 sunlamps) on five days a week for 24 weeks; further groups also received injections of 4 or 8 mg/kg bw *azathioprine* or 10 or 25 mg/kg bw *cyclosporine* three times a week. The mean latent period for tumour development was 16 weeks in the group receiving UV irradiation only and 12–13 weeks in the groups also receiving azathioprine or cyclosporine, indicating enhancement of photocarcinogenesis by both drugs (Nelson *et al.*, 1987).

Groups of female C3H/HeN(MTV$^-$) mice [initial numbers unspecified], four to six weeks of age, received grafts of fragments of an antigenic ('regressor') tumour (fibrosarcoma) previously induced in a host animal by UVB. Some animals received no further treatment; other groups received UVB irradiation (Westinghouse FS40; 5 kJ/m^2 per day on five days a week for four to six weeks), subcutaneous injections of 25 or 75 mg/kg bw *cyclosporine* once a day on eight consecutive days, or injections of 20 mg/kg bw *cyclophosphamide* 1, 3, 6, 9 and 13 days after tumour challenge. Tumours grew progressively in the groups treated with UVB or cyclosporine, but not in the groups receiving no further treatment or cyclophosphamide (Servilla *et al.*, 1987).

Groups of six female albino HRA/Skh-1 hairless mice, 10–12 weeks of age, were irradiated with UVA plus UVB (one Oliphant FL40SE tube and three Sylvania F4/350 BL tubes) on five days a week until death (about 35 weeks). During that time, they were also injected intraperitoneally with 15 mg/kg bw *azathioprine*, 20 mg/kg bw *prednisolone* or 15 mg/kg bw *cyclophosphamide* in 0.1 ml saline or given 60 mg/kg bw *cyclosporine* in 0.1 ml peanut oil by gavage or 0.1 ml vehicle alone. Azathioprine, cyclophosphamide and cyclosporine all significantly enhanced photocarcinogenesis with regard to median latent periods and tumour multiplicity. Prednisolone did not enhance this effect, nor did it interfere with the enhancement by other drugs when given in combination with them (Kelly *et al.*, 1987).

Groups of 15-32 female albino Skh-hr 1 hairless mice, 10-12 weeks of age, were irradiated with UVA plus UVB (250-700 nm; one Oliphant FL40SE tube, three Sylvania F40/350 BL tubes and two True-Lite [Duro-Test Corp] tubes) on five days per week for 12 weeks. Two weeks after the first irradiation, mice received intraperitoneal injections on five days a week of 15 mg/kg bw *azathioprine* or *6-mercaptopurine* in 0.1 ml saline or 0.1 ml vehicle alone. Both compounds significantly enhanced skin photocarcinogenesis with regard to median latent period, proportion of malignant:benign growths and tumour multiplicity (Kelly *et al.*, 1989).

3.10 Molecular genetics of animal skin tumours induced by ultraviolet radiation

Three skin papillomas and three skin carcinomas produced in female SENCAR mice after a single exposure to UVB (280-315 nm; Westinghouse FS20; 70 kJ/m^2) were examined for *ras* gene alterations. A five- to 10-fold increase in cHa-*ras* RNA gene expression associated with the gene amplification was found in papillomas and carcinomas, while DNA from carcinomas, but not from papillomas, induced foci in the NIH-3T3 cell transfection assay (Husain *et al.*, 1990).

4. Other Relevant Data

4.1 Transmission and absorption in biological tissues

UVR may be transmitted, reflected, scattered or absorbed by chromophores in any layer of tissue, such as the skin and eye. Absorption is strongly related to wavelength, as it depends on the properties of the responsible chromophore(s). Accordingly, transmission is also wavelength-dependent. Transmission and other optical properties are affected by changes in the structure of the tissue and, especially in the case of the lens of the eye, by ageing.

Absorption of radiation by a tissue chromophore is a prerequisite for any photochemical or photobiological effect; however, absorption does not necessarily have a biological consequence.

4.1.1 *Epidermis*

Since UVR-induced skin cancer is an epidermal phenomenon, this section focuses on epidermis and excludes the dermis.

The epidermis, a tissue with a high replication rate, can be divided functionally into two: an inner, living part (60–160-μm thick in humans) of cells at various stages of differentiation and the outermost, non-living, terminally differentiated stratum corneum (8–15-μm thick in humans). The dividing cell population is located in the innermost basal layer of the living epidermis. Optical properties have usually been studied using isolated strateum corneum or whole epidermis. Absorption and scattering of UVR by the stratum corneum afford some protection to the living part of the epidermis from UVR exposure.

Human and mouse epidermis have important structural differences. The living part and the stratum corneum of human epidermis have about 10 cell layers each. In mice, the living part has two to three cell layers and the stratum cornea one to two cell layers. The interphase of human epidermis and dermis is highly undulated (i.e., epidermal thickness varies), whereas in the mouse it is flat.

Skin contains sebaceous glands which secrete lipid-containing sebum, which forms a film on the stratum corneum.

(a) Humans

The optical properties of human skin have been reviewed (Anderson & Parrish, 1981, 1982).

Everett *et al.* (1966) used a variety of methods to obtain whole epidermal and stratum corneum preparations of human skin. Transmission characteristics (from 240 to 700 nm) were measured using a recording spectrophotometer *via* an integrating sphere which permits the measurement of forward scattered radiation. Transmission values of whole epidermis in white skin ranged from 1% at 250 nm to 44% at 320 nm, while transmission at 400 nm was about 50%.

Kaidbey *et al.* (1979) compared the optical properties (250–400 nm) of whole epidermis and stratum corneum from black and white skins. In general, the absorption spectra from the stratum corneum were similar in shape and magnitude; however, the absorption spectra for whole epidermis were clearly different: At about 300 nm, the absorbance (accounting for scattering) of black epidermis was twice that of white epidermis.

Anderson and Parrish (1981, 1982) presented data which show that epidermal transmission between 260 and 290 nm will be overestimated if no correction is made for tissue fluorescence (330–360 nm). This is most evident at about 280 nm and is consistent with tryptophan or tyrosine fluorescence.

Bruls *et al.* (1984a) measured transmission in whole human epidermis and stratum corneum of UVR between 248 and 546 nm, using a solar blind detector which corrects for fluorescence, and found results different from those of Everett, in particular, that UVC transmission was one to two magnitudes lower. The transmission spectra of whole epidermis and stratum corneum showed a similar general shape but with differences in minima and magnitude. The minimum for epidermis was 265 nm and that for stratum corneum was 275 nm, presumably reflecting different chromophores in those tissues. At 254 nm, transmission in stratum corneum was about two orders of magnitude greater than that in whole epidermis. At about 300 nm, this difference was only one order of magnitude. The transmission in stratum corneum from previously sun-exposed skin was about one order of magnitude less than that in unexposed epidermis at 254 nm. The difference was less at wavelengths > 290 nm. The minimal transmission in stratum corneum from previously sun-exposed skin was shifted from 275 to 265 nm. The authors also showed that the relationship between tissue thickness and transmission of UVR and visible light (log scale) is linear.

Bruls *et al.* (1984b) studied the relationship between the MED of UVB (filtered mercury arc) and UVC (germicidal lamp) and epidermal transmission. A clear linear (log–log) relationship was demonstrated; the MED increased with decreased transmission. Repeated exposure to UVB resulted in higher MEDs of UVB and UVC and decreased transmission of UVB (only epidermis measured) and UVC (epidermis and stratum corneum measured).

Beadle and Burton (1981) extracted skin lipids from human scalps and measured their transmission spectra in hexane. They estimated that lipid concentrations normally present on the skin surface of the forehead would reduce transmission at 300 nm by about 10%.

(b) Experimental systems

No data are available on transmission in the stratum corneum of mice. Sterenborg and van der Leun (1988) measured transmission of 246–365 nm in Skh-hr 1 mouse epidermis *in vitro*. Minimal transmission (about 2%) was observed at 254 nm and 270 nm; 10% was transmitted at 290 nm, 50% at 313 nm and 70% at 365 nm. Agin *et al.* (1981a) studied changes in optical properties of the epidermis of six to eight Skh-1 albino and Skh-2 pigmented (ears and tails) hairless mice irradiated dorsally with a single, 125-h exposure to a UVA source (GE F8T5-BL) with and without a 3-mm glass filter. When unfiltered, 1.4% of the radiation was < 320 nm and when filtered, 0.12% was < 320 nm. The mid-back (whole epidermis) was examined by forward scattering absorption spectroscopy (250–400 nm) at

48 h, 96 h, nine days and 23 days. With the filtered source, there was an increase in absorbance across the spectrum at 48 h, and the absorption spectrum was similar to that of control skin. Transmission returned to the control baseline by 23 days. With the unfiltered source, there was a smaller increase towards baseline absorbance at 48 h. With time, there was a general decrease in absorbance, except at 250–280 nm at which there was an increase at nine and 23 days. At 23 days, the spectrum had not returned to baseline level, despite a normal histological appearance.

de Gruijl and van der Leun (1982a) studied the effect of repeated exposure to UVR on epidermal transmission in Skh-hr 1 hairless albino mice. Groups of 11–40 mice were exposed to daily doses of UVR ranging from 0.11 to 1.9 kJ/m^2 from Westinghouse FS-40 sunlamps. Transmission measurements corrected for fluorescence of the epidermis were made at 313, 302 and 297 nm. After six weeks' exposure, the higher daily doses resulted in decreased transmission at all wavelengths. The optical density (the negative logarithm of transmission) ratios for the three wavelengths were fairly constant with each dose. There was a simple linear relationship between duration of treatment, increased optical density at 297 nm and epidermal thickness, measured microscopically from frozen sections, which indicates that increased optical density is a result of UVR-induced epidermal hyperplasia. These data show that UVR-induced changes in epidermal transmission may modify the UVR dose–response relationship for skin cancer.

(c) Epidermal chromophores

The influence of chromophores on the optical properties of the epidermis has been reviewed by Anderson and Parrish (1981). The main chromophores are urocanic acid (λ_{max}, 277 nm at pH 4.5), DNA (λ_{max}, 260 nm at pH 4.5), the aromatic amino acids tryptophan (λ_{max}, 280 nm at pH 7) and tyrosine (λ_{max}, 275 nm at pH 7), and melanins (Morrison, 1985).

Urocanic acid is the deamination product of histidine and is present in human and guinea-pig epidermis (mainly stratum corneum) at about 35 μg/cm^2 dry weight. It exists in two isomers, *trans* (E) and *cis* (Z); the *trans*-isomer is converted to the *cis*-isomer upon UV irradiation. The absorption spectra of the two isomers are virtually superimposable, but the extinction coefficient of the *cis* isomer at λ_{max} is 20% lower (Morrison, 1985). Norval *et al.* (1988) quantified urocanic acid isomers in mouse (C3Hf Bu/Kam) skin during development and after exposure to UVB radiation. Fetal dorsal mouse skin had a low total urocanic acid content, which increased in neonatal and older animals. Exposure to UVR increased the proportion of the *cis*-isomer within 16 h from 4.7% in nonirradiated mice to 31%, and this was maintained for days (16% after seven days). The photostationary state for in-vivo isomerization in guinea-pig skin is 45% *cis*-/55% *trans*-isomer (Baden & Pathak, 1967).

DNA is not present to any extent in the stratum corneum of guinea-pigs (Suzuki *et al.*, 1977). Bruls *et al.* (1984a) attributed the differences in transmission minima between whole epidermis (265 nm) and stratum corneum (275 nm) in humans to the lack of DNA. Absorption by protein occurs throughout the epidermis.

Melanins are stable protein polymers packaged in melanosomes, produced by melanocytes and transferred to keratinocytes. Melanins absorb broadly over the UV and visible spectrum although they are not neutral density filters of the skin. For example, 3,4-dihydroxyphenylalanine (dopa)-melanin shows a steady decline in optical density between 210 and 340 nm (Anderson & Parrish, 1981). There is no significant racial difference in the number of melanocytes/unit area of a given body site (Szabó et al., 1972), so that differences in the transmission properties of black and white skin are believed to be due to differences in melanin content and in the packaging and distribution of melanosomes in the epidermis (Kaidbey et al., 1979).

(d) *Enhancement of epidermal penetration of ultraviolet radiation*

Prolonged exposure of skin to water increases sensitivity to UVB. This effect is thought to be due to the removal of UVR-absorbing compounds, especially urocanic acid, from the stratum corneum (Anderson & Parrish, 1981).

Spectral remittance at 300–400 nm has been measured in normal and psoriatic white skin after the application of mineral oil. No effect was observed in normal skin, but remittance in psoriatic skin was reduced within seconds after application of oil, implying greater transmission (Anderson & Parrish, 1982). A similar enhancement of transmission was proposed to explain the observation that topically applied arachis oil enhances tumorigenesis by solar-simulated radiation in hairless albino mouse skin (Gibbs et al., 1985).

4.1.2 *Eye*

(a) *Humans*

Boettner and Wolter (1962) measured transmission of direct and forward scattering UVR (220–400 nm) in the cornea, aqueous humour, lens and vitreous humour from nine freshly enucleated normal eyes. There was no corneal transmission of < 300 nm, beyond which the transmission spectrum showed a very steep increase to about 80% transmission at 380 nm (the curve was almost vertical between 300 and 320 nm). Aqueous humour transmitted > 220 nm, with a steep rise to 90% transmission at 400 nm and no evidence of scattering. In a young (4.5-year-old) lens, transmission started at 300 nm with a peak at 320 nm, declining sharply to no measurable transmission between 370 and 390 nm; thereafter, it showed a steep increase. A similar but slower pattern was reported for two older lenses (53 and 75 years old), with greater light scattering. Transmission in the vitreous humour began at 300 nm with a steep increase to 80% transmission at 350 nm. Lerman (1988) showed that transmission of UV at 300–400 nm in normal human lenses decreases with age between three days and 82 years. A review by Sliney (1986) stated that 1% of incident radiant energy in the 300–315 nm range reaches the human retina early in life.

(b) *Experimental systems*

Kinsey (1948) measured transmission of direct UVR [no mention of instrumentation to detect scattering] in the corneal epithelium, whole cornea, aqueous humour, lens and vitreous humour of young adult albino rabbits. The cornea, aqueous and vitreous humor absorbed virtually all radiation at < 300 nm; the lens absorbed > 90% radiation at wavelengths < 370 nm.

Bachem (1956) measured absorption of UVR at 293–435 nm by the lens and cornea from rabbit eyes. Few technical details were given, but the author indicated that scattering was taken into account. The cornea absorbed all radiation at 293 nm, and the lens absorbed all radiation < 334 nm. Calculation of absorption by the lens *in situ* gave a maximum at 365 nm, with little or no absorption at > 400 and < 300 nm.

Ringvold (1980) studied the absorption of UVR at 200-330 nm by cornea from young adult albino rabbits, rats, guinea-pigs and domestic cats. In contrast to the results of other studies, the cornea did not completely absorb wavelengths < 300 nm; depending on the species, absorption at 300 nm ranged from about 30 to 80%. [The Working Group noted that this discrepancy cannot be explained by scattering, as presumed failure to take its effect into account would overestimate absorption.]

4.2 Adverse effects (other than cancer)

This section deals generally with adverse effects of UVR; however, beneficial effects also occur in humans. The vitamin D_3 precursor, previtamin D_3, is formed in the epidermis and dermis through the photochemical action of UVB (Holick *et al.*, 1980). The total daily requirement of vitamin D_3 (cholecalciferol) is supplied in most people by the combination of synthesis in the skin and contribution from dietary sources of animal origin. Older people are at particular risk for developing vitamin D_3 deficiency, partly because the capacity for its formation decreases with age (MacLaughlin & Holick, 1985). The sunscreen *para*-aminobenzoic acid efficiently blocks the photosynthesis of previtamin D_3 in the skin (Matsuoka *et al.*, 1987). It has been estimated that exposure of the cheeks for 10-15 min in the midday sun in Boston, USA, would be sufficient to provide the daily requirement of vitamin D.

4.2.1 *Epidermis*

(a) *Humans*

The most prominent acute effects of UVR on human skin are erythema ('sunburn') and pigmentation, with cellular and histological changes.

(i) *Erythema and pigmentation (sunburn and suntanning)*

Dose-response curves for erythema were constructed for four radiation wavelengths, 254, 280, 300 and 313 nm, by Farr and Diffey (1985); the erythemal response on the back was assessed quantitatively by a reflectance instrument. At 254 nm, erythema was maximal approximately 12 h after irradiation at doses up to about five times the MED. At higher doses, erythema was more persistent, with little change in intensity from about 12 h to at least 48 h after irradiation.

At 313 nm, with doses around the MED, the maximal response was seen 7 h after irradiation; with doses of two to three times the MED, the maximal response occurred at about 4 h. The MED at 254 and 280 nm was substantially lower than that at 300 and 313 nm; however, the slopes of the dose-response curves for erythema with 254 nm and 280 nm radiation were much flatter than those at 300 nm and 313 nm (Farr & Diffey, 1985).

The time-course of UVA erythema following irradiation with a high-intensity UVA source (predominantly 360-400 nm) was found to be biphasic. Erythema, which may be due to heat, was present immediately. It was minimal at about 4 h then increased between 6 and 24 h. The intensity of the early phase was dose-rate dependent, whereas the intensity in the latter phase depended on dose only. The slope of the log dose-erythema response to UVA at 24 h did not differ from that to UVB (Diffey *et al.*, 1987).

A number of variables affect the observation of erythema, including anatomical site, time of observation after irradiation, size of irradiated area, method of recording erythema and season (Diffey, 1982).

The pharmacological changes that may be responsible for erythema have been studied. Plummer *et al.* (1977) examined suction blisters raised on UVB-inflamed human abdominal skin. Bioassayable prostaglandin activity was elevated 6 and 24 h after irradiation, and levels of prostaglandin $F_{2\alpha}$, measured by radioimmunoassay, were elevated at 24 h; levels had returned to normal at 48 h, but erythema persisted. Greaves *et al.* (1978) extended these observations. Following UVC irradiation, arachidonic acid and prostaglandin E_2 and F_2 levels were elevated at 6 h, reached a maximum between 18 and 24 h, when erythema was most intense, but returned to control levels by 48 h, at which time the erythema had subsided. Indomethacin substantially reduced blood flow, with a good correlation between the reduction in visible erythema and prostaglandin E_2 and F_2 activity in irradiated skin. The results are compatible with the view that UVC-induced erythema is mediated by products of arachidonic acid metabolism. Changes in UVB-induced erythema were similar to those with UVC at 24 h, but by 48 h the levels of arachidonic acid and of metabolites had returned to normal, although erythema persisted. Further, although indomethacin suppressed prostaglandin formation, it altered blood flow only slightly, indicating that other factors must play an important role in inflammation following UVB irradiation. Elevated histamine levels have also been observed, but antihistamines have little effect in diminishing erythema (Gilchrest *et al.*, 1981).

Increased pigmentation of the skin by UVR occurs in two distinct phases: immediate pigmentation and delayed tanning (Hawk & Parrish, 1982; Gange, 1987). Immediate pigmentation, thought to result from oxidation and redistribution of melanin in the skin, begins during irradiation and is maximal immediately afterwards; it occurs following exposure to UVA and visible light and may fade within minutes or, after greater doses to people with darker skin, may last up to several days. Delayed tanning is induced maximally by exposure to UVB and becomes visible about 72 h after irradiation. It is associated with an increase in the number of melanocytes as well as with increased melanocytic activity, elongated dendrites, increased tyrosinase activity and increased transfer of melanosomes to keratinocytes. Small freckles may be formed, particularly in fair-skinned individuals.

Not all pigmentary changes induced by UVR are localized at the site of irradiation. Experimental exposures to UVB three times a week for eight exposures at the MED increased the number of melanocytes and produced larger, more dendritic melanocytes in both exposed skin and, to a much lesser extent, areas of skin shielded from the radiation. The increase in melanocyte number in both exposed and covered areas was greater in individuals whose melanocyte density was lower prior to exposure than in individuals with a high initial density (Stierner *et al.*, 1989).

The erythemal and tanning responses of human skin are genetically determined. Responses to a first seasonal exposure of about 30 min to the midday sun have been used as part of the basis for a skin type classification for white-skinned people ranging from Celtic to Mediterranean (Morison, 1983a; Pathak *et al.*, 1987):

Skin type I   Always burn, never tan
Skin type II   Usually burn, tan less than average (with difficulty)

Skin type III   Sometimes mild burn, tan about average
Skin type IV   Rarely burn, tan more than average (with ease)

UVA radiation produces immediate changes in melanocytes in white-skinned people. In individuals with type-II skin, multiple pinocytotic vesicles, larger vacuoles, swelling and partial-to-total dissolution of the inner membranes of mitochondria and numerous small vesicles associated with an enlarged Golgi apparatus were seen with doses that did not produce immediate pigment darkening (Beitner & Wennersten, 1983). In those with type-III skin, similar changes occurred but only with doses that produced immediate pigment darkening (Beitner, 1986).

Three Japanese skin types have been described on the basis of personal reactions to the sun (Kawada, 1986). Experimental exposure to monochromatic UVR showed that the MED correlated well with skin type. Immediate tanning occurred but was not related to skin type. After irradiation with the minimal dose that would produce immediate tanning, the tan faded within 3–15 min; after greater exposures, the tan remained longer but never for more than 60 min. The action spectrum for immediate tanning had a maximum at 320 nm and decreased gradually towards 400 nm. New pigment formation (delayed tanning) after exposure to 290 nm and 305 nm radiation began about 65 h after irradiation and increased until it reached a maximum at 124 h (with a dose four times the MED) or 151 h (with a dose eight times the MED). Following a dose three times the MED, some delayed tanning was still evident after two months. The minimal melanogenic dose (producing delayed tanning) was greater than the MED for all Japanese skin types, in contrast to findings in white Caucasians.

Parrish *et al.* (1981) showed that repeated daily exposure to doses of broad-band UVB and UVA lower than the MED lowered the threshold for both erythema and true melanogenesis for several subsequent days; the threshold for melanogenesis was decreased to a greater extent than that for erythema, a separation that was more pronounced for UVA than for UVB radiation.

(ii) Pigmented naevi

Exposure to the sun appears to stimulate the occurrence and behaviour of acquired pigmented naevi. Kopf *et al.* (1985) showed, in 80 consecutive patients with dysplastic naevus syndrome, that the concentration of naevi on areas of the thorax protected relatively well from the sun was substantially lower than that on areas exposed to the sun. Augustsson *et al.* (1990) showed that, in melanoma cases as well as in controls, the concentration of common naevi was higher on the sun-exposed skin of the back than on the protected skin of the buttocks. An Australian study compared naevi excised in summer to those excised in winter in Western Australia. Inflammation, regression, mitotic activity and lymphocytic infiltration were significantly more prevalent in naevi excised in summer than in winter (Holman *et al.*, 1983b; Armstrong *et al.*, 1984). [The Working Group noted that these observations may be confounded by the site of the naevi.]

In an Australian cross-sectional study of 511 people, the presence of palpable naevi on the forearm was associated with female sex, young age, not having southern European grandparents, being born in Australia and intermediate categories of variables indicating sun exposure (Armstrong *et al.*, 1986).

Gallagher *et al.* (1990a,b) studied risk factors for common naevi in school children in Vancouver, British Columbia, Canada. The number of naevi increased with age (from six to 18 years). Naevi occurred most commonly on intermittently than on constantly exposed parts of the body and less commonly in skin that was rarely exposed. Light and freckled skin, propensity to burn rather than tan upon exposure to the sun and a history of frequent or severe sunburn were associated with a large number of naevi.

Green *et al.* (1988b) compared the prevalence of melanocytic naevi (benign pigmented moles) in children aged 8–9 in Kiddermister, United Kingdom, and Brisbane, Australia. Regardless of skin colour, the mean number of naevi was at least five times larger in the Australian children than in the British children. In both populations, naevi were more prevalent in children with fair skin.

(iii) *Ultrastructural changes*

Jones, S.K. *et al.* (1987) and Roth *et al.* (1989) each described a patient who developed many freckle-like lesions on all exposed sites following repeated exposure to high-dose UVA from a home sunbed for tanning the skin. Biopsy showed increased numbers of large melanocytes in the basal layers.

Rosario *et al.* (1979) examined the sequential histological changes produced by single exposures to UVA, UVB and UVC radiation on untanned skin of the lower back. Exposures were designed to cause approximately equal degrees of erythema. Following UVB and UVC, dyskeratotic cells ('sunburn cells') were scattered throughout the malpighian layer of the epidermis at 24 and 48 h. By 72 h and seven days, they formed a continuous band in the upper malpighian layer or the stratum corneum. Epidermal hyperkeratosis, parakeratosis and acanthosis appeared concurrently at 72 h. The granular layer was focally absent at 24 and 48 h and had increased focally at 72 h and seven days. There was a minimal-to-moderate lymphocytic infiltrate in the dermis which was most pronounced after 48–72 h. Infrequent mitotic figures were observed in keratinocytes. UVA caused fewer dyskeratotic cells at all time intervals, and these never coalesced into a band. UVA, however, elicited the greatest degree of inflammation at 24, 48 and 72 h in terms of both quantity and depth of cellular infiltrate. Endothelial cell swelling, nuclear dust and extravasation of red blood cells were generally observed together. These dermal findings were more pronounced at 72 h. Neither epidermal hyperkeratosis, parakeratosis nor acanthosis was observed. Intracellular oedema of moderate degree was noted with all wavebands at all time intervals. The authors considered that the production of more prominent dermal changes by UVA than by UVB and UVC might be related to greater penetration of longer wavelengths. The histological changes returned to normal earliest after UVB and latest after UVA irradiation.

Pearse *et al.* (1987) examined the effects of repeated irradiation with UVB (0.5, 1 and 2 times the MED three times a week for six weeks) and UVA (6 J/cm^2 [60 kJ/m^2] three times a week for three weeks). UVB irradiation at twice the MED led to significant increases in epidermal thickness, stratum corneum thickness and keratinocyte height, as did UVA irradiation. Both UVA and UVB significantly increased glucose-6-phosphate dehydrogenase activity and decreased succinic dehydrogenase activity throughout the epidermis. The autoradiographic labelling index was significantly increased following the highest dose of UVB.

The benign skin changes attributed to sunlight and seen on physical examination include wrinkles, atrophy, cutis rhomboidalis nuchae (thick, yellow, furrowed skin, particularly on the back of the neck), yellow papules and plaques on the face, colloid milium (firm, small, yellow, translucent papules on the face, forearms and hands), telangiectasia, diffuse erythema, diffuse brown pigmentation, ecchymoses in sun-damaged areas, freckles, actinic lentigo (large, irregular, brown areas), Favre–Racouchot syndrome (yellow, thick comedones and follicular cysts of the periorbital, malar and nasal areas) and reticulated pigmented poikiloderma (reddish-brown reticulated pigmentation with telangiectasia and atrophy and prominent hair follicles on exposed chest and neck) (Goldberg & Altman, 1984). Although most commonly seen in fair-skinned Caucasians, these changes may also be seen in Chinese heavily exposed to the sun (Giam, 1987). A visual system using facial photographs has been developed to enable grading of the degree of elastosis (Cameron et al., 1988).

Holman et al. (1984a,b) made silicone rubber moulds of the microtopography of the skin of the hands of 1216 subjects and developed a grading system to describe alterations in skin surface characteristics observed under a low-power microscope. Using multivariate analysis, independent risk factors for topographic evidence of actinic skin damage were: male sex, age, tendency to burn upon exposure to the sun and outdoor occupation. Similar results were reported by Green (1991).

Everett et al. (1970) reported ultrastructural changes in the epidermis of six elderly, fair-skinned, freckled, blue-eyed, Caucasian male farmers with a history of multiple actinic keratoses and skin cancers. Light microscopy showed effacement of epidermal rete ridges and an irregular decrease in epidermal thickness in areas of skin exposed to sunlight. Three groups of changes were apparent upon transmission electron microscopic examination: firstly, local areas of degeneration involving groups of adjacent cells, with degenerative changes resembling dyskeratosis in both the basal and the spinous layers of the epidermis; secondly, disturbed cellular cohesion, with variable numbers, distribution and degrees of maturity; and thirdly, changes in epidermal pigment—with the melanin concentration varying from none to excessive—and melanosome complexes that were often abnormally large.

Kligman (1969) described the changes in elastic tissue (elastic hyperplasia or actinic elastosis) seen in the dermis of sun-exposed Caucasian facial skin. Such changes were quite advanced before the extent of the damage became visible clinically. Some elastic hyperplasia was seen in elderly blacks over the age of 70, but the changes were markedly less extensive than those seen in whites.

Bouissou et al. (1988) studied elastic fibres in protected skin and skin highly exposed to the sun from normal Caucasians of different ages, using light and electron microscopy. In skin exposed to the sun, there was elastotic degeneration in the reticular dermis and progressive thickening and curling of the elastic fibres in the upper dermis. Altered fibres progressively formed thick, irregular masses, with clumps of amorphous, granular, elastotic material and large areas of uneven staining appearing frequently thereafter. Electron microscopy revealed that normal collagen and elastotic material were often contiguous but never continuous.

(iv) *Keratosis*

The occurrence of keratosis, a benign but probably premalignant squamous neoplasm of the skin (Marks *et al.*, 1988), has been studied in relation to exposure to sunlight in several cross-sectional studies.

Chronic solar damage (assessed by cutaneous microtopographs and paraocular photographs) was associated with keratosis, after adjustment for age, in a study of 1216 people in Busselton, Australia (Holman *et al.*, 1984a). A similar association between cutaneous microtopography and prevalence of keratosis was observed by Green (1991) in a study of 1539 people in Nambour, Australia.

Vitasa *et al.* (1990) conducted a study of 808 white watermen in Maryland, USA. The prevalence of keratosis was 25%. The risk factors for this condition were found in a multivariate analysis to be age, individually estimated cumulative exposure to sunlight, blue eyes, childhood freckling and a tendency to sunburn.

Marks *et al.* (1983) studied 2113 adults in Maryborough, Australia. The prevalence of keratosis was 56.9%. Adjusted for age, the prevalence of keratosis was significantly associated with being born in Australia, with a tendency to sunburn and not tan and with blue eye colour. In another survey by these authors, of 2000 adult in-patients from a hospital in Melbourne, Australia, the prevalence of keratosis on the light-exposed areas of the head and neck, forearms and back of hands was 37.7%. Prevalence of keratosis was significantly associated with age and with being born in Australia and, among men, with outdoor occupation (Goodman *et al.*, 1984). The Melbourne and Maryborough populations were compared further by Marks and Selwood (1985), who attributed the higher prevalence of keratosis in Maryborough to the fact that this population had a 14.2% higher erythemal UVR level.

Foley *et al.* (1986) studied 766 consecutive patients with keratosis. Lesions on the hands and forearms in men were seen more often on the right side than on the left, which the authors attributed to the higher exposure of the right side while driving an automobile. In women, more lesions of the head and neck were on the left side.

(v) *Photosensitivity disorders*

Abnormal reactions to solar radiation, termed photosensitivity disorders, occur in a relatively small number of exposed individuals; these have been reviewed comprehensively (Harber & Bickers, 1981; Bernhard *et al.*, 1987). Genetic and metabolic diseases that may be associated with photosensitivity include xeroderma pigmentosum, phenylketonuria, Bloom's syndrome, Cockayne's syndrome, Rothmund–Thomson syndrome, certain porphyrias, Hartnup syndrome and pseudoporphyria cutanea tarda. The excision repair disorders are discussed on pp. 191–194. Defects in pigmentation due to an absence of melanocytes (vitiligo) and defective functioning of melanocytes (albinism) also confer susceptibility to UVR because of failure to develop photoprotection through tanning responses.

In idiopathic photodermatoses, the primary abnormality is an acquired alteration in reaction to sunlight. The commonest form is polymorphous light eruption, in which individuals who previously tolerated sun exposure develop itchy papules, vesicles or erythematous patches or plaques on exposed areas after moderate exposure to the sun (Bernhard *et al.*, 1987). Other photosensitivity conditions include solar urticaria (Armstrong, 1986), hydroa vacciniforme (hydroa aestivale) (Halasz et al., 1983) and actinic reticuloid (Bernhard et al., 1987).

Photoaggravated dermatoses are conditions that may occur in the absence of exposure to sunlight but can be induced or exacerbated by such exposure. The commonest is recurrences of herpes simplex viral eruptions, usually on the upper lip; this viral infection has been reproduced by exposure to artificial sources of UVR (Spruance, 1985).

Other skin diseases reported to be photoaggravated include lupus erythematosus, Darier's disease, acne vulgaris, atopic dermatitis, bullous pemphigoid, disseminated superficial actinic porokeratosis, erythema multiforme, lichen planus, pellagra, pemphigus, pityriasis alba, pityriasis rubra pilaris, psoriasis, acne rosacea, seborrheic dermatitis and transient acantholytic dermatitis (Grover's disease) (Bernhard et al., 1987).

(b) Experimental systems

Agin et al. (1981b) found that single exposures to UVA plus UVB caused thickening of the whole epidermis and stratum corneum in pigmented and albino hairless mice. Sterenborg et al. (1986) found similar changes after repeated exposures to mainly UVB in hairless albino mice.

C57Bl mice irradiated with UVB daily for 10 days had a four-fold increase in the number of epidermal melanocytes, with increased pigmentation and local thickening of the epidermis (Rosdahl, 1979). A gradual, delayed, three-fold increase in the number of melanocytes also occurred in shielded contralateral ears, without increased pigmentation or epidermal thickening.

Generally consistent observations have been reported on chronic changes (photoageing) in hairless mice (Bissett et al., 1987, 1989; Kligman, 1989). Bissett et al. (1987) described the progression of chronic UV damage to the skin in albino hairless Skh:Hr-1 mice irradiated with UVB or UVB plus UVA three times a week for 16 weeks, with a 17-week recovery period. UVB and a combination of UVA and UVB produced similar changes. An early increase in transepidermal water loss was seen, with a doubling of skin thickness and changes in the microtopography of the skin surface with visible skin wrinkling. Dose-dependent histological changes were seen, with thickening and hyperplasia of the epidermis. Dermal elastic fibres thickened and proliferated throughout the upper dermis, and there was a proliferation of fibroblasts, sebaceous cysts and dermal cysts in the upper dermis. By week 16, the skin was clearly elastotic, with thick, tangled masses of elastic fibres in the dermis. Use of a broad-spectrum sunscreen product with a claimed SPF (skin protector factor) of 15 retarded but did not completely prevent the effects of UVB and of UVB plus UVA radiation. Animals exposed to UVB and then allowed to recover for 12 weeks exhibited a zone of clearance of all abnormal elastin from the dermal–epidermal junction to mid-way down the dermis.

Animals exposed to UVA alone for 33 weeks with a recovery period of 18 weeks (Bissett et al., 1987) exhibited a different pattern of changes. Epidermal thickening occurred at a slower rate, there was no increase in water loss; and sagging rather than wrinkling of the skin occurred. There was a very gradual increase in cellularity; focal areas of collagen damage and absence of elastic fibres were seen; the size and number of dermal cysts increased; and there was only slight evidence of recovery after 18 weeks. UVA appeared to accelerate several changes similar to those that occur with chronological ageing in mice. Using a dual grating monochromator, Bissett et al. (1989) examined the action spectra for these changes. Most were similar and occurred in the UVB waveband: wrinkling, glycosaminoglycan increase, collagen damage, elastosis, epidermal thickening, dermal cellularity and dermal inflammatory cell increase. In contrast, the spectrum for skin sagging was very broad, with a maximum near 340 nm. These results suggest that more than one chromophore is involved in UV-induced chronic skin changes.

High doses of UVA (cumulative dose, 3000 J/cm^2) were reported to produce severe elastic fibre hyperplasia, but no large aggregates of elastosis or destruction of collagen, in female Skh-hr 1 albino mice (Kligman et al., 1985; Kligman, 1989). A dose of 13 000 J/cm^2 from a filtered (50% cutoff at about $\leq$ 345 nm) UVA source, however, produced only insignificant changes. Dose–response studies with another UVA source, filtered to remove all radiation below 340 nm, produced some elastin thickening at a total dose of 8000 J/cm^2 as well as increased epidermal proliferation and increased and enlarged dermal cysts (Kligman et al., 1987).

Kligman and Sayre (1991) found that the action spectrum for elastosis in albino hairless mice was similar to that for erythema, except that longer UVA wavelengths (> 330 nm) were less effective for elastosis.

The chronic effect of repeated UV irradiation was also investigated in naked albino Ng/- mice using high total doses (> 20 000 J/cm^2) from a predominantly UVA source (but containing some UVB) administered for 16 h daily for 8.5 months (Berger et al., 1980a). Dermal changes similar to those seen in human actinic elastosis were observed. There was endothelial swelling of dilated small capillary vessels and slight perivascular infiltration. Particularly in the upper dermis, collagen was replaced with an amorphous material that stained faintly with haematoxylin–eosin. Mast cells and a relatively increased number of spindle-shaped fibroblasts were found in the middle and lower dermis. Large aggregates of numerous tangled, thickened fibres with the staining properties of elastic tissue were seen. Electron microscopy showed that elastic fibres were increased in number and size and there was splitting of collagen fibres. Most small blood vessels were dilated, with multiple basal lamina. The elastic tissue changes showed no signs of regression 2.5 months after irradiation had been discontinued, although the epithelial changes regressed over this period.

Similar changes in elastic tissue (Berger et al., 1980b) were found after exposure to a filtered UVA source which contained no UVB, but no alteration of collagen was observed and inflammatory changes were absent. Electron microscopy showed changes similar to those observed in actinic elastosis.

In female, lightly pigmented, hairless Oslo/Bom mice, UVB alone produced moderate elastosis, UVB and UVA together produced a slightly reduced degree of elastosis, but UVB followed by large doses of UVA produced severe elastosis; UVA alone was reported to have no effect (Poulsen et al., 1984). In Skh:Hr 1 albino hairless mice, a combination of UVA and UVB had additive effects (Kligman et al., 1985).

(c) Comparison of humans and animals

No direct comparison has been reported of the optical properties of whole human and mouse epidermis; however, the available data suggest that the absorption/transmission spectra are of a similar general shape but have marked quantitative differences. For example, a comparison of data on a graph of effects on human epidermis not previously exposed to UVR (Bruls et al., 1984a) with tabulated data on mouse epidermis not previously exposed (Sterenborg & van der Leun, 1988), generated in the same laboratory, showed that transmission in the mouse was two orders of magnitude greater in the UVC region and one order of magnitude greater in the UVB and UVA regions than in humans. In human and mouse epidermis, prior exposure to UVR resulted in marked decreases in UVR transmission. No study has been reported on mouse stratum corneum.

4.2.2 Immune response

Exposure to solar radiation and UVR can alter immune function in experimental animals and humans. This area of research is known as photoimmunology and has recently been reviewed (Daynes et al., 1983; Parrish, 1983; Parrish et al., 1983; Bergstresser, 1986; Roberts et al., 1986; Krutmann & Elmets, 1988; Morison, 1989).

(a) Humans

  (i) *Contact hypersensitivity (allergy)*

Exposure of normal subjects to radiation in a tanning solarium which emitted mainly UVA but also UVB radiation reduced allergic reactions to 2,4-dinitrochlorobenzene (Hersey et al., 1983a). Halprin et al. (1981) and Nusbaum et al. (1983) found that UVB radiation partially suppressed the development of contact allergy to nitrogen mustard in patients with mycosis fungoides and psoriasis. Exposure to UVB was begun prior to treatment with mustard, and the field of exposure to the chemical was included in the area exposed to radiation, so that both a local and systemic effect may have been measured. In both studies, the proportion of patients sensitized to mustard gas was reduced by exposure to UVB radiation, and sensitization, when it did occur, was delayed. [The Working Group noted that the presence of diseases known to influence the immune system makes the findings difficult to interpret.]

Response to 2,4-dinitrochlorobenzene was diminished in sun-damaged skin in subjects previously sensitized to the allergen (Kocsard & Ofner, 1964; O'Dell et al., 1980). UVB-induced suppression of contact allergy to nickel and other allergens (e.g., cobalt) has also been reported (Mørk & Austad, 1982; Sjövall & Christensen, 1986).

Studies on the possible mechanism of suppression have focused mainly on the effects on antigen presentation in the skin. At low doses of UVB ($\leq$ 15 mJ/cm^2), Langerhans' cells are the only epidermal cells to be altered morphologically (Aberer et al., 1981). Depletion of Langerhans' cells after a few exposures to UVB radiation is transient (Tjernlund & Juhlin, 1982; Scheibner et al., 1986a); however, chronic exposure to sunlight appears to result in a sustained reduction, since fewer Langerhans' cells are found in exposed than in unexposed skin of older adults but not of young adults (Gilchrest et al., 1982; Scheibner et al., 1983; Thiers et al., 1984; Czernielewski et al., 1988). Pigmentation does not seem to protect Langerhans' cells, since exposure to UVB plus UVA radiation (simulating natural UVR) produced similar degrees of depletion of these cells in dark-skinned Australian aboriginals and in fair-skinned people of Celtic descent (Hollis & Scheibner, 1988); Langerhans' cells were equally affected in fair-skinned and dark-skinned people after multiple exposures to sunlight (Scheibner et al., 1986b).

The antigen-presenting function of Langerhans' cells is also diminished after irradiation *in vivo* with UVB (Cooper et al., 1985; Räsänen et al., 1989). The function returns to the epidermis within 24 h, owing to the appearance of two cell populations that are distinct and different from Langerhans' cells (Cooper et al., 1986). Both populations have receptors for the monoclonal OKM5 antibody; one also has receptors for the OKM1 antibody and is possibly a dendritic cell from blood, while the other is OKM1⁻ and is related to a subset of blood monocytes. These cells can activate T cells in the absence of exogenous antigen and lead to the generation of T-suppressor cells which can inhibit various immune responses. Baadsgaard et al. (1988) showed that epidermal cells from UVB-irradiated skin can stimulate suppressor/cytotoxic lymphocytes. This may occur *via* at least two pathways: activation of T-suppressor/inducer cells or induction of interleukin-2 production. These observations suggest that UV-induced immune suppression is more closely related to the appearance of OKM5$^+$ cells in the epidermis than to the disappearance of Langerhans' cells.

Systemic suppression of contact allergy may also result from exposure to UVR. Granstein and Sauder (1987) exposed subjects to a MED of mainly UVB radiation and measured levels of serum interleukin-1 activity that peaked 1–4 h after exposure and returned to baseline by 8 h. This activity may originate from the skin, in which increased levels have been detected after UVB irradiation (Kupper et al., 1987; Oxholm et al., 1988; Räsänen et al., 1989).

A recent study (Yoshikawa et al., 1990) showed that suppression of UVB-induced contact allergy may be a risk factor for nonmelanocytic skin cancer. Approximately 60% of normal subjects were sensitized by application of 2,4-dinitrochlorobenzene to UVB-irradiated skin compared to 8% of patients with a history of skin cancer. Many skin cancer patients were also immunologically tolerant to this allergen; this was not observed in normal subjects.

Pigmentation does not protect against UV-induced immunosuppression, since it occurs in the same proportion of black and white people (Vermeer et al., 1991).

(ii) *Lymphocytes*

A single, whole-body exposure to UVB radiation which produced painful erythema produced a transient decrease in the proportion of circulating E rosette-forming cells and in the response of lymphocytes to a mitogen (Morison et al., 1979a). McGrath et al. (1986) found a decrease in the proportion of circulating suppressor cells following exposure to half the MED of UVB, although the total number of T lymphocytes was not altered. Exposure of normal subjects to sunlight daily for two weeks, however, produced different effects: The total proportion of T lymphocytes was diminished owing to a pronounced drop in the proportion of helper/inducer cells associated with an increase in the proportion of suppressor cells in the peripheral blood (Hersey et al., 1983b). Similar changes occurred after exposure of normal subjects to UVA plus UVB radiation (Hersey et al., 1983a). When UVB radiation was removed by a Mylar filter (Hersey et al., 1988) or a sunscreen (Hersey et al., 1987), most of the effect was removed. The numbers of circulating T cells and helper-T cells were significantly reduced by exposure of normal subjects to solar lamps containing UVA (with minimal UVB) and to fluorescent tubes emitting mainly visible light, which contained small quantities of UVB, but the number of T-suppressor cells was only slightly reduced. These effects were considered to be due to the UVB radiation (Rivers et al., 1989).

(iii) *Infectious diseases*

Recurrent infections due to herpes simplex virus types 1 and 2 can be induced by exposure to UVB radiation (Wheeler, 1975; Spruance, 1985; Klein & Linnemann, 1986; Perna et al., 1987). Presumably, local alterations of immunity, associated with extensive UV-induced tissue damage, are responsible for this reactivation.

(iv) *Photosensitive disease*

An interaction between solar radiation and the immune system was first postulated on the basis of observations that the pathogenesis of several diseases is characterized by photosensitivity. Solar urticaria, photoallergy and lupus erythematosus are the main examples (for reviews, see Morison, 1983b,c; Morison & Kochevar, 1983).

(b) Experimental systems

(i) *Contact hypersensitivity*

The first report of UV-induced suppression of contact hypersensitivity was in guinea-pigs that received applications of a sensitizing chemical through UV-irradiated skin (Haniszko & Suskind, 1963). This effect has since been termed local suppression of contact hypersensitivity. Later, in studies of UV-induced tumour susceptibility in mice, it was found that UVR could also induce systemic suppression of contact hypersensitivity when the sensitizer is applied through unexposed skin only (Kripke et al., 1977). This occurred during chronic treatment of mice, was transient and appeared to be due to failure of an effector mechanism (efferent block) of the immune response (Jessup et al., 1978). These two phenomena, local and systemic suppression of contact hypersensitivity, are probably mediated by different mechanisms.

Local suppression of contact hypersensitivity: Pretreatment of mice with low doses of UVB radiation (100–700 J/m^2 fluorescent sunlamp radiation daily for four days) suppressed the development of contact hypersensitivity to sensitizing chemicals (e.g., 2,4-dinitrofluorobenzene) applied subsequently to irradiated skin (Toews et al., 1980; Elmets et al., 1983). This effect was associated with generation of hapten-specific LyT-1^+ T cells which suppress the induction phase of the immune response (Elmets et al., 1983). The most effective wavelengths are < 300 nm (Elmets et al., 1985). Local suppression of contact hypersensitivity by UVB radiation also occurs in hamsters (Streilein & Bergstresser, 1981).

Several hypotheses have been explored to explain the mechanism of local suppression. Multiple exposures to sunlight result in a striking reduction in the number of Langerhans' cells in guinea-pigs, as detected by ultrastructural examination (Fan et al., 1959). UV-induced alterations occur in Ia^+ Langerhans' cells (Streilein et al., 1980; Perry & Greene, 1982; Gurish et al., 1983; Stingl et al., 1983), but alterations in other cells may be involved.

Thy-1^+ dendritic epidermal cells (identified by antibodies to surface markers on lymphocytes), found in mouse but not reported in human skin, are bone marrow-derived lymphocytes which down-regulate contact hypersensitivity. They are not affected by low-dose UVR, and hapten-conjugated Thy-1$^+$ dendritic epidermal cells can induce tolerance on subcutaneous injection into the footpad or after intravenous injection (Welsh & Kripke, 1990). This finding is supported by the observations (Okamoto & Kripke, 1987) that (i) the draining lymph nodes of mice treated with low doses of UVR contained these hapten-conjugated cells after exposure to a contact sensitizer, (ii) injection of these cells into other syngeneic mice resulted in the generation of suppressor cells, and (iii) removal of these cells from the lymph node cells abolished the suppression.

I-J$^+$, Thy-1$^-$, Ia$^-$ antigen-presenting cells, which are also resistant to low doses of UVB radiation and preferentially generate a suppressor cell pathway, may also be involved in local suppression (Granstein et al., 1984; Granstein, 1985; Granstein et al., 1987; Okamoto & Kripke, 1987).

Keratinocytes may also be involved through the production of epidermal cell-derived thymocyte-activating factor (ETAF), which is functionally and biochemically very similar to interleukin-1, a nonspecific helper factor necessary for activation of T cells by antigen. Interleukin-1 can reduce expression of contact hypersensitivity in mice (Robertson et al., 1987). Studies by several workers have suggested that exposure to UVR inhibits the production of ETAF (Sauder et al., 1983) or decreases its activity (Stingl et al., 1983). When antigen-presenting cells are exposed to UVR, their ability to activate T cells is markedly inhibited (Tominaga et al., 1983). UV irradiation of mice induces the release of a specific interleukin-1 inhibitor, keratinocyte-derived, EC-contra IL 1 (Schwarz et al., 1988). Other workers (Ansel et al., 1983; Gahring et al., 1984) have found increased production of ETAF. [The Working Group noted that differences in the radiation sources and model systems could explain the discrepancies between the results of these studies.]

Systemic suppression of contact hypersensitivity: Systemic suppression of contact hypersensitivity in mice requires a higher exposure dose (40–50 KJ/m^2) than local suppression (Kripke & Morison, 1986a). A dose of 8.2 kJ/m^2 at 320 nm produced nearly 50% systemic suppression, and 100 kJ/m^2 produced 80% suppression (Noonan et al., 1984). Like local suppression, systemic suppression is associated with the generation of suppressor Lyt-1$^+$ T lymphocytes (Noonan et al., 1981a; Ullrich & Kripke, 1984). The pathways leading to the appearance of these lymphocytes are, however, probably different. Systemic suppression has also been induced in guinea-pigs (Morison & Kripke, 1984) and in the South American opossum, *Monodelphis domestica* (Applegate et al., 1989). Artificial sources of UVB radiation and sunlight, but not UVA, induce systemic suppression of contact allergy in mice and guinea-pigs (Morison et al., 1985).

Determination of an action spectrum for systemic suppression of contact hypersensitivity in mice revealed peak activity in the 260–270 nm region, which is consistent with a superficial location of the chromophore in the epidermis (De Fabo & Noonan, 1983; Noonan & De Fabo, 1985). Two candidate molecules, urocanic acid and DNA, have been suggested.

Several lines of evidence indicate that abnormalities in Langerhans' cells are not involved in systemic suppression, in contrast to local suppression (Lynch et al., 1983; Morison et al., 1984; Noonan et al., 1984), and that a defect of antigen presentation is not an initial step (Kripke & McClendon, 1986). Soluble mediators are released from irradiated skin and may generate suppressor cells in a distant organ. Serum collected from UV-exposed mice and epidermal cells exposed to UVR *in vitro* contain factors that can induce systemic suppression

(Schwarz et al., 1986). The situation is far from straightforward, however, since a recent study indicated that multiple suppressive factors, with different immunosuppressive properties, may be released by different wavelengths of UVR (Kim et al., 1990). Indomethacin blocks the development of suppression (Chung et al., 1986; Jun et al., 1988), indicating that prostaglandins may also be involved in the pathway.

Several properties of the suppressor cells have been defined: (i) they suppress primary proliferative responses but not a secondary response *in vitro* (this is consistent with the idea that they suppress induction of sensitization but not with the proposal that they elicit a response in a previously sensitized animal) (Ullrich, 1985); (ii) their action is limited to T-dependent antigens (Ullrich, 1987); and (iii) they can modulate other immunological pathways, such as formation of anti-hapten antibodies and cytotoxic-T lymphocytes (Ullrich et al., 1986a).

(ii) *Delayed hypersensitivity to injected antigens*

Systemic suppression of delayed hypersensitivity was induced by UVB irradiation of mice following injection of 2,4-dinitrochlorobenzene into the footpad (Jessup et al., 1978), of hapten-coupled spleen cells into the footpad (Greene et al., 1979) or the ear (Noonan et al., 1981b) or of erythrocytes and soluble protein antigens into the footpad (Ullrich et al., 1986b) and is associated with the generation of antigen-specific T lymphocytes. This suppression differs from the suppression of contact hypersensitivity to topically applied allergens because delayed hypersensitivity can be restored in UV-irradiated mice by injection of hapten-coupled spleen cells from normal mice (Noonan et al., 1981b; Kripke & Morison, 1985, 1986b). Furthermore, systemic injection of methylprednisolone before immunization prevented suppression of delayed hypersensitivity but had no effect on the suppression of contact hypersensitivity (Kripke & Morison, 1986b).

Systemic depression of splenic antigen-presenting cell function was demonstrated in UVB-exposed mice (Letvin et al., 1980a,b; Gurish et al., 1982). Two explanations have been advanced: a transient redistribution of antigen-presenting cells to peripheral lymphoid tissues in response to UV-induced inflammation (Gurish et al., 1982; Spangrude et al., 1983) or direct damage to blood monocytes or other precursors of splenic antigen-presenting cells as they circulate through the skin (Spangrude et al., 1983). The latter theory is supported by the observation that immunization with hapten-conjugated splenic antigen-presenting cells or epidermal cells exposed *in vitro* to UVR can induce hapten-specific T-suppressor cells (Fox et al., 1981; Sauder et al., 1981).

The role of one of the proposed chromophores, urocanic acid, has been explored. UV-irradiated urocanic acid (containing 74% *cis*-urocanic acid after 4 h) suppresses delayed hypersensitivity to HSV-1 when injected subcutaneously or applied to the skin of mice (Ross et al., 1986), and is thus similar to UVB radiation (Ross et al., 1987). In both instances, phenotypically similar suppressor cells were induced (Howie et al., 1986a; Ross et al., 1987). In addition, intravenous administration of *cis*-urocanic acid impairs antigen-presenting cell function in splenic dendritic cells. These observations suggest that *trans*-urocanic acid is the photoreceptor for UVB-induced systemic suppression of delayed hypersensitivity and that *cis*-urocanic acid acts as an immunomodulator (Noonan et al., 1988).

(iii) *Immunology of ultraviolet-induced skin cancer*

Most UV-induced tumours in mice are highly antigenic and are rejected upon transplantation into normal syngeneic recipients; however, they grow progressively in immunosuppressed recipients (Kripke, 1974). The specific immunological rejection of these transplanted tumours is mediated by cytolytic-T lymphocytes aided by natural killer and cytotoxic-T cells (Fortner & Kripke, 1977; Fortner & Lill, 1985; Streeter & Fortner, 1988a,b). Tumours grow in UV-irradiated recipients or primary hosts because T-suppressor lymphocytes induced by the exposure to UVR block the normal immunological surveillance system (Fisher & Kripke, 1977; Spellman *et al.*, 1977; Fisher & Kripke, 1978; Spellman & Daynes, 1978). The function of these suppressor cells is specific in that, whereas they prevent development of UVR-induced tumours, they do not alter the growth of chemically induced tumours or skin allografts (Kripke & Fisher, 1976; Fisher & Kripke, 1978).

The phenotype of the suppressor cells is $LyT1^+ 2^-$, Ia^- (antibodies to surface markers on lymphocytes), similar to that of other UV-induced suppressor cells (Ullrich & Kripke, 1984). These suppressor cells are important in the development of primary neoplasms. de Gruijl and van der Leun (1982b, 1983) found accelerated development of UVR-induced tumours in hairless mice that had been exposed previously to UVR at a separate site. Fisher and Kripke (1982) observed that, if suppressor cells were present from the time of commencement of exposure to UVR, the latent period for development of tumours was shortened and the tumour yield was increased. Thus, photocarcinogenesis in mice appears to involve at least two UVR-induced alterations: (i) an alteration in DNA leading to transformation of cells (see pp. 188–189) and (ii) a specific systemic immunological alteration that permits expression of the tumour (Fisher & Kripke, 1977).

Suppressor cells can be induced by doses of 40–50 kJ/m^2 of radiation from fluorescent sunlamps (see Fig. 9c, p. 64) (Kripke & Morison, 1986a), and susceptibility to transplanted tumours is evident long before the de-novo appearance of tumours (Fisher & Kripke, 1977). Suppressor cells can be induced by exposure to UVC (from low-pressure mercury discharge lamps) (Lill, 1983), UVB (De Fabo & Kripke, 1980), large doses of UVA (Morison, 1986) and sunlight (Morison & Kelley, 1985). Wiskemann *et al.* (1986) described an effect of neutral white fluorescent bulbs. [The Working Group considered that this effect may have been due to low levels of UVB from this source.]

(iv) *Transplantation immunity*

The immune responses in graft rejection and graft-*versus*-host disease are complex and directed against class I antigens of the major histocompatibility complex which are expressed on all nucleated cells and class II Ia antigens which are expressed normally on lymphocytes and macrophages. Lindahl-Kiessling and Säfwenberg (1971) demonstrated that UV irradiation of stimulator cells could abrogate the proliferation of responder cells in a mixed-lymphocyte reaction. Subsequent studies (Alter *et al.*, 1973; Bach *et al.*, 1977) indicated that this effect was due to alteration of class II Ia antigens on the cells bearing them. These initial observations have been extended to various systems.

Pre-transplant, donor-specific blood transfusions have been used to reduce the need for post-transplant immunosuppression, with varying success. The basis for this effect is thought to be generation of donor-specific T-suppressor lymphocytes in the host. Lau *et al.* (1983)

found that exposure of the blood to UVB radiation prior to transfusion greatly enhanced this effect and permitted long-term survival of allografts of islets of Langerhans across a major histocompatibility barrier in rats. The effect was shown to be due to inactivation of lymphocytes by radiation, resulting in cancellation of a signal from Ia antigen-positive cells and permitting the generation of donor-specific T-suppressor cells. A similar effect was demonstrated with rat heart allografts (Balshi *et al.*, 1985).

Deletion of Ia antigens or inactivation of cells bearing them may explain prolonged graft survival in other systems. Exposure of mouse tail skin to UVB radiation *in vitro* prolonged its survival as a graft when I-region differences only were present, but UVB had no effect in the case of complete H-2 differences (Claas *et al.*, 1985). Similarly, mouse corneal allograft survival was prolonged by exposure to UVB radiation *in vitro* (Ray-Keil & Chandler, 1986). Prolonged survival as grafts of rat islets of Langerhans exposed to UVB radiation *in vitro* was apparently due to inactivation of dendritic cells bearing Ia antigens (Lau *et al.*, 1984).

The model of UVR-induced systemic suppression of delayed hypersensitivity has been extended to transplantation studies, because of the considerable potential for manipulating the immune system in transplantation. Sensitization of mice with allogeneic spleen cells after a single exposure to UVB radiation suppressed the delayed hypersensitivity response to these cells and proliferation of lymphocytes from the irradiated mice in a mixed-lymphocyte reaction; these effects are due to generation of suppressor cells specific for donor antigens (Ullrich, 1986). Interestingly, exposure of the mice to radiation need not precede exposure to the antigen but can be delayed up to five days after first contact with the antigen, unlike other forms of suppression of delayed hypersensitivity (Magee *et al.*, 1989a). Similar observations have been made in rats, but suppressor cells were not demonstrated in the spleen (Magee *et al.*, 1989b). Subcutaneous injection of epidermal cells that have been exposed to UVB radiation *in vitro* can similarly cancel a delayed hypersensitivity response in mice; this effect is associated with prolongation of skin allograft survival (Tamaki & Iijima, 1989).

Graft-*versus*-host disease can also be reversed by UVR. Two rat models have been studied. Pretreatment of donor bone marrow with UVB radiation did not increase the failure of grafts, but it prevented graft-*versus*-host disease in most instances (Pepino *et al.*, 1989). Pre-irradiation of rat skin with UVB prevented subsequent development of cutaneous graft-*versus*-host disease at the site of exposure (Glazier *et al.*, 1984). In both of these studies, an alteration of Ia-bearing cells was postulated as the mechanism.

(v) *Infectious diseases*

Classic delayed hypersensitivity to complex protein antigens (correlated with resistance to a number of infections) can be suppressed by exposure to UVB radiation (Ullrich *et al.*, 1986b).

Exposure of mice to low doses (1.3–3.4 kJ/m^2) of UVB (less than a human MED) at the site of intradermal infection with herpes simplex type 2 virus increased the severity of the disease. Unirradiated mice developed only a single vesicle at the site of inoculation, whereas irradiated mice developed zosteriform lesions which healed slowly and, at the highest dose of radiation, were lethal. At doses that increased the severity of the infections, systemic suppression of delayed hypersensitivity to the virus due to generation of antigen-specific T-suppressor lymphocytes was observed (Yasumoto *et al.*, 1987). In-vitro assays showed

UVB-induced impairment of antigen presentation, which may have been due to the presence of suppressor factors in the supernatant (Hayashi & Aurelian, 1986). Similar results were found in a model of herpes simplex virus type 1 infections in mice (Howie *et al.*, 1986a,b,c; Otani & Mori, 1987). [The Working Group considered that these experiments have not demonstrated clearly that the effect of radiation on the induction of immunity is local, since the possibility of an indirect systemic effect has not been explored.]

Exposure to low doses of UVB radiation prevented the development of delayed hypersensitivity to the protozoan, leishmania, and reduced the number and severity of skin lesions when leishmania was inoculated at the site of exposure. Exposure to radiation did not, however, alter the viability of the organisms or the degree of their dissemination to distant sites—the spleen, lymph nodes and skin. Furthermore, the irradiated mice reacted to a second, distant inoculation as if it were a primary infection, presumably because they lacked the cell-mediated immunity that would be needed to control this second attack of the organism (Giannini, 1986).

Exposure of mice to UVB radiation also caused systemic suppression of delayed hypersensitivity to the yeast *Candida albicans* (Denkins *et al.*, 1989), through two possible mechanisms: one mediated by suppressor cells (detected in the spleen) triggered by exposure to radiation prior to contact with the antigen and another which did not involve splenic suppressor cells and was triggered by exposure to radiation following exposure to the antigen.

(vi) *Human lymphocytes* in vitro

Lymphocytes are highly sensitive to low doses of UVR. UVC was approximately 10 times more effective than UVB and 10^5 times more effective than UVA on mononuclear peripheral blood cells *in vitro* (Morison *et al.*, 1979b). Cripps *et al.* (1978) found that UVC was preferentially toxic to T lymphocytes, but that T and B lymphocytes were similarly susceptible to UVB. UVA did not appear to kill T or B cells. Exposure of mononuclear peripheral blood cells to UVB radiation inhibited both natural killer cell activity and the response of these cells to stimulation by a mitogen (phytohaemagglutinin) (Schacter *et al.*, 1983), in the absence of any apparent change in viability. The effect on natural killer cell activity occurred selectively at the post-binding stage of lysis (Elmets *et al.*, 1987) and could be virtually reversed by the addition of interleukin-2 and superoxide dismutase (Toda *et al.*, 1986).

(c) *Comparison of humans and animals*

Firstly, most observations have been made in experimental systems and few studies have involved humans, and it can be only assumed that results of studies in mice can be extrapolated to humans. Furthermore, in no instance have parallel studies in an experimental system and in humans been performed to test this assumption. Secondly, while most investigations of photoimmunology have focused on the effects of 'UVB' radiation, in most studies this term refers to the emission spectrum of a fluorescent sunlamp (see Fig. 9c, p. 64) which contains both UVC and UVA, as well as UVB radiation, besides having little in common with the spectrum of sunlight. Fortunately, in the few studies in which the effects of fluorescent sunlamps and sunlight have been compared in experimental systems, similar alterations in immunity have been observed. Finally, with few exceptions, the effect of exposure to UVR is to suppress immunity highly selectively, at least in experimental animals. Thus, in mice, certain cell-mediated immune responses are suppressed by UVR, whereas humoral immunity is largely unaffected. The selective nature of UVR-induced immunosuppression has not been established in humans, but no evidence exists to suggest that it does not apply. The importance of such selectivity is that it differs from the forms of immunosuppression seen most commonly in humans, namely viral and drug-induced suppression, which affect most functions of the immune system. Exposure of humans to UVR is unlikely to cause paralysis of immune function but probably selectively negates a few immune responses.

4.2.3 *Eye*

(a) *Humans*

(i) *Anterior eye (cornea, conjunctiva)*

The cornea absorbs UVC and UVB radiation (Sliney & Wolbarsht, 1980). Sunlight has been implicated as causing nodular band keratinopathies (spheroidal degeneration and climatic droplet keratopathy), pinguecula, pterygium, photokeratitis and photokeratoconjunctivitis (Wittenberg, 1986). Artificial sources of UVR, including welding arcs and germicidal lamps, cause photokeratoconjunctivitis and photokeratitis (Sliney, 1986). A study by Taylor *et al.* (1989) of the association between exposure to broad-band UVR and corneal disease in 838 fishermen in Chesapeake Bay, Maryland, USA, reported a significant association with pterygium and climatic droplet keratopathy but a weak association with pinguecula.

(ii) *Lens*

The lens absorbs radiation between 305 and 400 nm (Wittenberg, 1986). UVR produces substantial photodamage to both the structural proteins and key enzymes of the lens (for review, see Andley, 1987).

Taylor *et al.* (1988) studied the two major types of senile cataract (nuclear and cortical cataracts) in 838 Maryland fishermen for each of whom mean annual and cumulative UVB exposure had been assessed. High cumulative exposure to UVB and high annual exposure to UVB were both associated with increased risk of cortical cataract, but no association was seen with nuclear cataracts. The association between exposure to solar radiation and cataract is also supported by studies of cataract in northern India and China and in aborigines in Australia and by an analysis of data from the US National Health and Nutritional Examination Survey. These studies were reviewed by Wittenberg (1986).

It has been claimed that the presence of low levels of photosensitizing compounds in lens tissue may contribute to cataractogenesis (Lerman, 1988).

(iii) *Posterior eye*

The posterior eye is composed of the vitreous humour and the retina (Lerman, 1980). In the normal eye, solar radiation in the visible and near infrared regions (400–1400 nm) reaches these structures. Refraction of this waveband by the cornea and lens greatly increases the irradiance between the surface of the cornea and the retina (Sliney & Wolbarsht, 1980).

Permanent retinal damage was observed after direct viewing of the sun and viewing of solar eclipses and in aircraft spotters during the Second World War, but no epidemiological study has associated retinal pathology with routine environmental exposure to sunlight (Wittenberg, 1986). The suggestion that senile macular degeneration is related to solar exposure was not supported by a large study of fishermen in Maryland (West et al., 1989).

(b) *Experimental systems*

(i) *Anterior eye*

Pitts et al. (1977) and Cullen (1980) studied the effects of exposure to UVR at 295 nm on the corneas of pigmented rabbit eyes. The threshold dose for corneal damage was 0.05 J/cm^2. Changes observed with a slit lamp biomicroscope included discharge, corneal debris, haziness, granular change, epithelial exfoliation, stromal opacities and stromal haze.

Applegate and Ley (1991) showed that UVR-induced corneal opacification and neovascularization of the cornea of the South American opossum *M. domestica* was due to DNA damage, as these effects could be delayed by subsequent illumination with photoreactivation light, which specifically monomerizes pyrimidine dimers.

(ii) *Lens*

Cataracts have been produced in pigmented rabbit eyes by exposure to UVB radiation (Pitts et al., 1977). Cataracts were produced in young albino mice 60 weeks after irradiation with a black light (predominantly UVA) (Zigman & Vaughan, 1974; Zigman et al., 1974). Albino mice developed anterior lens opacities after daily exposure for one to two months to a UVB plus UVA source (290–400 nm), but not after the source was filtered to remove radiation < 320 nm (Jose & Pitts, 1985).

(iii) *Posterior eye*

The effects of solar radiation on the posterior eye have been reviewed (Wittenberg, 1986; Andley, 1987). Irradiation of calf vitreous humour *in vitro* with visible radiation in the presence of photosensitizers resulted in partial liquefaction, suggesting that photogenerated active species of oxygen may damage the vitreous structure. In rabbits *in vivo*, however, little liquefaction was seen, suggesting a protective mechanism in the intact organ (Pitts et al., 1977).

Damage to the retina by exposure to sunlight may also be due to thermal effects at high irradiances or to photochemical effects at lower irradiances. In various animals, continuous exposure to sunlight produces a photochemical lesion involving the entire retina and affecting both rods and cones (Young, 1988). The photopigment, rhodopsin, is the chromophore for damage to the rods, while the three cone pigments are the chromophores for cones. In monkeys, blue-light damage caused by exposure to the 400–500 nm waveband affected the macular or paramacular region of the retinal pigment epithelium. The chromophore involved has been postulated to be melanin; active species of oxygen appear to act as mediators of the photochemistry (Lerman, 1980; Andley, 1987).

(c) *Comparison of humans and animals*

The limited data available indicate that the optical properties of the components of human and animal eye are broadly similar.

4.3 Photoproduct formation

4.3.1 *DNA photoproducts*

A multitude of photoproducts are formed in cellular DNA by solar UVR, many of which were first recognized after their induction by non-solar radiation at a wavelength of 254 nm. The ratio of the different photoproducts changes markedly with wavelength. A brief description of the photoproducts is given below, together with a note on the wavelength dependence of formation and susceptibility to repair. Substantial information on biological consequences is available only for cyclobutane-type pyrimidine dimers and pyrimidine-pyrimidone (6-4) photoproducts.

(a) Cyclobutane-type pyrimidine dimers

Shortly after the observation that thymine compounds irradiated with UVC in the frozen state rapidly lose their absorption (Beukers *et al.*, 1958), a dimer of thymine was shown to be responsible for this effect, the two molecules being linked by a cyclobutane ring involving the 5 and 6 carbon atoms (Beukers & Berends, 1960; Wulff & Fraenkel, 1961). Continued irradiation leads to a wavelength-dependent equilibrium between dimer formation and dimer splitting to reform the monomer. Dimer formation is favoured when the ratio of dimer to monomer absorbance is relatively small (wavelengths > 260 nm), whereas monomerization is favoured at shorter wavelengths (around 240 nm), when the ratio is larger (Johns *et al.*, 1962). Although several isomers of the cyclobutane-type thymidine dimer have been isolated from irradiated thymine oligomers, only the *cis*-syn isomer appears to predominate in biological systems (Ben-Hur & Ben-Ishai, 1968; Varghese & Patrick, 1969; Banerjee *et al.*, 1988).

Cytosine–thymine (cyt↔thy), thymine↔thymine (thy↔thy) and cytosine–cytosine (cyt↔cyt) cyclobutane-type dimers are also formed in irradiated *Escherichia coli* DNA but deaminate to uracil↔thymine (ura↔thy) and uracil–uracil dimers after the acid hydrolysis usually used in chromatographic analysis (Setlow & Carrier, 1966). Cytosine moieties in dimers are also deaminated at a slower rate under physiological conditions that produce uracil residues (Fix, 1986), and recent evidence obtained in bacteria suggests that the rate may be more significant than was previously thought (Tessman & Kennedy, 1991). After treatment at 254 nm, thy↔thy, cyt↔thy and cyt↔cyt appear in irradiated DNA at a ratio of 2:1:1 (Unrau *et al.*, 1973), but this ratio changes quite markedly at longer wavelengths, e.g., to 5:4:1 at 265 nm (Setlow & Carrier, 1966). At 254 nm, the relative proportion of cyclobutane dimers was: 5'-thy↔thy, 0.68; 5'-cyt↔thy, 0.17; 5'-cyt↔thy, 0.08; and 5'-cyt↔cyt, 0.07 (Kraemer *et al.*, 1988). Ellison and Childs (1981) showed in *E. coli* that the ratio of cyt↔thy:thy↔thy increases from 0.75 at 254 nm to 1.5 at 313 nm then decreases to 0.8 at 320 nm, the longest wavelength tested. At 365 nm, the longest wavelength at which dimers have been detected, the ratio of thy↔thy:ura↔thy was 5–6:1 (Tyrrell, 1973). The proportion of cyt↔cyt:thy↔thy increased up to 300 nm, but cyt↔cyt was undetectable at longer wavelengths (Ellison & Childs, 1981). On the basis of these data, the latter authors argued that the predominant dimer species formed in *E. coli* by exposure to sunlight are likely to be mixed dimers of cyt↔thy rather than thy↔thy (cyt↔thy:thy↔thy, 1.2:1). The ratio of formation of thy↔thy:ura↔thy dimers in bacterial DNA at 254 and 365 nm is approximately 7×10^5 nm

(Tyrrell, 1973). A similar ratio of total dimer product formation was found in cultured human skin fibroblasts irradiated at 254–265 nm (Enninga et al., 1986).

Fisher and Johns (1976) described the photochemistry and mechanism of formation of cyclobutane-type pyrimidine dimers in considerable detail. The mechanism of dimer formation in the UVB region almost certainly involves direct absorption, since the action spectrum for induction closely resembles that for the appropriate monomer for wavelengths as long as 313 nm (Ellison & Childs, 1981). The mechanism of formation by longer wavelengths (e.g., 365 nm) has not been clarified.

Cyclobutane-type dimers can be removed from the DNA of both prokaryotic and eukaryotic cells by the powerful excision repair mechanism that is deficient in cells from most sun-sensitive, skin cancer-prone patients with the hereditary disease, xeroderma pigmentosum (see Friedburg, 1984; Cleaver & Kraemer, 1989). Photoreactivation is specific for pyr↔pyr (pyrimidine dimers) and monomerizes them *in situ* via a photolyase. Many microorganisms and higher eukaryotes contain a photolyase, but the proteins and light-activation spectra differ from species to species. The specificity of this process has proved a powerful tool in analysing the role of pyr↔pyr in biological effects. For example, the potential photoreactivation of pyr↔pyr has been studied in a set of experiments to demonstrate that the presence of UVC-induced pyr↔pyr in fish can be a precarcinogenic lesion (Setlow, 1975). More recently, the small opossum, *M. domestica*, has been used by Ley and coworkers as an animal model in studies on the effects of UVR, predominantly UVB, mainly because cells of the skin of this animal, unlike that of the mouse, contain a photoreactivating enzyme(s). They showed that several biological effects, including decreased hair growth, erythema and tumour formation, were suppressed by exposure to longer wavelengths (photoreactivating light) (Ley & Applegate, 1989; Ley et al., 1991).

Considerable evidence, including the fact that photoreactivation prevents formation of the majority of mutations induced in bacteria by UVC, shows that the argument that pyr↔pyr is a major premutagenic lesion is overwhelming (Doudney, 1976). Recognition that UV-induced mutagenesis in bacteria is an inducible process (see Witkin, 1976), however, complicates this argument, since, assuming that a structure involving pyr↔pyr constitutes the inducing event, its elimination by photoreactivation would preclude error-prone repair at the site of any premutagenic lesion. When all inducible functions relevant to mutagenesis are turned on, the photoreversibility of UVC mutagenesis at several pyr↔pyr sites disappears (Bridges & Brown, 1992); e.g., UV-induced mutagenesis to his^+ in certain $recA441$ $lexA51$ bacteria was not photoreversible, indicating that pyrimidine dimers are not target lesions (Ruiz-Rubio et al., 1986). This suggests that non-photoreversible photoproducts (such as the pyrimidine–pyrimidone 6-4 photoproduct) are the principal premutagenic lesions at dithymine sequences and that cyclobutane-type thymine dimers are weakly mutagenic. This conclusion is consistent with the results of other studies with single-stranded vector DNA containing cyclobutane-type (6-4) thy↔thy photoproducts at specific sites (Banerjee et al., 1988, 1990; LeClerc et al., 1991).

(b) Pyrimidine–pyrimidone (6-4) photoproducts

The most extensively studied non-dimer photoproduct is that formed from thymine and cytosine. Indirect evidence (Varghese & Patrick, 1969) suggests that this structure is the in-vivo precursor of the compound 6-4′-[pyrimidin-2′-one]thymine (thy(6-4)pyo), originally found in acid hydrolysates of UV-irradiated DNA (Varghese & Wang, 1967; Wang & Varghese, 1967). Some years later, a type of UV-induced photoproduct, the pyrimidine nucleoside–cytidine lesion, was recognized in highly reiterated sequences of human DNA (Lippke et al., 1981); this is also probably a precursor of the thy(6-4)pyo product (Brash & Haseltine, 1982; Franklin et al., 1982). Using DNA sequencing analysis, UV photoproducts were more frequent at the 3′ end of pyrimidine runs. Although the overall ratio of 6-4 photoproducts to dimers was 15% at certain 5′-thy↔cyt sequences, 6-4 photoproducts occurred at approximately the same frequency as that of the cyclobutane dimer (Kraemer et al., 1988).

Patrick (1977) originally reported that the action spectrum for (6-4) photoproduct formation resembles that for cyclobutane dimer formation, although the quantum yields are two and ten times lower than that of cyt↔thy and thy↔thy formation, respectively. Using irradiation at wavelengths as long as 334 nm, Chan et al. (1986) found that the action spectrum for induction of hot alkali sites (presumably the thy(6-4)pyo hydrolysis product) was also similar to that for pyr↔pyr formation. The action spectra for the induction of thymine dimers and (6-4) photoproducts were similar from 180 to 300 nm, whereas the action spectrum values for thymine dimer induction were about nine and 1.4 times higher or more than the values for (6-4) photoproduct induction below 160 nm and above 313 nm, respectively (Matsunaga et al., 1991).

Most xeroderma pigmentosum patients are defective in the excision of (6-4) photoproducts (Mitchell et al., 1985) and cyclobutane pyrimidine dimers (Cleaver & Kraemer, 1989). In addition, a group of patients with trichothiodystrophy (type 3) showed a marked reduction in the repair of (6-4) photoproducts (Broughton et al., 1990).

Glickman et al. (1986) demonstrated in *E. coli* that the cytosine–cytosine pyrimidine–pyrimidone (6-4) photoproduct is highly mutagenic; however, in other studies (e.g., Hutchinson et al., 1988), cyclobutane dimers were shown to be responsible for the majority of observed mutations. Assessment of the relative contributions to mutagenesis of all dipyrimidine photoproducts will require comprehensive studies in different biological systems with specifically designed sequences containing the appropriate photoproducts. Both pyrimidine dimers and pyrimidine–pyrimidone (6-4) photoproducts appear to be important in inducing cytotoxic and mutagenic lesions in human cells, although the relative contributions of each type remain controversial (Mitchell, 1988).

(c) *Thymine glycols*

A group of monomeric ring-saturated lesions of the 5,6-dihydroxydihydrothymine type (thymine glycols) have been detected by alkaline–acid degradation in the DNA of UV-irradiated human cells (Hariharan & Cerutti, 1976, 1977). Alkaline–acid degradation (see Cerutti, 1981) can be used to detect a class of structurally related lesions rather than a single lesion, with a yield that has been estimated to be approximately 20% of the total of ring-saturated thymine products (t_{sat}).

Two aspects of this class of UV photoproduct are of particular interest: firstly, they are closely related to a class of ionizing radiation products and are believed to arise through a similar mechanism, i.e., indirectly *via* the action of hydroxyl radicals; secondly, their yield (relative to that of other UV-induced base damage) increases with exposures in the UVB region. Measurements in HeLa cells showed that at 265 nm the ratio of thy↔thy to t_{sat} was 21, whereas at 313 nm the ratio decreased to 1.3 (Cerutti & Netrawali, 1979). The saturated thymine damage induced by UVA and UVB radiation may thus be due to the effects of active oxygen species generated *via* endogenous cell components. There is little evidence pertaining to the lethal or other biological consequences of such lesions in mammalian cells, although a glycosylase capable of repairing these lesions has been isolated from human cells (Higgins *et al.*, 1987).

(d) Cytosine damage

The photochemical induction of pyrimidine hydrates has been reviewed (Fisher & Johns, 1976). Significant levels of hydrates are probably formed initially by UVR; however, their instability hampers measurement of their induction and removal in cells, and it has not been possible to establish a cause-and-effect relationship between photohydrate induction and biological effects *in vivo*. Using sequencing techniques, Gallagher *et al.* (1989) observed incision by human endonucleases of unidentified cytosine photoproducts that were neither cyclobutane-type nor (6-4) pyrimidine dimers. The frequency of these two photoproducts was two orders of magnitude lower than that of pyrimidine dimers, and the optimal wavelengths for induction were between 270 and 295 nm.

(e) Purine damage

Purine damage has been studied less frequently than pyrimidine damage, since the quantum yields are at least one order of magnitude lower; however, the development of sequencing techniques has made their detection easier (Kumar *et al.*, 1991). Incisions (endonuclease V) are detected at unidentified purine or purine–pyrimidine moieties after broad-spectrum UV irradiation (Gallagher & Duker, 1986). Such damage appears to be induced maximally in the wavelength region of 260–300 nm (Gallagher & Duker, 1989). Although the overall yield is much lower than that of pyr↔pyr, similar yields occur at certain loci.

(f) DNA strand breaks

UVC radiation induces a lower proportion of single-strand breaks than of other photoproducts. In contrast, strand breaks are the commonest initial lesion induced by ionizing radiation. Although strand breaks form only a minority of lesions after irradiation at wavelengths up to 365 nm, they become increasingly important at longer wavelengths in the solar UV region (290–400 nm). At 313 nm, the ratio of DNA strand breakage to pyr↔pyr induction in intact *E. coli* was 1:44 (Miguel & Tyrrell, 1983), whereas at 365 nm one strand break was formed for approximately every two pyrimidine dimers (Tyrrell *et al.*, 1974). An action spectrum for break induction in *Bacillus subtilis* DNA *in vivo* is available (Peak & Peak, 1982). More recently, an action spectrum for single-strand breaks in human skin cells has been determined which shows that irradiation in the presence of deuterium (which enhances singlet oxygen lifetime) increases the number of strand breaks observed at 365 and 405 nm. At wavelengths of 405 nm and longer, strand breaks and DNA–protein cross-links are the only forms of photochemical damage that have been determined (Peak *et al.*, 1987). Between 10 and 20% of the breaks induced at 365 nm are not frank breaks but rather alkali-labile bonds which presumably include apurinic and apyrimidinic sites (Ley et al., 1978; Peak & Peak, 1982). The formation of breaks is strongly dependent upon oxygen at both 313 (Miguel & Tyrrell, 1983) and 365 nm (Tyrrell et al., 1974; Peak & Peak, 1982). Their formation *in vitro* at 365 nm is also quenched by free-radical scavengers. Strand breaks are repaired rapidly by a variety of cellular mechanisms in both prokaryotes and eukaryotes. The role of these lesions in the biological action of solar radiation is not well understood (Tyrrell et al., 1974).

(g) DNA–protein cross-links

The photochemical addition of nucleic acids to amino acids and proteins both *in vitro* and *in vivo* has been the subject of several reviews (Smith, 1976; Shetlar, 1980). Of the 22 common amino acids, 11 undergo photochemical addition to labelled uracil, the most reactive of which is cysteine, and several heterophotoproducts involving cysteine have been isolated and characterized.

Several prokaryotic and eukaryotic proteins have been cross-linked photochemically to DNA *in vitro*, including DNA polymerase, RNA polymerase, helix destabilizing protein and mixtures of proteins (Shetlar, 1980).

There is evidence that DNA–protein cross-links are formed in mammalian cells in significant yields by wavelengths longer than 345 nm (Bradley et al., 1979; Peak & Peak, 1991). Action spectra for the formation of DNA–protein cross-links in human cells have now been obtained. Two peaks of induction are observed: one at 254–290 nm, corresponding to the peak of DNA absorption, and a second at 405 nm, presumably resulting from a photosensitization reaction (Peak et al., 1985). [The Working Group noted that DNA–protein cross-links are likely to have important consequences for cells, but no data are available to allow evaluation of their effects in eukaryotic cells.]

4.3.2 Other chromophores and targets

In addition to DNA, many other cellular components absorb and/or are damaged by solar UVR and may influence the biological outcome of exposure. Both informational and transfer RNA molecules are susceptible to photomodification. Studies in insects indicate that damage to messenger RNA may be relevant to embryonic development, but the relevance of these results to mammalian systems is unclear (Kalthoff & Jäckle, 1982). Detailed results of bacterial studies on the photolability of certain components of transfer RNA (Jagger, 1981) are almost certainly not relevant to mammalian cells. Damage to proteins could lead to modification of the level of persistent primary damage in DNA, such that cellular DNA repair and antioxidant pathways are compromised (Tyrrell, 1991). There is also evidence that components of electron transport and oxidative phosphorylation, as well as membranes and membrane transport systems, can be damaged by solar wavelengths (Jagger, 1985). Non-DNA chromophores and targets become particularly relevant at longer wavelengths.

(a) Chromophores

Both nucleic acids and proteins weakly absorb UVA, and, although direct photochemical events may occur, it appears likely that the initial event in the biological effects of UVA radiation is absorption by a non-DNA chromophore which results in generation of active oxygen species or energy transfer to the critical target molecules. As a consequence, at long UV wavelengths, the range of targets is extended to all critical molecules that are susceptible to active intermediates generated by chromophores.

Most of the knowledge on relevant chromophores has been obtained from in-vitro experiments or from studies in bacteria (Eisenstark, 1987). Indirect evidence indicates that porphyrins play a role in the inactivation of *Propionibacterium acnes* by UVA (Kjeldstad & Johnsson, 1986). It has also been shown that *E. coli* mutants defective in the synthesis of δ-aminolaevulinic acid are resistant to inactivation by UVA (Tuveson & Sammartano, 1986), which strongly suggests that porphyrin components of the respiratory chain act as endogenous photosensitizers. This conclusion is supported by the finding that strains that overproduce cytochrome were sensitive to broad-band UVA radiation (Sammartano & Tuveson, 1987). Porphyrins are also essential to human cellular metabolism, and overproduction of iron-free porphyrins in erythropoietic or hepatic tissues is the underlying cause of the photodestruction of the skin seen in the group of diseases known as porphyrias. Although direct evidence is lacking, free porphyrins and proteins containing haem (such as catalase, peroxidases and cytochromes) are also potentially important chromophores in skin cells from normal individuals. Many other cellular compounds which contain unsaturated bonds, such as flavins, steroids and quinones, should also be considered potential chromophores. Although normal levels of catalase (which contains haem) and alkyl hydroperoxide reductase (which contains FAD) would be expected to exert a protective role in bacteria (see below), overproduction of these enzymes is correlated with an increase in sensitivity to UVA radiation in bacteria (Kramer & Ames, 1987).

Porphyrins are an important class of photodynamic sensitizers which are believed to exert their biological action *via* the generation of singlet oxygen. Recent experiments have shown that deuterium oxide (which prolongs the lifetime of singlet oxygen) sensitizes human fibroblast cell populations to the lethal action of UVA radiation, while sodium azide (which destroys singlet oxygen) protects them (Tyrrell & Pidoux, 1989). Although this finding is consistent with the involvement of porphyrins in the lethality of UVA, other cellular compounds may also generate singlet oxygen. It is also important to consider active oxygen species that may be generated intracellularly. Not only can hydrogen peroxide be generated by UVA irradiation of tryptophan (McCormick *et al.*, 1976), but both superoxide anion and hydrogen peroxide can be generated by photo-oxidation of NADH and NADPH (Czochralska *et al.*, 1984; Cunningham *et al.*, 1985).

The presence of chromophores (such as psoralens) in the diet may also influence susceptibility to damage, but this reaction is clearly subject to enormous individual variability. Accidental and deliberate application of chemical agents (such as sunscreens and drugs) to the skin may also introduce potentially damaging chromophores.

(b) Membranes

The lipid membrane is readily susceptible to attack by active oxygen intermediates. Many reports (e.g., Desai *et al.*, 1964; Roshchupkin *et al.*, 1975; Putvinsky *et al.*, 1979; Azizova *et al.*, 1980) have shown that UVR can induce peroxidation of membrane lipids. In-vitro studies with lecithin microvesicles have shown UVR-induced changes in the microviscosity of membrane bilayers (Dearden *et al.*, 1981) which are correlated with the degree of unsaturation of fatty acid chains (Dearden *et al.*, 1985). UVC and UVA radiation and sunlight have been shown to cause lipid peroxidation in the liposomal membrane (Mandal & Chatterjee, 1980). Haem proteins such as cytochrome *c* and catalase are known to catalyse lipid peroxidation and peroxidative breakdown of membranes (e.g., Brown & Wüthrich, 1977; Goñi *et al.*, 1985; Szebeni & Tollin, 1988). A dose-dependent, linear increase in lipid peroxidation of liposomal membranes was induced by UVA radiation, which was inhibited to a large extent by butylated hydroxytoluene, a nonspecific scavenger of lipid-free radicals. Since both sodium azide and L-histidine (quenchers of singlet oxygen) led to 40–50% inhibition of peroxidation, the authors suggested that singlet oxygen is involved in initiation of the reaction (Bose *et al.*, 1989).

UVA irradiation of liposomes leads to lipid peroxidation in the absence of photosensitizer molecules, so that singlet oxygen may arise through direct stimulation of molecular oxygen (Bose *et al.*, 1989). Biological membranes are, however, rich in endogenous photosensitizer molecules, such as those involved in electron transport, and these may contribute to the peroxidation of lipids observed in biological systems (see Jagger, 1985). Membrane damage has long been implicated in the lethality of UVA in bacteria (Hollaender, 1943) and almost certainly contributes to the sensitivity of UVA-treated populations plated on minimal medium—a phenomenon which is highly dependent on oxygen (Moss & Smith, 1981). Sensitivity to UVA has been related to levels of unsaturated fat in membranes (Klamen & Tuveson, 1982; Chamberlain & Moss, 1987). Furthermore, the presence of deuterium oxide enhances the levels of membrane damage, sensitivity to UVA and lipid peroxidation (Chamberlain & Moss, 1987), suggesting that singlet oxygen plays a role in all three processes. Leakage experiments have also been used to assess UVA-induced membrane damage in yeast: again, changes in permeability correlated well with lethality and were highly oxygen dependent (Ito & Ito, 1983). UVA irradiation of cultured human and mouse fibroblasts led to the release of arachidonate metabolites from the membrane in a dose-dependent fashion. The release was also dependent on the presence of both oxygen and calcium ion and may be related to the induction of cutaneous erythema, which is also oxygen dependent (Hanson & DeLeo, 1989). Studies of the effects of UVR on membrane transport have been undertaken in prokaryotes (Jagger, 1985), but no information was available on the effects of UVR on eukaryotic membrane transport.

4.4 Human excision repair disorders

4.4.1 *Xeroderma pigmentosum*

The commonest, most characteristic photoproducts produced in DNA by UVB and UVC radiation involve adjacent pyrimidines. Evidence summarized above argues strongly that these products give rise to a wide variety of alterations in DNA sequence and gene expression. Like many other types of DNA damage, these photoproducts may be excised, and the resulting gap in one strand can be resynthesized accurately using the undamaged strand as a template. How this is accomplished is best understood in the bacterium *E. coli*, in which a multiprotein complex including the products of the *uvr*A, B and C genes excises an oligonucleotide 12 or 13 bases in length containing the photoproduct. The resulting gap is filled by a DNA polymerase (usually III), and the final ligase link to the adjacent DNA is effected by polynucleotide ligase (Bridges *et al.*, 1987; Bridges, 1988; Bridges & Bates, 1990). Other gene products are involved in the process, and a more comprehensive discussion is given by Sancar and Rupp (1983). Bacteria that have defects in the *uvr*A or B genes cannot excise UV photoproducts and are 10–20 times more sensitive to killing and the induction of mutations by UVC. They are also more sensitive to UVB and (under certain conditions) UVA (Webb, 1977). It can be concluded that the function of excision repair is to minimize the deleterious consequences of DNA damage, such as the persistence of UV photoproducts.

A similar process takes place in humans. Although much less is known about the mechanism, many genes have been shown to be involved, and these are being cloned and the role of their products is being elucidated (Hoeijmakers & Bootsma, 1990; Bootsma & Hoeijmakers, 1991). Like bacteria, humans can also be deficient in aspects of excision repair. The prototypic example is the genetic disorder xeroderma pigmentosum, which is actually a complex of disorders comprising at least 10 different forms of DNA repair defect (nine excision defective complementation groups and one excision repair proficient variant group) (Kraemer *et al.*, 1987; Cleaver & Kraemer, 1989). The sensitivity of fibroblasts and lymphocytes from excision-defective individuals with xeroderma pigmentosum to mutation and lethality by UVC is up to 10 times greater than that of cells from normal individuals (Arlett *et al.*, 1992) and for UVR from a solar simulator (Patton *et al.*, 1984). The pigmentary abnormalities are confined to sun-exposed portions of the skin.

The incidences of tumours of the skin, anterior eye and tip of the tongue in these individuals are much higher than those in unaffected populations (Kraemer *et al.*, 1987), and the median age of patients at onset of skin cancers appears to be much younger than that of the general population. Multiple primary skin cancers are common, which arise predominantly on sunlight-exposed areas of the body (Kraemer *et al.*, 1987); there is anecdotal information that they are largely prevented if protection against exposure to sunlight is afforded early in life (Kraemer & Slor, 1984). Studies of patients with excision-defective xeroderma pigmentosum provide the strongest evidence that sunlight-induced photoproducts can result (in the absence of repair) in the genesis of basal-cell carcinomas, squamous-cell carcinomas and melanomas and strongly support the contention that they can also do so in normal individuals in whom repair is more efficient (although probably never complete). The photoproducts that fail to be excised in xeroderma patients are known to be produced in human skin, not only by UVC (used in most laboratory experiments with cells) but also by UVB, particularly by wavelengths around 300 nm (Bridges, 1990; Athas *et al.*, 1991). Action spectra show that the difference in the cytotoxic action of UVB on cultured cells from normal and xeroderma pigmentosum patients is similar to that of UVC, whereas the differences in the response to UVA are only slight (Keyse *et al.*, 1983). The studies on xeroderma pigmentosum illustrate that DNA repair is a major defence of the human skin against the carcinogenic action of sunlight.

4.4.2 *Trichothiodystrophy*

The conclusions derived from studies of xeroderma pigmentosum have become more complex with the availability of information on two related excision disorders. Trichothiodystrophy is a rare disease in which patients generally have skin judged to be sun-sensitive by erythemal response but no indication of the pronounced freckling or elevated incidence of early skin tumours associated with xeroderma pigmentosum (Bridges, 1990). In the majority of cases studied, trichothiodystrophy is associated with a deficiency in the ability to repair UV-induced damage in cellular DNA.

Three categories of response to UVR have been identified. In type 1, the response is completely normal, whereas type-2 cells are deficient in excision repair, with properties indistinguishable from those of xeroderma pigmentosum complementation group D. Type-3 cells survive normally after UV irradiation, and the rates of removal of cyclobutane pyrimidine dimer sites are also normal (Broughton et al., 1990). In xeroderma pigmentosum diploid fibroblast lines, catalase activity was decreased on average by a factor of five as compared to controls, while heterozygotic lines exhibited intermediary responses. All trichothiodystrophy lines tested were deficient in UV-induced lesion repair and exhibited a high level of catalase activity; however, molecular analysis of catalase transcription showed no difference between normal, xeroderma and trichothiodystrophy cell lines. UV irradiation induces five times more hydrogen peroxide production in xeroderma lines than in trichothiodystrophy lines and three times more than in controls. These striking differences indicate that UVR, directly or indirectly, together with defective oxidative metabolism may increase the initiation and/or the progression steps in patients with xeroderma pigmentosum to a greater degree than in people with trichothiodystrophy, which may partly explain the different tumoral phenotypes in the two diseases (Vuillaume et al., 1992).

Five patients with trichothiodystrophy type 2 appeared to be in one of the xeroderma pigmentosum complementation groups: Fibroblasts from these individuals were indistinguishable from xeroderma fibroblasts in the same complementation group and were equally sensitive to the lethal and mutagenic effects of UVC (Stefanini et al., 1986; Lehmann et al., 1988). Two other trichothiodystrophy patients (type 3) had cells markedly defective in the removal of (6-4) pyrimidine photoproducts but not cyclobutane-type dimers (Broughton et al., 1990).

4.4.3 Cockayne's syndrome

A third sun-sensitive excision repair disorder is Cockayne's syndrome. Patients with this condition have fibroblasts which undergo normal excision repair in the overall genome but which are defective in the excision of dimers from DNA strands undergoing active transcription (Mayne et al., 1988). Cockayne's syndrome cells are sensitive to both killing and mutation induction by UVC (Arlett & Harcourt, 1983) and have reduced repair of cyclobutane dimers; they show, however, normal repair of non-dimer photoproducts in a UV-treated shuttle vector plasmid. Like patients with trichothiodystrophy, those with Cockayne's syndrome do not have pronounced freckling or enhanced early incidence of skin cancers (Barrett et al., 1991).

4.4.4 Role of immunosuppression

If it is assumed that UV-induced DNA damage sustained by patients with trichothiodystrophy type 2 results in the same photo-induced mutations in their skin cells (including mutations associated with the initiation of cancer) as is seen in xeroderma pigmentosum patients of the same complementation group (D) (Bridges, 1990; Broughton et al., 1990), something other than unrepaired DNA damage and an elevated frequency of mutations must be needed to trigger initiated cells into clonal expansion and early tumours, as is seen in xeroderma pigmentosum. The assumed latency of initiated cells in such trichothiodystrophy patients may be related to the latency seen in epidemiological studies of skin cancer in the normal population (see section 2).

The nature of the circumstances that allow initiated skin cells to develop into tumours in xeroderma pigmentosum patients, and perhaps later in life in other individuals, is unclear. Burnet (1971) first suggested that individuals with this disorder might be deficient in some immunosurveillance step. Bridges (1990) proposed that they were also hypersensitive to both the immunosuppressive and the mutagenic action of UVR, so that the elevated skin cancer rate in individuals with xeroderma pigmentosum would not accurately reflect the actual increase in mutation frequency in exposed skin but would exaggerate it greatly.

4.5 Genetic and related effects

Any cell that is UV-irradiated can be expected to sustain DNA damage. The nature of this damage is wavelength-dependent, and the major photoproducts of short-wavelength UV irradiation are various types of dipyrimidine photoproducts, while DNA strand breakage and DNA–protein cross-linkage occur relatively more frequently after irradiation with long-wavelength UVR. As the wavelength is increased above 290 nm, the efficiency of formation of pyrimidine dimers and other DNA photoproducts decreases greatly. This wavelength-dependency of response presents a fundamental problem for the quantitative interpretation of the genetic activities of different regions of the UV spectrum. In most experimental studies with UVA and UVB irradiation and, of course, simulated solar radiation, monochromatic radiation was not used. Also, the characteristics of the radiation emitted from the source are variable over time and from source to source. Because of these practical considerations, comparisons of the effects seen in different studies in terms of dose are commonly invalid: Photoproduct yield is dependent on the energy contributions from the different wavelengths within the spectrum used, but incident doses (fluences) are measured only as energy fluxes over the whole spectrum emitted from the source. The problem of dosimetry within experimental systems is compounded by the fact that absorbed dose is determined by the geometry of the system and the position of the target within it: absorption by one layer (e.g., the medium or a layer of cells) will affect the fluence incident upon the layer beneath. The fluence absorbed may thus differ substantially from the incident fluence of the system. For these reasons, it was considered inappropriate to compile quantitative genetic profiles as is customary in these monographs.

Given the generally significant responses in many different tests for the genetic activity of UVR in a wide range of organisms and cultured cells, the simple qualitative questions appear to have been answered in abundance. The main issues of outstanding interest are: identification of the types of damage induced by the various portions of the UV spectrum; the mechanisms by which damage is translated into mutation or other genetic changes; and the dose characteristics of these responses.

4.5.1 *Humans*

The portions of the body that receive most exposure to UVR are the skin, anterior eye and lip. Because dermal capillaries approach the skin surface, it can be anticipated that blood will be exposed to the portion of UVR (see Kraemer & Weinstein, 1977; Morison et al., 1979a; Larcom et al., 1991) that penetrates the dermis. The biological consequences of this exposure are unknown.

DNA damage in skin cells has been studied using three methods that are sensitive enough to detect DNA damage after exposure to doses of UVR too low to induce erythema:

(i) use of antibodies specific for UV-altered DNA, followed by immunofluorescence. This method can be used with immunoperoxidase staining and a secondary antibody (Eggset et al., 1983, 1986) or without them (Tan & Stoughton, 1969);

(ii) autoradiography after tritiated thymidine incorporation (Epstein et al., 1969, 1970; Hönigsmann et al., 1987; Wolf et al., 1988); and

(iii) treatment of extracted DNA with *Micrococcus luteus* cyclobutyl pyrimidine dimer site-specific endonuclease, followed by alkaline agarose gel electrophoresis of the single-stranded DNA fragmented at the dimer sites (Sutherland et al., 1980; D'Ambrosio et al., 1981; Gange et al., 1985; Freeman et al., 1986, 1987, 1989; Alcalay et al., 1990). This method suffers the disadvantage that damage cannot be localized to particular layers of the skin, but dimer yield can be calculated. Methods for the study of resolved genetic damage have not been pursued.

(a) Epidermis

(i) *Broad-spectrum ultraviolet radiation, including solar simulation*

Effects on DNA synthesis were demonstrated in human skin *in vivo* which had been exposed to three times the MED of UVR (< 320 nm; mercury arc lamp [Fig. 9a, p. 64]) and then injected intradermally with tritiated thymidine (8–41 × 10^6 ergs/cm^2 [8–41 kJ/m^2]) in the irradiated area immediately and at 0.25, 3, 5 and 24 h subsequently. S Phase was suppressed in cells of the basal layer at 3-h and 5-h sampling times, but not at 24 h. Sparsely labelled cells (indicating DNA repair) occurred in greatly variable proportions from person to person in the basal, malpighian and granular layers at 0, 0.25, 3 and 5 h, but not at 24 h, indicating that repair was complete by 24 h (Epstein et al., 1969). DNA repair was also reduced in the skin cells of three patients with xeroderma pigmentosum in comparison to eight normal controls (Epstein et al., 1970).

Sutherland et al. (1980) demonstrated a dose-related response for the induction of pyrimidine dimers after exposure to a Westinghouse sun lamp (Fig. 9c, p. 64), with 50% energy < 320 nm, at 0, 970, 1940 and 3880 J/m^2. In one subject, 0.5 of the MED of sun-lamp exposure resulted in about 6 ± 0.6 dimers per 10^8 Da.

D'Ambrosio et al. (1981) reported that approximately 12.8 and 23.6 dimers per 10^8 Da were induced in skin DNA *in vivo* following irradiation with a mercury arc lamp (200–450 nm) at 150 and 300 J/m^2, respectively. Repair or removal of dimers was measured 0–24 h following exposure. About 50% of the dimers were lost 58 min after irradiation, and less than 10% remained at 24 h. In an experiment with patients with lupus erythematosus, D'Ambrosio et al. (1983) obtained results similar to those found in the skin of normal individuals.

Strickland et al. (1988) measured the induction of cyclobutane dithymidine photoproducts in human skin samples after exposure to simulated solar radiation. Tissue samples from three non-pigmented (white) individuals were exposed to 18 or 36 kJ/m^2 UVR (0.5–1 MED), and those from three constitutively pigmented (black) individuals were exposed to 72 and 144 kJ/m². Constitutively pigmented skin required doses of UVR two to four times higher than non-pigmented skin to produce roughly equivalent levels of thymine dimers. [The Working Group noted the small number of people studied.]

(ii) *UVA radiation*

Freeman *et al.* (1987) showed in two subjects that similar pyrimidine dimer yields were produced in skin by a broad-band UVA source (UVASUN 2000), by broadband UVA filtered to remove all light of wavelengths < 340 nm and by narrow-band radiation centred at 365 nm (xenon–mercury compact arc), indicating that UVA radiation and not stray shorter wavelength radiation was responsible. Dimer production was observed following exposures to 5×10^5 J/m². Since exposure to a UVA-emitting tanning lamp results in a dose of about 5×10^5 J/m², UVA exposure for cosmetic purposes could result in measurable levels of DNA damage.

(iii) *UVB radiation*

The efficiency of UVA- and UVB-induced tans in protecting against erythema and the formation of dimers induced by UVB was studied in five subjects by Gange *et al.* (1985). The radiation sources were a UVASUN 2000 lamp (UVA; Fig. 8d, p. 61) and an FS36 Elder fluorescent sunlamp (UVB). UVB-induced tanning protected against erythema produced by subsequent UVB exposure two to three times better than UVA-induced tanning; however, tanning with either UVA or UVB was associated with a similar reduction in yield of endonuclease-sensitive sites in epidermal DNA (about 50%).

Eggset *et al.* (1983) observed DNA damage in both epidermis and dermis following exposure to a Westinghouse FS-20 sunlamp (Fig. 9c, p. 64) at 0.5–2 MED (2 MED, 900 J/m²). The outer layers were more heavily damaged after small doses than the basal layer, which may be better protected by its deeper location and shielding by melanin. The authors claimed that DNA repair was well under way after 4–5 h and was apparently nearly complete at 24 h, as judged by immunofluorescence and immunoperoxidase staining. Repair was faster in the presence of visible light than when irradiated skin was shielded with thick black plastic. [The Working Group noted the absence of quantitative data.]

In a study of two volunteers (Eggset *et al.*, 1986), tanning was shown to protect against DNA damage in skin (induced in a UVB solarium), but the conclusions were based solely on observations of immunofluorescence. [The Working Group noted the absence of quantitative data.]

Freeman *et al.* (1986) measured UVB-induced DNA damage in the skin of seven individuals with different sensitivities to UVB irradiation, as measured by the MED, with irradiation from an FS36 Elder fluorescent sunlamp (280–320 nm). The production of dimers was correlated inversely with the MED. The slopes of the dose–response curves for the most UVB-sensitive individual (MED, 240 J/m²) and for the least sensitive individual (MED, 1460 J/m²) were 11.5×10^{-4} and 2.6×10^{-4} dimer sites per 1000 bases per mJ/cm² [10 J/m²], respectively.

Hönigsmann *et al.* (1987) studied unscheduled DNA synthesis in epidermal cells in the skin of 25 male volunteers (four with skin type II and 21 with skin type III; see pp. 168–169) after exposure to doses of UVB of 0.06–6 MED, from a 6-kW xenon arc lamp (292–304 nm). The MED values ranged from 140 to 550 J/m². The dose–response curve showed a significant increase in unscheduled DNA synthesis between 0.06 and 1 MED but no difference between 1 and 6 MED, suggesting a saturation of excision repair *in vivo*.

Freeman (1988) studied interindividual variability in 17 healthy volunteers in the repair of pyrimidine dimers induced following exposure to 0.25–1.5 MED from a Westinghouse FS-40 sunlamp (see Fig. 9c, p. 64). Removal of dimers was detected within 6 h of irradiation. The average half-time for removal of dimers was 11.0 ± 4.3 (SD) h (range, 5.5–21.1 h). [The Working Group noted that the spectra and doses used in this study were different from those used by D'Ambrosio *et al.* (1981). It is not clear if the interindividual variability is greater than the experimental error.]

Interindividual variability in the repair of UVB-induced pyrimidine dimers was also studied by Alcalay *et al.* (1990) in 22 patients aged 31–84 with at least one basal-cell carcinoma. The control group consisted of 19 cancer-free volunteers aged 25–61. Both groups were given one MED of radiation from a 150-W xenon arc solar UV-simulated lamp equipped with a 50-cm liquid light guide and a filter eliminating wavelengths below 295 nm. Dimers were measured immediately and after 6 h. The two groups were similar at time 0, but after 6 h, $22 \pm 4\%$ (range about 8–64) of the dimers were removed in the cancer group compared to $33 \pm 4\%$ (range about 4–64) in the control group. Of the cancer patients, 23% had repaired more than 30% of the DNA damage, compared to 53% of the control group. [The Working Group noted that it is not clear if the interindividual variability is greater than the experimental error.]

Wolf *et al.* (1988) observed measurable amounts of unscheduled DNA synthesis in the skin of 23 volunteers exposed to 0.5 MED UVB irradiation from a high-pressure mercury lamp [spectral emission not given]. Administration of carotenoids (to reduce light sensitivity in patients with erythropoietic protoporphyria) at a dose of 150 mg per day for 30 days did not significantly alter the amount of unscheduled DNA synthesis (6 ± 1.2 grains/cell before and 8 ± 2 grains/cell after carotenoid treatment; seven subjects). The same investigation showed no significant protection by carotenoids against UVA-, UVB- or PUVA-induced erythema, on the basis of pre- and post-carotenoid MED or minimal phototoxic dose.

In 30 volunteers, it was demonstrated that the action spectrum for the frequency of pyrimidine dimer formation in human skin DNA for a given fluence (incident dose) has its maximum near 300 nm and decreases sharply on either side of this wavelength (Fig. 12). The decrease at < 300 nm is probably due to absorption in the upper layers of skin. These data were used to estimate that, at a solar angle of 40°, a reduction in the thickness of the stratospheric ozone layer from 0.32 cm down to 0.16 cm would be expected to result in a 2.5-fold increase in dimer formation (Freeman *et al.*, 1989).

A dose–response for the formation of thymine dimers in epidermal cells isolated from human skin irradiated with UVB *in vitro* was determined by Roza *et al.* (1988) using a monoclonal antibody.

(iv) UVC radiation

Exposure of human skin, from which the stratum corneum had been removed, to either a germicidal (UVC) or a Hanovia hot quartz lamp *in vivo* resulted in DNA damage demonstrable by immunofluorescence (Tan & Stoughton, 1969). When the stratum corneum was intact, DNA damage was detected only after exposure to the germicidal lamp. [The

Fig. 12. Action spectrum for pyrimidine dimer formation in human skin (•) and solar spectra at the surface of the Earth for stratospheric ozone levels of 0.32 cm (dotted line) and 0.16 cm (solid line). Each point in the action spectrum represents the slope of the dose–response line (dimer yields at three exposures) for one volunteer at one wavelength, obtained from triplicate independent determinations. Thirty points occur at 302 nm, although some points overlie other values; five points occur at each other wavelength: points at 290 and 334 nm are circled to indicate that identical dimer yields were recorded for two volunteers. ph, photon; ESS, endonuclease-sensitive site

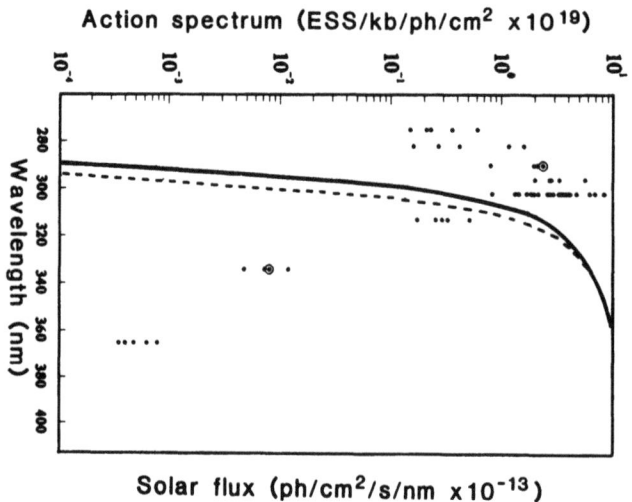

From Freeman *et al.* (1989)

Working Group noted that more sensitive analytical techniques for DNA damage are now available.]

(b) *Lymphocytes*

(i) *Broad-spectrum ultraviolet radiation*

In addition to cells of the skin, white blood cells are also subject to exposure to UVB and UVA, partly because some are temporarily resident in the skin and partly because it has been estimated that the equivalent of the total blood volume circulates through the dermal capillaries approximately every 11 min (Kraemer & Weinstein, 1977). Detecting effects, e.g., on lymphocytes, is likely to be extremely difficult owing to the fact that they are continually moving between the blood and other tissues; indeed, 90% of the lymphocyte population at any given time is resident outside the blood. Thus, the concentration in the blood of any lymphocytes irradiated while passing through the skin may fall substantially over time after irradiation ends as they are diluted in the whole body lymphocyte pool. Extravascular lymphocytes resident in the skin may also receive higher doses of UVR. Nevertheless, studies have been reported of genetic or related effects on lymphocytes sampled from peripheral blood.

Larcom et al. (1991) examined the capacity for DNA synthesis of lymphocytes from eight subjects exposed in two commercial tanning salons. Blood was taken immediately before tanning and again 24 h after tanning. System I used a sunlamp with a UVB:UVR ratio of 0.02% for 280–300 nm and 1.4% for 300–315 nm; the output of system II (Solana Voltarc lamp) was not indicated. There was a 24–84% (average, 53%) decrease in phytohaemagglutinin-induced DNA synthesis with system I and a 8–58% (average, 30%) decrease with system II.

(ii) *UVA radiation*

Seven of 13 psoriasis patients receiving oral 8-methoxypsoralen and high-intensity, long-wave UVA radiation had reduced leukocyte DNA synthesis; this did not occur in any of 10 controls (Kraemer & Weinstein, 1977). These results indicate that UVA reduces the incorporation of tritiated thymidine in lymphocytes circulating through the skin.

(iii) *UVB radiation*

In normal, fair-skinned subjects given whole-body exposure to 1.5–3 × MED doses of UVB from a sunlamp (280–380 nm), a dose-dependent decrease was seen in the incorporation of tritiated thymidine into DNA following stimulation by photohaemagglutinin; the proportion of circulating lymphocytes was decreased and the proportion of null cells was increased (Morison et al., 1979a).

These studies indicate that leukocytes should be included in any inventory of human cells potentially exposed to solar radiation or artificial UVR.

4.5.2 *Experimental systems* [see Tables 32–35, in which exposures are separated according to type of UVR]

(a) *DNA damage*

Inhibition of DNA synthesis has been induced in hairless albino mouse epidermis at wavelengths of 260–320 nm, with a maximal effect at 290 nm. Inhibition was not detected at 335 nm (Kaidbey, 1988). The action spectrum was similar to that for formation of cyclobutane-type pyrimidine dimers (Cooke & Johnson, 1978; Ley et al., 1983) and pyrimidine–pyrimidone (6-4) photoproducts in mouse skin (Olsen et al., 1989). Pyrimidine dimers (measured as endonuclease-sensitive sites) have been measured in the corneal DNA of the marsupial, *M. domestica*, following exposure to a sunlamp (280–400 nm) (Ley et al., 1988).

While DNA is the main photochromophore for UVC, there is evidence that active oxygen intermediates are involved in the production of DNA damage by UVA (Tyrrell, 1991). The production of several types of photolesions is oxygen dependent (Tyrrell, 1984, 1991). In addition, the irradiation lethality of both cultured bacterial (Webb, 1977) and mammalian (Danpure & Tyrrell, 1976) cells is dependent on the presence of oxygen; this observation was later linked with the production of singlet oxygen (Tyrrell & Pidoux, 1989). It has also been observed that irradiation of cultured human skin cells with UVB (302 nm, 313 nm), UVA (334 nm, 365 nm) and visible (405 nm) radiation is strongly enhanced in glutathione-depleted cells (Tyrrell & Pidoux, 1986, 1988). This apparent protection by glutathione appears to be due to its radical scavanging properties at the stated wavelength but may be due to induction of a more specific pathway (such as its essential role as a hydrogen donor for glutathione peroxidase) at longer wavelengths. Francis and Giannelli (1991) found that the abnormally high yield of single-stranded DNA breaks produced by UVA in six UVA-sensitive human fibroblasts (three from actinic reticuloid patients, two from sisters with familial actinic keratoses and internal malignancies and one from a patient with an abnormally high incidence of basal-cell carcinomas) could be reduced if sensitive cells were co-cultivated with normal fibroblasts or with radical scavengers. They suggested that the UVA-sensitive cells had deficits of small-molecular-weight scavengers of active oxygen species and that intercellular cooperation allows the transfer of these substances from resistant to sensitive cells. The presence of non-DNA chromophores that generate active oxygen species can also occur with UVC. Melanin, normally regarded as a solar screen, has also been associated with the formation of oxidative DNA damage, such as thymine glycols in mouse cells that vary in melanin content (Huselton & Hill, 1990). A slight increase in pyrimidine dimer yield was seen in human melanocytes as compared to keratinocytes following exposure to UVR at 254, 297, 302 and 312 nm but was significant only at 297 nm (Schothorst et al., 1991).

(b) Mutagenicity

Numerous reports show that sunlight or solar-simulated radiation induces mutations in bacteria, plants, Chinese hamster ovary (CHO) and lung (V79) cells, mouse lymphoma cells and human skin fibroblasts.

Studies in bacteria exposed to radiation throughout the solar UV spectrum (reviewed by Webb, 1977) demonstrate mutagenic activity unambiguously. The effects of sunlight on mammalian cells have been reviewed (Kantor, 1985). UVA (320–400 nm) is mutagenic to yeast and cultured mammalian cells, UVB (290–320 nm) to bacteria and cultured mammalian cells and UVC (200–290 nm) to bacteria, fungi, plants, cultured mammalian cells, including CHO and V79 cells, and human lymphoblasts, lymphocytes and fibroblasts. Since wavelengths in the UVC range do not reach the surface of the Earth, they are of no significance as a source of damage in natural sunlight.

A characteristic of all of these studies is that UVA appears to be relatively inefficient as a mutagen in comparison with UVB and UVC when activity is expressed per unit of energy fluence, but not necessarily so when expressed per DNA photoproduct (see Tyrrell, 1984). Webb (1977) compiled action spectra for the introduction of mutations in bacteria, as did Coohill et al. (1987) for mutagenesis in human epithelial cells. In both *Salmonella* and human cells, wavelengths > 320 nm were at least 10^3 times less effective than those between 270 and 290 nm.

A comparison of the mutagenicity of various UV-containing light sources towards a set of *S. typhimurium* strains was reported by De Flora et al. (1990). The approach did not involve measurement of cytotoxicity, and mutagenicity was compared at roughly equitoxic doses rather than as a function of fluence. Halogen lamps were as mutagenic as 254-nm UVC and more mutagenic than fluorescent sunlamps or sunlight. The mutagenicity of halogen lamps was attributed to their UVC component, in contrast to sunlight which produced mutagenic effects over a wide UV spectrum. The mutagenicity of halogen lamps, fluorescent lamps and sunlight was partially inhibited by catalase, suggesting that peroxides may be involved in this in-vitro system. It is also relevant that pretreatment of *E. coli* with hydrogen peroxide results in an increase in both UVA resistance and hydrogen peroxide scavenging ability (Moss, S.H., quoted by Tyrrell, 1985; Sammartano & Tuveson, 1985; Tyrrell, 1985).

Further evidence for the complexity of responses to the UVR region comes from Schothorst *et al.* (1987b), who examined the mutational response of human skin fibroblasts to 12 lamps differing widely in their emission characteristics. Surprisingly, they found that, whatever the light source, mutation induction per MED was similar with UVC, UVB and solar radiation; with UVA (only one data point), mutation induction per MED was much greater. The authors emphasized that these conclusions hold only if it is valid to calculate the mutagenicity of a light source by adding the effects of the contributing wavelengths; however, the data of Coohill *et al.* (1987) argue against this assumption.

The inevitable consequence of the absorption spectrum maximum of DNA is that there is a considerable body of data on mutagenicity toward microorganisms of UVC, which is usually delivered by radiation from germicidal lamps with more than 90% of their output at 254 nm. The types of mutations that are induced by UVC and the mechanisms of their induction have been reviewed (Witkin, 1976; Hall & Mount, 1981; Walker, 1984; Hutchinson & Wood, 1986; Bridges *et al.*, 1987; Hutchinson, 1987). Specific cellular proteins, including the products of *rec*A and *umu*C genes, together with a cleaved derivative of the *umu*D gene product, must be present for mutations to result from most types of DNA damage. These proteins are themselves part of an inducible response to DNA damage, and their intracellular level increases dramatically when photoproducts or other lesions are detected in DNA. It is not yet clear to what extent inducible systems are involved in UV mutagenesis in higher eukaryotes.

Current evidence suggests that all photoproducts are likely to be potentially mutagenic, although with greatly different specificities and potencies. The major UV photoproducts, cyclobutane-type thymine–thymine dimers, are, for example, relatively weakly mutagenic (Banerjee *et al.*, 1988, 1990), owing in part to the propensity of polymerases to insert adenine when the template instruction is unclear or missing (Sagher & Strauss, 1983; Schaaper *et al.*, 1983; Kunkel, 1984). The relatively minor (6-4) thymine–thymine photoproduct is, in contrast, highly mutagenic, the dominant mutation being a 3′ T→C transition (LeClerc *et al.*, 1991). By far the most frequent UVC-induced change in human cells is the transition from G:C to A:T (Bredberg *et al.*, 1986; Seetharam *et al.*, 1987; Hsia *et al.*, 1989; Dorado *et al.*, 1991). A number of investigators have noted the production of tandem transitions from G:C,G:C to A:T,A:T. Although this is not the most frequent change, it seems to be particularly characteristic for UVC mutagenesis in human cells. The frequency of mutation per lethal event at the *hprt* locus (which detects a broad spectrum of mutations) is approximately the same at 254 nm and 313 nm in human lymphoblastoid cells; however, the mutation frequency per lethal event at the Na^+/K^+ ATPase locus (which detects point mutations) is considerably higher at 313 nm. This finding may indicate a difference in types of premutagenic lesions and/or rates of mutation between the two wavelength regions (Tyrrell, 1984).

Two bacterial studies provide positive evidence for the mutagenic activity of fluorescent lamps. De Flora *et al.* (1990) employed Sylvania 36 W cool white tubes with *E. coli* and *Salmonella* strains. [The Working Group had difficulty in evaluating these data because they are presented in a highly transformed format.] Hartman *et al.* (1991) used General Electric F15T8CW lamps; a lowest effective dose of 5500 J/m^2 can be estimated from the results with *Salmonella* tester strains. Filters that block wavelengths < 370 nm effectively eliminated mutagenesis, while radical scavengers such as superoxide dismutase or catalase stimulated mutagenesis.

Hsie *et al.* (1977) irradiated the *hprt* CHO system with Westinghouse white light F40CW lamps. The minimal effective dose was 3.96×10^6 J/m^2. Putting lids on the petri dishes reduced mutant frequency by 30%. [The Working Group noted that the results were based on a single dose point in a single experiment.] Jacobson *et al.* (1978) exposed mouse lymphoma L5178Y $tk^{+/-}$ cells to Sylvania F18T8 cool white lamps. The estimated lowest effective dose was 2×10^4 J/m^2. [The Working Group noted that the selective agent used, BUdR, is regarded as inefficient and has been superseded by trichlorothymidine, so these results require confirmation.]

(c) Chromosomal effects

Sunlamps have been shown to produce sister chromatid exchange in amphibian cells (Chao & Rosenstein, 1985) and in human fibroblasts (Bielfeld *et al.*, 1989; Roser *et al.*, 1989). Fibroblasts from a panel of cutaneous malignant melanoma patients (Roser *et al.*, 1989) and heterozygotes of xeroderma pigmentosum (Bielfeld *et al.*, 1989) were more susceptible to the induction of both sister chromatid exchange and micronuclei than those from normal donors. Micronuclei were also induced in mouse splenocytes by exposure to sunlamps *in vitro* (Dreosti *et al.*, 1990).

A study with CHO cells provided evidence for a dose-related increase in the induction of sister chromatid exchange by UVA, but the increased induction of chromosomal aberrations showed no dose-response relationship (Lundgren & Wulf, 1988).

UVB induced sister chromatid exchange in CHO cells (Rasmussen *et al.*, 1989) and chromosomal aberrations in frog ICR 2A cells (Rosenstein & Rosenstein, 1985). In the latter study, photoreactivation reduced the number of chromosomal aberrations more effectively at 265, 289 and 302 than at 313 nm, suggesting that non-cyclobutane dimer photoproducts are more important primary lesions at the higher wavelength.

For UVC, more extensive data are available. Sister chromatid exchange was induced in Chinese hamster V79 (Nishi *et al.*, 1984) and CHO (Rasmussen *et al.*, 1989) cells. Chromatid exchange was also recorded in cultured fetal fibroblasts from New Zealand black mice, which proved to be more sensitive than BALB/c cells (Reddy *et al.*, 1978). The induction of chromosomal aberrations in Chinese hamster cells has been reported on a number of occasions (Chu, 1965a,b; Trosko & Brewen, 1967; Bender *et al.*, 1973; Griggs & Bender, 1973; Ikushima & Wolff, 1974).

Exposure of frog ICR 2A cells to 254 or 265 nm radiation induced both sister chromatid exchange (Chao & Rosenstein, 1985) and chromosomal aberrations, while photoreactivating light significantly reduced the frequency of chromosomal aberrations, which implies a role for pyrimidine dimers in their genesis (Rosenstein & Rosenstein, 1985). Chromosomal aberrations were also seen with *Xenopus* cell cultures (Griggs & Bender, 1973). The frequencies of sister chromatid exchange and chromosomal aberrations induced by UVC were reduced by photoreactivating light in chicken embryo fibroblasts (Natarajan *et al.*, 1980), lending further support to the concept that the cyclobutane pyrimidine dimer represents a primary lesion in these two end-points.

Parshad *et al.* (1980a) reported the induction of chromosomal damage in human IMR-90 fibroblasts following treatment with 4.6 W/m^2 over 20 h (331 kJ/m^2) from F15T8-CW tubes. Shielding and radical scavengers reduced the level of damage.

Extensive data are available on the induction of sister chromatid exchange in fibroblasts from patients with Bloom's syndrome (Krepinsky *et al.*, 1980), xeroderma pigmentosum (De Weerd-Kastelein *et al.*, 1977; Fujiwara *et al.*, 1981) or Cockayne's syndrome (Marshall *et al.*, 1980; Fujiwara *et al.*, 1981), as well as from normal individuals. In comparison with normal individuals, more sister chromatid exchanges were induced per lethal lesion in fibroblasts from excision-competent Bloom's syndrome (Kurihara *et al.*, 1987) and Cockayne's syndrome (Marshall *et al.*, 1980) patients. No such increase in sister chromatid exchange was seen in fibroblasts from excision-defective xeroderma pigmentosum patients or from an individual defective in the ligation step of repair (Henderson *et al.*, 1985).

The induction of sister chromatid exchange by UV irradiation has also been studied in human lymphocytes, with conflicting results. In one study, they were reported to be less responsive than either human fibroblasts or CHO cells (Perticone *et al.*, 1986), while another report, in which chromosomal aberrations were also studied, suggested that lymphocytes were more sensitive than fibroblasts in their response at both end-points (Murthy *et al.*, 1982). These results may have implications for the interpretation of the effect of UV on the immune system.

Fibroblasts from xeroderma pigmentosum patients are more sensitive to the induction of chromosomal aberrations than cells from normal donors (Parrington *et al.*, 1971; Parrington, 1972; Marshall & Scott, 1976). Seguin *et al.* (1988) showed that lymphoblastoid cells from five Cockayne's syndrome patients were similarly hypersensitive to UVC-induced chromosomal aberrations. The induction of micronuclei in two normal and three Bloom's syndrome-derived fibroblast cell cultures was reported by Krepinsky *et al.* (1980). One culture from a Bloom's syndrome patient, GM1492, proved to be exceptionally sensitive to the induction of micronuclei; the other two were indistinguishable from normal cells. This result emphasizes the potential importance of heterogeneity in response among patients with rare genetic syndromes.

(d) Transformation

Morphological transformation of mammalian cells has been induced by solar radiation, unshielded fluorescent tubes, solar simulators, UVA, UVB and, most extensively, UVC. There is weak evidence (Baturay *et al.*, 1985) for the induction of transformation by predominantly UVA radiation (20T12BLB bulbs) in BALB/c 3T3 cells. In the same report, UVA was shown to have promoting activity following initiation with β-propiolactone. The most effective wavelength for Syrian hamster embryo cells (Doniger *et al.*, 1981) and human embryonic fibroblasts (Sutherland *et al.*, 1981) appears to be in the UVC range at about 265 nm. Transformation of human cells can be enhanced by delivering the dose on a number of separate occasions (Sutherland et al., 1988). It has also been reported that excision repair-defective xeroderma pigmentosum cells can be transformed to the anchorage-independent phenotype at lower doses than those required for cells from normal individuals (Maher et al., 1982). Fisher and Cifone (1981) showed enhanced metastatic potential of mouse fibrosarcoma cells. Plasmids containing the human N-*ras* gene which were irradiated with UVR (254 nm) *in vitro* acquired the ability to transform cultured rat-2 cells after transfection; photoreactivation of irradiated plasmids eliminated their transforming ability (van der Lubbe et al., 1988). In another study, UVB irradiation activated the human Ha-*ras* gene on a plasmid in a transformation assay with mouse NIH-3T3 cells (Pierceall & Ananthaswamy (1991).

An investigation of chromosomal breaks and malignant transformation in embryonic mouse cells (Sanford et al., 1979; Parshad et al., 1980b) revealed that exposure of cultured cells to fluorescent lamps induced malignant transformation, as measured by tumour formation following implantation into syngeneic hosts. The potential importance of active oxygen species was revealed by experiments in which the partial pressure of oxygen in cultures was increased, resulting in increased malignant transformation and correlated chromosomal breakage.

Kennedy et al. (1980) reported induction of transformation in C3H 10T½ mouse embryonic cell cultures by light from General Electric F18T8 lamps. The lowest effective dose was estimated at 2×10^5 J/m^2, and use of petri dish lids was effective in reducing transformation.

(e) *Effects on cellular and viral gene expression*

A number of cellular oncogenes and other genes involved in the regulation of growth are implicated in the process of carcinogenesis, as they are subject to both gene mutation and alteration in expression due to chromosomal rearrangement. Many of these genes also show transient alterations in expression following DNA damage, which has led to the suspicion that such transient changes are involved, either directly or indirectly, in the carcinogenic process.

UVC radiation was found to increase transiently the expression of various cellular genes, including those that code for collagenase (Stein et al., 1989), the fos protein (Hollander & Fornace, 1989; Stein et al., 1989), the jun protein (Ronai et al., 1990), metallothioneins I and II (Fornace et al., 1988) and human plasminogen activator (Miskin & Ben-Ishai, 1981). UVA radiation enhanced expression of the genes that code for the fos protein (Hollander & Fornace, 1989), and UVB radiation increased the level of ornithine decarboxylase (Verma et al., 1979). Different levels of cytotoxicity were seen in these experiments. UVA radiation at doses that inactivate a small fraction of the fibroblast cell population induced expression of the haem oxygenase gene (Keyse & Tyrrell, 1989) by a transient enhancement in transcription rate (Keyse et al., 1990). *cis*-Acting enhancer elements have been shown to be involved in activation of the collagenase and c-*fos*, as well as human immunodeficiency promoter (Stein et al., 1989). In both rat fibroblasts and human keratinocyte cell lines, exposure to UVR increased the levels of c-*fos* RNA within 10 min and of c-*myc* RNA after about 1 h. The levels peaked at 30 min and 7 h and returned to normal within 1 h and 24 h, respectively. The order of effectiveness was UVC > UVB > UVA

(Ronai *et al.*, 1990). Elevated levels of p53 protein were observed in mouse cells treated with UVR; the increase was due to post-translation activation or stabilization (Maltzman & Czyzyk, 1984). In human keratinocytes exposed to UVA, increased levels of human epidermal growth factor receptor RNA (HER-1) were found (Yang *et al.*, 1988).

The mechanisms that mediate these transient and immediate inducible responses are largely unknown. Some of them, however, overlap with those seen in response to tumour promoters, and it is significant that natural sunlight has been reported to enhance the expression of protein kinase C in cultured human epithelial P3 cells (Peak *et al.*, 1991a). For reviews of this general area, see Ananthaswamy and Pierceall (1990) and Ronai *et al.* (1990).

Other transient responses to UVR have been noted at somewhat later times (12–48 h). Methotrexate resistance due to gene amplification was reported in 3T6 mouse cells (Tlsty *et al.*, 1984). Another selective DNA amplification response is induction by UVR of viral DNA synthesis, e.g., of polyoma virus in rat fibroblasts. UVC was more effective than UVB, and UVA was ineffective (Ronai *et al.*, 1987). In Chinese hamster embryo cells, UVC irradiation increased DNA binding to the early domain of the SV40 minimal origin, resulting in SV40 DNA amplification (Lücke-Huhle *et al.*, 1989). The induction of asynchronous viral replication is mediated by cellular proteins that bind to specific sequences in the DNA of polyoma (Ronai & Weinstein, 1988) and SV40 viruses (Lücke-Huhle *et al.*, 1989).

Exposure to UVR can activate viruses. This phenomenon has been known for herpes simplex virus for a long time (for a recent report, see Rooney *et al.*, 1991). It was reported recently that UVC can activate the gene promoters of the human immunodeficiency virus (HIV) (Valerie *et al.*, 1988) and Moloney murine sarcoma virus (Lin *et al.*, 1990). Furthermore, activation of complete HIV grown in cells pre-exposed to UVC radiation was observed (Valerie *et al.*, 1988). HIV activation may contribute to faster development of AIDS, which in turn may facilitate development of malignancies. Further studies showed that the HIV promoter and HIV are activated by UVC and UVB, but not UVA radiation even at very high exposures (Stanley *et al.*, 1989; Beer *et al.*, 1991 [abstract]; Lightfoote *et al.*, 1992). There are indications that pyrimidine dimers (Stein *et al.*, 1989) or chromatin damage (Valerie & Rosenberg, 1990) play a role in the initiation of HIV activation by UVR. The in-vitro observations have been verified for UVC, UVB and UVA in experiments with transgenic mice carrying the HIV promoter/reporter gene constructs (Cavard *et al.*, 1990; Frucht *et al.*, 1991; Vogel *et al.*, 1992). For reviews on the activation HIV by UVR, see Zmudzka and Beer (1990) and Beer and Zmudzka (1991).

Table 32. Genetic and related effects of solar, simulated solar and sunlamp (UVA and UVB) irradiation

Test system	Result[a]	Reference
BS?, *Bacillus subtilis*, mutation	+	Munakata (1989)
SSB, *Saccharomyces cerevisiae* D7, DNA damage	+	Hannan et al. (1984)
PLM, Wheat mutation	+	Morgun et al. (1988)
DIA, DNA damage, ICR 2A frog cells in vitro	+	Chao & Rosenstein (1986)
DIA, DNA damage, ICR 2A frog cells in vitro	+	Rosenstein et al. (1989)
DIA, DNA strand breaks, Chinese hamster V79 cells	+	Elkind & Han (1978)
DIA, DNA damage, Chinese hamster V79 cells in vitro	+	Suzuki et al. (1981)
DIA, DNA damage, C3H 10T½ mouse cells in vitro	+	Suzuki et al. (1981)
GCO, Gene mutation, Chinese hamster ovary cells in vitro	+	Hsie et al. (1977)
G9H, Gene mutation, Chinese hamster V79 lung cells in vitro, 6-TG[r]	+	Zölzer et al. (1988)
G5T, Gene mutation, mouse lymphoma L5178Y cells in vitro	+	Jacobson et al. (1978)
G9H, Gene mutation, Chinese hamster V79 lung cells in vitro, 6-TG[r]	+	Bradley & Sharkey (1977)
GCO, Gene mutation, Chinese hamster ovary cells in vitro	+	Burki & Lam (1978)
G9H, Gene mutation, Chinese hamster V79 lung cells in vitro, 6-TG[r]	+	Suzuki et al. (1981)
SIA, Sister chromatid exchange, ICR 2A frog cells in vitro	+	Chao & Rosenstein (1985)
MIA, Micronucleus test, mouse splenocytes in vitro	+	Dreosti et al. (1990)
TBM, Cell transformation, BALB/c 3T3 mouse cells in vitro	+	Withrow et al. (1980)
TBM, Cell transformation, BALB/c mouse epidermal cells in vitro	+	Ananthaswamy & Kripke (1981)
TCM, Cell transformation, C3H 10T½ mouse embryo cells in vitro	+	Kennedy et al. (1980)
TCM, Cell transformation, C3H 10T½ mouse cells in vitro	+	Suzuki et al. (1981)
TCL, Cell transformation, mouse fibrosarcoma cells in vitro	+	Fisher & Cifone (1981)
TCL, Cell transformation, 10T½ mouse skin fibroblasts in vitro	+	Ananthaswamy (1984a)
DIA, DNA damage, fish in vitro	+	Applegate & Ley (1988)
DIH, DNA damage, human skin fibroblasts in vitro	+	Rosenstein et al. (1985)
DIH, DNA damage, human skin fibroblasts in vitro	+	Chao & Rosenstein (1986)
DIH, DNA damage, human skin fibroblasts in vitro	+	Rosenstein (1988)
DIH, DNA damage, human skin fibroblasts in vitro	+	Rosenstein & Mitchell (1991)
DIH, DNA damage, human HeLa cells in vitro	+	Elkind & Han (1978)
GIH, Gene mutation, human xeroderma pigmentosum fibroblasts in vitro	+	Patton et al. (1984)
SHF, Sister chromatid exchange, human[b] fibroblasts in vitro	+	Knees-Matzen et al. (1991)

Table 32 (contd)

Test system	Result[a]	Reference
SIH, Sister chromatid exchange, human xeroderma pigmentosum fibroblasts	+	Bielfeld et al. (1989)
SIH, Sister chromatid exchange, human malignant melanoma cells	+	Roser et al. (1989)
MIH, Micronucleus test, human xeroderma pigmentosum fibroblasts	+	Bielfeld et al. (1989)
MIH, Micronucleus test, human malignant melanoma cells	+	Roser et al. (1989)
DVA, DNA damage, BALB/c mouse skin cells in vivo	+	Ananthaswamy & Fisher (1981)
DVA, DNA damage, marsupial corneal cells in vivo	+	Freeman et al. (1988a)
DVA, DNA damage, marsupial corneal cells in vivo	+	Ley et al. (1988)
TVI, Cell transformation, 10T½ mouse skin fibroblasts treated in vivo scored in vitro	+	Ananthaswamy (1984b)
DVH, DNA damage, human skin cells in vivo	+	Eggset et al. (1983)
DVH, DNA damage, human skin cells in vivo	+	Freeman et al. (1988b)

[a] +, positive
[b] First-degree relatives of melanoma patients

Table 33. Genetic and related effects of predominantly UVA irradiation (near UV)

Test system	Result[a]	Reference
SA9, *Salmonella typhimurium* TA98, reverse mutation	+	Calkins *et al.* (1987)
ECW, *Escherichia coli* WP2 *uvrA*, reverse mutation	+	Tyrrell (1982)
EC2, *Escherichia coli* WP2 hcr–, reverse mutation	+	Kubitschek (1967)
ECR, *Escherichia coli* B/tr/l, *trp*, reverse mutation	+	Webb & Malina (1970)
ECR, *Escherichia coli* WP2 *recA*, reverse mutation	+	Tyrrell (1982)
ECR, *Escherichia coli* WP2 *uvrA recA*, reverse mutation	+	Tyrrell (1982)
ECR, *Escherichia coli* B/r *uvrA trp thy*, reverse mutation	+	Tyrrell (1982)
ECR, *Escherichia coli* wild type, reverse mutation	+	Tyrrell (1982)
ECR, *Escherichia coli*, mutation	+	Wood *et al.* (1984)
SSB, *Saccharomyces cerevisiae* wild type, DNA damage	+	Zölzer & Kiefer (1983)
SSB, *Saccharomyces cerevisiae* excision-deficient, DNA damage	+	Zölzer & Kiefer (1983)
SSB, *Saccharomyces cerevisiae* D7, DNA damage	+	Hannan *et al.* (1984)
DIA, DNA damage, Chinese hamster ovary cells *in vitro*	+	Zelle *et al.* (1980)
DIA, DNA strand breaks, Chinese hamster ovary cells *in vitro*	+	Churchill *et al.* (1991)
GCO, Gene mutation, Chinese hamster ovary cells *in vitro*	+	Zelle *et al.* (1980)
GCO, Gene mutation, Chinese hamster ovary cells *in vitro*	+	Singh & Gupta (1982)
GCO, Gene mutation, Chinese hamster ovary cells *in vitro*	+	Lundgren & Wulf (1988)
G9H, Gene mutation, Chinese hamster lung V79 cells, *hprt* locus	+	Wells & Han (1984)
G9O, Gene mutation, Chinese hamster lung V79 cells, 6-TG[r]	+	Wells & Han (1984)
G5T, Gene mutation, mouse lymphoma L5178Y cells, *tk* locus	+	Hitchins *et al.* (1987)
SIC, Sister chromatid exchange, Chinese hamster ovary cells *in vitro*	+	Lundgren & Wulf (1988)
CIC, Chromosomal aberrations, Chinese hamster ovary cells *in vitro*	(+)	Lundgren & Wulf (1988)
TCL, Cell transformation, Syrian hamster embryo cells *in vitro* (neoplastic transformation)	+	Barrett *et al.* (1978)
TCL, Cell transformation, Syrian hamster embryo cells *in vitro* (morphological transformation)	–	Barrett *et al.* (1978)
DIH, DNA strand breaks, human fibroblasts *in vitro*	+	Rosenstein & Ducore (1983)
DIH, DNA strand breaks, human teratoma cells *in vitro*	+	Peak *et al.* (1987)
DIH, DNA double strand breaks, human teratocarcinoma cells *in vitro*	+	Peak & Peak (1990)
DIH, DNA strand breaks, human fibroblasts *in vitro*	+	Francis & Giannelli (1991)
DIH, DNA–protein cross-links, human teratocarcinoma cells *in vitro*	+	Peak & Peak (1991)
DIH, DNA strand breaks, human epithelial P3 cells *in vitro*	+	Peak *et al.* (1991b)
DIH, Pyrimidine dimer formation, human skin fibroblasts *in vitro*	+	Enninga *et al.* (1986)

Table 33 (contd)

Test system	Result[a]	Reference
DIH, Pyrimidine dimer formation, human skin fibroblasts *in vitro*	+	Rosenstein & Mitchell (1987)
GIH, Gene mutation, human lymphoblastoid cell line *in vitro*	−	Tyrrell (1984)
GIH, Gene mutation, human skin fibroblasts *in vitro*	+	Enninga *et al.* (1986)
GIH, Gene mutation, human epithelial cells *in vitro*	+[b]	Jones *et al.* (1987)
DVH, Pyrimidine dimer formation, human skin *in vivo*	+	Freeman *et al.* (1989)

[a]+, positive; (+), weakly positive; −, negative
[b]Positive result with 365 nm but not with 334 nm at same fluence

Table 34. Genetic and related effects of predominantly UVB irradiation

Test system	Result[a]	Reference
SA9, *Salmonella typhimurium* TA98, reverse mutation	+	Calkins et al. (1987)
EC2, *Escherichia coli* WP2, reverse mutation	+	Peak et al. (1984)
TSC, *Tradescantia*, chromosomal aberrations	+	Kirby-Smith & Craig (1957)
DIA, DNA damage, Chinese hamster ovary cells *in vitro*	+	Zelle et al. (1980)
DIA, DNA strand breaks, Chinese hamster V79 cells	+	Matsumoto et al. (1991)
DIA, DNA-protein cross-links, Chinese hamster V79 cells	+	Matsumoto et al. (1991)
GCO, Gene mutation, Chinese hamster ovary cells *in vitro*	+	Zelle et al. (1980)
GCO, Gene mutation, Chinese hamster ovary cells *in vitro*	+	Rasmussen et al. (1989)
G9H, Gene mutation, Chinese hamster V79 lung cells, *hprt* locus	+	Wells & Han (1984)
G9H, Gene mutation, Chinese hamster V79 lung cells, *hprt* locus	+	Zölzer & Kiefer (1984)
G9O, Gene mutation, Chinese hamster V79 lung cells, ouabain[r]	+	Wells & Han (1984)
G9H, Gene mutation, Chinese hamster V79 lung cells *in vitro*, 6TG[r]	+	Colella et al. (1986)
G51, Gene mutation, mouse lymphoma L5178Y cells *in vitro*	+	Jacobson et al. (1981)
SIC, Sister chromatid exchange, Chinese hamster ovary cells *in vitro*	+	Rasmussen et al. (1989)
CIA, Chromosomal aberrations, ICR 2A frog cells *in vitro*	+	Rosenstein & Rosenstein (1985)
TCS, Cell transformation, Syrian hamster embryo cells *in vitro*	+	Doniger et al. (1981)
DIH, DNA strand breaks, human skin fibroblasts *in vitro*	+	Rosenstein & Ducore (1983)
DIH, Pyrimidine dimer formation, human skin fibroblasts *in vitro*	+	Enninga et al. (1986)
DIH, Pyrimidine dimer formation, human skin fibroblasts *in vitro*	+	Rosenstein & Mitchell (1987)
DIH, DNA strand breaks, human teratoma *in vitro*	+	Peak et al. (1987)
DIH, DNA double strand breaks, human teratocarcinoma *in vitro*	+	Peak & Peak (1990)
DIH, DNA-protein cross-links, human teratocarcinoma *in vitro*	+	Peak & Peak (1991)
DIH, Pyrimidine dimer formation in human skin keratinocytes *in vitro*	+	Schothorst et al. (1991)
DIH, Thymine dimer formation, human fibroblasts *in vitro*	+	Roza et al. (1988)
GIH, Gene mutation, human lymphoblastoid cell line *in vitro*	–	Tyrrell (1984)
GIH, Gene mutation, human skin fibroblasts *in vitro*	+	Enninga et al. (1986)
GIH, Gene mutation, human epithelial cells *in vitro*	+	Jones, C.A. et al. (1987)
TIH, Cell transformation, human fibroblasts *in vitro*	+	Sutherland et al. (1981)
DVA, Cyclobutane dimers in SV40 plasmid DNA in human fibroblasts *in vivo*	+	Mitchell et al. (1991)
DVA, Cytosine photohydrates in SV40 plasmid DNA in human fibroblasts *in vivo*	+	Mitchell et al. (1991)

Table 34 (contd)

Test system	Result[a]	Reference
DVA, Pyrimidine dimer induction, mouse skin *in vivo*	+	Cooke & Johnson (1978)
DVA, Pyrimidine dimer formation, mouse skin *in vivo*	+	Ley et al. (1983)
DVA, (6–4) Photoproduct formation, mouse epidermis *in vivo*	+	Olsen et al. (1989)
DVH, Pyridime dimer formation, human skin *in vivo*	+	Freeman et al. (1989)
UVH, Unscheduled DNA synthesis, human cornea *in vivo*[b]	+	Grabner & Brenner (1981)

[a] +, positive; –, negative
[b] From people who had been dead for 15 min

Table 35. Genetic and related effects of UVC irradiation

Test system	Result[a]	Reference
ECB, *Escherichia coli*, thymine dimer formation	+	Setlow *et al.* (1963)
ECB, *Escherichia coli*, photoproduct formation	+	Setlow (1968)
ECB, *Escherichia coli*, thymine photoadduct formation	+	Smith (1964)
ECB, *Escherichia coli*, pyrimidine dimers	+	Brash & Haseltine (1982)
ECB, *Escherichia coli*, (6-4) photoproducts	+	Brash & Haseltine (1982)
ECF, *Escherichia coli*, miscellaneous strains, forward mutation	+	Miller (1985)
ECR, *Escherichia coli*, mutation	+	Witkin (1976)
ECR, *Escherichia coli*, mutation	+	Walker (1984)
ECR, *Escherichia coli*, mutation	+	Franklin & Haseltine (1986)
ECR, *Escherichia coli*, mutation	+	Bridges *et al.* (1987)
ECR, *Escherichia coli*, mutation	+	Schaaper *et al.* (1987)
SSB, *Saccharomyces cerevisiae*, pyrimidine dimer formation	+	Wheatcroft *et al.* (1975)
SSB, *Saccharomyces cerevisiae*, pyrimidine dimer formation	+	Resnick *et al.* (1987)
SCN, *Saccharomyces cerevisiae*, aneuploidy	+	Parry *et al.* (1979)
SCF, *Saccharomyces cerevisiae*, forward mutation	+	Lee *et al.* (1988)
SCR, *Saccharomyces cerevisiae*, reverse mutation	+	Siede & Eckardt (1986)
PLU, Plants, DNA damage	+	McLennan (1987)
PLU, *Nicotiana tabacum*, unscheduled DNA synthesis	+	Cieminis *et al.* (1987)
PLU, *Chlamydomonas reinhardtii*, pyrimidine dimer formation	+	Vlček *et al.* (1987)
PLM, *Chlamydomonas reinhardtii*, mutation	+	Vlček *et al.* (1987)
TSC, *Tradescantia*, chromosomal aberrations	+	Kirby-Smith & Craig (1957)
DM?, *Drosophila melanogaster* embryo cells *in vitro*, DNA damage	+	Koval (1987)
DIA, DNA damage, ICR 2A frog cells *in vitro*	+	Chao & Rosenstein (1986)
DIA, DNA strand breaks, Chinese hamster V79 cells	+	Elkind & Han (1978)
DIA, DNA damage, Chinese hamster ovary cells *in vitro*	+	Zelle *et al.* (1980)
GCO, Gene mutation, Chinese hamster ovary cells *in vitro*	+	Zelle *et al.* (1980)
GCO, Gene mutation, Chinese hamster ovary cells *in vitro*	+	Rasmussen *et al.* (1989)
GCO, Gene mutation, Chinese hamster ovary cells *in vitro*	+	Drobetsky & Glickman (1990)
G9H, Gene mutation, Chinese hamster V79 lung cells *in vitro*	+	Colella *et al.* (1986)
G9H, Gene mutation, Chinese hamster V79 lung cells, *hprt* locus	+	Suzuki *et al.* (1981)
G9H, Gene mutation, Chinese hamster V79 lung cells, *hprt* locus	+	Zölzer & Kiefer (1984)
G9O, Gene mutation, Chinese hamster V79 lung cells, ouabain[r]	+	Suzuki *et al.* (1981)

Table 35 (contd)

Test system	Result[a]	Reference
G51, Gene mutation, mouse lymphoma L5178Y cells *in vitro*	+	Jacobson *et al.* (1981)
SIC, Sister chromatid exchange, Chinese hamster V79 cells *in vitro*	+	Nishi *et al.* (1984)
SIC, Sister chromatid exchange, Chinese hamster ovary cells *in vitro*	+	Rasmussen *et al.* (1989)
SIA, Sister chromatid exchange, ICR 2A frog cells *in vitro*	+	Chao & Rosenstein (1985)
SIA, Sister chromatid exchange, chick embryo fibroblasts *in vitro*	+	Natarajan *et al.* (1980)
CIC, Chromosomal aberrations, Chinese hamster fibroblasts *in vitro*	+	Chu (1965a)
CIC, Chromosomal aberrations, Chinese hamster fibroblasts *in vitro*	+	Chu (1965b)
CIC, Chromosomal aberrations, Chinese hamster V79 cells *in vitro*	+	Bender *et al.* (1973)
CIC, Chromosomal aberrations, Chinese hamster V79 cells *in vitro*	+	Griggs & Bender (1973)
CIC, Chromosomal aberrations, Chinese hamster ovary cells *in vitro*	+	Ikushima & Wolff (1974)
CIC, Chromosomal aberrations, Chinese hamster CHEF-125 cells *in vitro*	+	Trosko & Brewen (1967)
CIA, Chromosomal aberrations, chick embryo fibroblasts *in vitro*	+	Natarajan *et al.* (1980)
CIA, Chromosomal aberrations, A8W243 *Xenopus* cells *in vitro*	+	Griggs & Bender (1973)
CIA, Chromosomal aberrations, ICR 2A frog cells *in vitro*	+	Rosenstein & Rosenstein (1985)
CIA, Chromosomal aberrations, New Zealand black mouse fetal fibroblasts	+	Reddy *et al.* (1978)
TBM, Cell transformation, BALB/c 3T3 mouse cells	+	Withrow *et al.* (1980)
TCM, Cell transformation, C3H 10T½ mouse cells	+	Chan & Little (1976)
TCM, Cell transformation, C3H 10T½ mouse cells	+	Mondal & Heidelberger (1976)
TCM, Cell transformation, C3H 10T½ mouse cells	+	Chan & Little (1979)
TCM, Cell transformation, C3H 10T½ mouse cells	+	Suzuki *et al.* (1981)
TCM, Cell transformation, C3H 10T½ mouse cells	+	Borek *et al.* (1989)
TCS, Cell transformation, Syrian hamster embryo cells	+	DiPaolo & Donovan (1976)
TCS, Cell transformation, Syrian hamster embryo cells	+	Doniger *et al.* (1981)
TCS, Cell transformation, Syrian hamster embryo cells	+	Borek *et al.* (1989)
TEV, Cell transformation, SV-40/BALB/c 3T3 mouse cells	+	Withrow *et al.* (1980)
DIH, DNA strand breaks, human skin fibroblasts *in vitro*	+	Rosenstein & Ducore (1983)
DIH, DNA damage, human skin fibroblasts *in vitro*	+	Rosenstein *et al.* (1985)
DIH, Pyrimidine dimer formation, human skin fibroblasts *in vitro*	+	Enninga *et al.* (1986)
DIH, Pyrimidine dimer formation, human skin fibroblasts *in vitro*	+	Rosenstein & Mitchell (1987)
DIH, DNA strand breaks, human teratoma cells *in vitro*	+	Peak *et al.* (1987)
DIH, Thymine dimer formation, human skin fibroblasts *in vitro*	+	Roza *et al.* (1988)
DIH, DNA damage, human skin fibroblasts *in vitro*	+	Chao & Rosenstein (1986)

Table 35 (contd)

Test system	Result[a]	Reference
DIH, DNA strand breaks, human fibroblasts *in vitro*	+	Lai & Rosenstein (1990)
DIH, DNA-protein cross-links, human fibroblasts *in vitro*	+	Lai & Rosenstein (1990)
DIH, DNA double strand breaks, human teratocarcinoma cells *in vitro*	+	Peak & Peak (1990)
DIH, DNA-protein cross-links, human teratocarcinoma cells *in vitro*	+	Peak & Peak (1991)
DIH, Pyrimidine dimer formation, human skin keratinocytes and melanocytes *in vitro*	+	Schothorst *et al.* (1991)
GIH, Gene mutation, human fibroblasts *in vitro*	+	Maher *et al.* (1979)
GIH, Gene mutation, human fibroblasts *in vitro*	+	Myhr *et al.* (1979)
GIH, Gene mutation, human lymphocytes *in vitro*	+	Sanderson *et al.* (1984)
GIH, Gene mutation, human lymphoblastoid cell line *in vitro*	+	Tyrrell (1984)
GIH, Gene mutation, human skin fibroblasts *in vitro*	+	Enninga *et al.* (1986)
GIH, Gene mutation, human epithelial cells *in vitro*	+	Jones, C.A. *et al.* (1987)
GIH, Gene mutation, human HeLa cells *in vitro*	+	Musk *et al.* (1989)
GIH, Gene mutation, human lymphocytes *in vitro*	+	Norimura *et al.* (1990)
GIH, Gene mutation, human fibroblasts *in vitro*	+	Dorado *et al.* (1991)
GIH, Gene mutation, human fibroblasts *in vitro*	+	McGregor *et al.* (1991)
GIH, Gene mutation, human melanoma cells *in vitro*	+	Musk *et al.* (1989)
SHF, Sister chromatid exchange, human fibroblasts *in vitro*	+	Fujiwara *et al.* (1981)
SHF, Sister chromatid exchange, human fibroblasts *in vitro*	+	Kurihara *et al.* (1987)
SHL, Sister chromatid exchange, human lymphocytes *in vitro*	+	Murthy *et al.* (1982)
SHL, Sister chromatid exchange, human lymphocytes *in vitro*	+	Perticone *et al.* (1986)
SHF, Sister chromatid exchange, human skin fibroblasts	+	De Weerd-Kastelein *et al.* (1977)
SHF, Sister chromatid exchange, human skin fibroblasts	+	Krepinsky *et al.* (1980)
SHF, Sister chromatid exchange, human skin fibroblasts	+	Marshall *et al.* (1980)
SHF, Sister chromatid exchange, human skin fibroblasts	+	Henderson *et al.* (1985)
SHF, Sister chromatid exchange, human skin fibroblasts	+	Krepinsky *et al.* (1980)
MIH, Micronucleus test, human skin fibroblasts *in vitro*	+	Parrington (1972)
CHF, Chromosomal aberrations, human skin fibroblasts	+	Parrington *et al.* (1971)
CHF, Chromosomal aberrations, human skin fibroblasts	+	Marshall & Scott (1976)
CHF, Chromosomal aberrations, human skin fibroblasts	+	Murthy *et al.* (1982)
CHL, Chromosomal aberrations, human lymphocytes *in vitro*	+	Holmberg & Gumauskas (1990)
CHL, Chromosome exchanges, human lymphocytes *in vitro*	+	Sutherland *et al.* (1981)
TIH, Cell transformation, human fibroblasts *in vitro*	+	Maher *et al.* (1982)
TIH, Cell transformation, human fibroblasts *in vitro*	+	

Table 35 (contd)

Test system	Result[a]	Reference
TIH, Cell transformation, human fibroblasts *in vitro*	+	Sutherland *et al.* (1988)
???, Cyclobutane dimers in SV40 plasmid DNA in human skin fibroblasts *in vitro* and *in vivo*	+	Mitchell *et al.* (1991)
???, Cytosine photohydrates in SV40 plasmid DNA in human skin fibroblasts *in vitro* and *in vivo*	+	Mitchell *et al.* (1991)
DVA, Pyrimidine dimer formation, mouse skin *in vivo*	+	Bowden *et al.* (1975)

[a] +, positive

5. Summary of Data Reported and Evaluation

5.1 Exposure data

Terrestrial life is dependent on radiant energy from the sun. Approximately 5% of solar terrestrial radiation is ultraviolet radiation (UVR), and solar radiation is the major source of human exposure to UVR. Before the beginning of this century, the sun was essentially the only source of UVR, but with the advent of artificial sources the opportunity for additional exposure has increased.

UVR spans the wavelengths from 100 to 400 nm. The biological effects of UVR vary enormously with wavelength; by convention, the ultraviolet spectrum has been further subdivided into three regions: UVC (100–280 nm), UVB (280–315 nm) and UVA (315–400 nm).

Solar UVR that reaches the Earth's surface comprises approximately 95% UVA and 5% UVB: UVC is completely filtered out by the Earth's atmosphere. The amount of solar UVR measured at the Earth's surface depends upon a number of factors, which include solar zenith angle (time of day, season and geographical latitude), stratospheric ozone, atmospheric pollutants, weather, ground reflectance and altitude.

Exposed skin surface is irradiated differently depending on cultural and social behaviour, clothing, the position of the sun in the sky and the relative position of the body. Exposure to UVB of the most exposed skin surfaces, such as nose, tops of the ears and forehead, relative to that of the lesser exposed areas, such as underneath the chin, normally ranges over an order of magnitude. Ground reflectance plays a major role in exposure to UVB of the eye and shaded skin surfaces, particularly with highly reflective surfaces such as snow.

In cutaneous photobiology, radiant exposure is frequently expressed as 'exposure dose' in units of J/cm^2 (or J/m^2). 'Biologically effective dose', derived from radiant exposure weighted by an action spectrum, is expressed in units of J/cm^2 (effective) or as multiples of 'minimal erythema dose' (MED). In cellular photobiology, the term 'fluence' is often used incorrectly as equivalent to radiant exposure.

The cumulative annual exposure dose of solar UVR varies widely among individuals in a given population, depending to a large extent on occupation and extent of outdoor activities. For example, it has been estimated that indoor workers in mid-latitudes (40–60 °N) receive an annual exposure dose of solar UVR to the face of about 40–160 times the MED, depending upon propensity for outdoor activities, whereas the annual solar exposure dose for outdoor workers is typically around 250 times the MED. Because few actual measurements have been reported of personal exposures, these estimates should be considered to be very approximate and subject to differences in cultural and social behaviour, clothing, occupation and outdoor activities.

Cumulative annual outdoor exposures may be augmented by exposures to articial sources of UVR. For example, the use of cosmetic tanning appliances increased in popularity in the 1980s. The majority of users are young women, and the median annual exposure dose is probably 20–30 times the MED. Currently used appliances emit primarily UVA radiation; prior to the 1980s, tanning lamps emitted higher proportions of UVB and UVC.

UVR has been used for several decades to treat skin diseases, notably psoriasis. A variety of sources of UVR are employed, and nearly all emit a broad spectrum of radiation. A typical dose in a single course of UVB phototherapy might lie between 200 and 300 times the MED.

UVR is used in many different industries, yet there is a paucity of data concerning human exposure from these applications, probably because in normal practice sources are well-contained and exposure doses are expected to be low. Acute reactions to overexposure are common among electric arc welders. Staff in hospitals who work with unenclosed phototherapy equipment are at potential risk of overexposure unless protective measures are taken. Individuals exposed to lighting from fluorescent lamps may typically receive annual exposure doses of UVR ranging from 0 to 30 times the MED, depending on illuminance levels and whether or not the lamps are housed behind plastic diffusers. There is increasing use of tungsten–halogen lamps, which also emit UVR, for general lighting.

5.2 Human carcinogenicity data

5.2.1 *Solar radiation*

Subjects with the inherited condition xeroderma pigmentosum appear to have frequencies of nonmelanocytic skin cancer and melanoma that are much higher than expected. Some evidence suggests that the greatest excess occurs on the head and neck.

(a) Nonmelanocytic skin cancer

The results of descriptive epidemiological studies suggest that exposure to sunlight increases the risk of nonmelanocytic skin cancer. These tumours occur predominantly on the skin of the face and neck, which is most commonly exposed to sunlight, although the distribution of basal-cell carcinomas is not as closely related to the distribution of exposure to the sun as is that of squamous-cell carcinomas. There is a strong inverse relationship between latitude and incidence of or mortality from skin cancer and, conversely, a positive relationship between incidence or mortality and measured or estimated ambient UVR. Migrants to Australia from the British Isles have lower incidence of and mortality from nonmelanocytic skin cancer than the Australian-born population. People who work primarily outdoors have higher mortality from these cancers, and there is some evidence that outdoor workers have higher incidence.

In several cross-sectional studies, positive associations have been seen between measures of solar skin damage and the prevalence of basal- and squamous-cell carcinomas. Measures of actual exposure to the sun have been less strongly associated with these cancers, possibly because of errors in measurement and inadequate control for potential confounding variables. In a study of US fishermen, estimates of individual annual and cumulative exposure to UVB were positively associated with the occurrence of squamous-cell carcinoma but not with the occurrence of basal-cell carcinoma.

Only two population-based case–control studies have been conducted. In one of these, from Canada, the response rate was low and the measures of exposure were crude. In the other study, from Australia, facial telangiectasia and solar elastosis of the neck were strongly associated with the risk for squamous-cell carcinoma, and cutaneous microtopography and solar elastosis of the neck were strongly associated with risk for basal-cell carcinoma. Migrants to Australia had a lower risk of squamous-cell carcinoma than did native-born Australians, and migrants who arrived after childhood had a lower risk for basal-cell carcinoma.

The hospital-based case–control studies that have been conducted suffer from methodological deficiencies, including choice of controls, measurement of exposure and confounding by reaction to sunlight, and are therefore difficult to interpret.

In a cohort study of nurses in the USA, those who spent more than 8 h per week outside without sunscreens had a similar incidence rate of basal-cell carcinoma to those who spent fewer than 8 h per week outdoors. In a cohort study from Victoria, Australia, the rates of both types of skin cancer were increased in outdoor workers, but the effect was not significant after adjustment for reaction to sunlight.

(b) Cancer of the lip

Cancer of the lip has been related to outdoor occupation in a number of descriptive studies. Migrants to Australia and Israel have lower risks than native-born residents.

Three case–control studies provide useful information about the association between outdoor work, taken as a proxy measure for exposure to UVR, and cancer of the lip. All of them showed a significantly increased risk, although potential confounding by tobacco use was not controlled adequately in any of the studies.

Assessment of the carcinogenicity of solar radiation for the lip is complicated by the fact that carcinoma of the lip as actually diagnosed is a mixture of cancers of the external lip and cancers of the buccal membranes. Use of alcohol and tobacco are known causes of the latter tumours.

(c) Malignant melanoma of the skin

Descriptive studies in whites in North America, Australia and several other countries show a positive association between incidence of and mortality from melanoma and residence at lower latitudes. Studies of migrants suggest that the risk of melanoma is related to solar radiant exposure at the place of residence in early life. The body site distribution of melanoma shows lower rates per unit area on sites usually unexposed to the sun than on usually or regularly exposed sites.

A large number of case–control studies are pertinent to the relationship between melanoma and exposure to the sun. These include large, carefully conducted population-based studies carried out in Western Australia, Queensland, western Canada and Denmark. Their results are generally consistent with positive associations with residence in sunny environments throughout life, in early life and even for short periods in early adult life. Positive associations are generally seen between measurements of cumulative sun damage expressed biologically as microtopographical changes or history of keratoses or nonmelanocytic skin cancer.

In contrast, the associations with total exposure to the sun over a lifetime or in recent years, as assessed by questionnaire, are inconsistent. This inconsistency may be due to differences in the effects of chronic and intermittent exposure. Chronic exposure, as assessed through occupational exposure, appeared to reduce melanoma risk in three of the large studies, particularly in men; this observation is consistent with the descriptive epidemiology of the condition, which shows lower risks in groups that work outdoors. Several other studies, which were generally smaller or had less detailed methods of exposure assessment, show either no effect or an increased risk associated with occupational exposures.

Assessment of intermittent exposure is complex; nonetheless, most studies show positive associations with measure of intermittent exposure, such as particular sun-intensive activities, outdoor recreation or vacations.

Most studies show positive associations with a history of sunburn; however, this association cannot be easily interpreted, because while it might accurately reflect sunburn it could just as well reflect either the tendency to sunburn, if exposed, or intermittent exposure more generally.

(d) *Melanoma of the eye*

There is no latitude gradient among white populations of the incidence of ocular neoplasms, some 80% of which are likely to be ocular melanomas. No effect of southern US birthplace was seen in the two descriptive studies in the USA that examined this aspect.

Four case–control studies, from western Canada and from Philadelphia, San Francisco and Boston, USA, provided information on the association between exposure to solar radiation and ocular melanoma. All of these studies demonstrate an increased risk of ocular melanoma in people with light skin, light eye colour or light hair colour. Two of the studies compared effect of southern US birthplace with birth elsewhere in the USA; a significant difference was seen in the Philadelphia study.

Past residence south of 40 °N latitude was positively associated with ocular melanoma in the Boston study but was not significant in the Philadelphia study after control for southern birthplace. Although several outdoor activities, such as gardening and sunbathing, were associated in the Philadelphia study with ocular melanoma, participation in outdoor activities did not increase risk significantly in Boston or San Francisco.

The lack of consistency of the results of these studies makes their interpretation difficult.

(e) *Other cancers*

No adequate study was available to evaluate the role of solar radiation in cancers at other body sites.

5.2.2 *Artificial sources of ultraviolet radiation*

No adequate study was available on nonmelanocytic skin cancer in relation to exposure to artificial sources of UVR.

Two case–control studies, one from Scotland and one from Ontario, with detailed information on use of sunbeds and sunlamps showed positive relationships between duration of use and risk of melanoma of the skin. Several other studies with limited information showed no association.

One case–control study from Sydney, Australia, showed a positive relationship between melanoma of the skin and exposure to fluorescent lights at work among women, but the measurement of exposure was crude and among exposed cases there was a relative excess of melanoma on the trunk, a site likely to be covered at work. A more detailed study from Australia showed no consistent association between cumulative exposure or rate of exposure to fluorescent lights and melanoma. Two other studies had detailed information on exposure. One, from Scotland, showed no such association, while the other, from England, had inconsistent effects depending on the method of ascertainment of information. Another study, from New York, with limited information also showed inconsistent effects depending on the source of information.

Two case–control studies, from Boston and Philadelphia, USA, showed significant positive associations between use of sunlamps and melanoma of the eye. Another case–control study, from San Francisco, showed an increased risk for exposure to 'UV or black light', although the nature of the exposure was not specified.

Two studies, from Philadelphia and Montréal, showed significant positive associations between welding and melanoma of the eye.

5.2.3 *Molecular genetics of human skin cancers*

Base substitutions in a tumour suppressor gene, p53, found in human squamous-cell skin carcinomas that had developed at sites exposed to the sun were similar to those found in experimental systems exposed to UVR, and especially to UVB.

5.3 Carcinogenicity in experimental animals

Solar radiation was tested for carcinogenicity in a series of exceptional studies in mice and rats. Large numbers of animals were studied, and well-characterized benign and malignant skin tumours developed in most of the surviving animals. Although the reports are deficient in quantitative details, the results provide convincing evidence that sunlight is carcinogenic for the skin of animals.

Broad-spectrum UVR (solar-simulated radiation and ultraviolet lamps emitting mainly UVB) was tested for carcinogenicity in many studies in mice, to a lesser extent in rats and in a few experiments in hamsters, guinea-pigs, opossums and fish. Benign and malignant skin tumours were induced in all of these species except guinea-pigs, and tumours of the cornea and conjunctiva were induced in rats, mice and hamsters.

The predominant type of tumours induced by UVR in mice is squamous-cell carcinoma. Basal-cell carcinomas have been observed occasionally in athymic nude mice and rats exposed to UVR. Melanocytic neoplasms of the skin were shown to develop following exposure of opossums and hybrid fish to broad-spectrum UVR.

Studies in hairless mice demonstrated the carcinogenicity of exposures to UVR in the wavelength ranges 315–400 nm (UVA), 280–315 nm (UVB) and $\leq$ 280 nm (UVC), UVB radiation being the most effective, followed by UVC and UVA. UVB radiation is three to four orders of magnitude more effective than UVA. Both short-wavelength UVA (315–340 nm) and long-wavelength UVA (340–400 nm) induced skin cancer in hairless mice. The carcinogenic effectiveness of the latter waveband is known only as an average value over the entire range; the uncertainty of this average is about one order of magnitude. In none of the experiments involving UVC was it possible to exclude completely a contribution of UVB, but the size of the effects observed indicate that they cannot be due to UVB alone.

No experimental data were available on the carcinogenicity to animals of radiation from general lighting fixtures, including fluorescent and quartz halogen lamps.

UVR has been studied in protocols involving two-stage chemical carcinogenesis (substituting UVR for the chemical initiator or for the chemical promoter or giving it in addition to both). UVR has been reported to exert many effects on the carcinogenic process, including initiation, promotion, cocarcinogenicity and even tumour inhibition. Chemical immunosuppressive agents have been shown to enhance the probability of developing UVR-induced tumours in mice.

5.4 Other relevant data

5.4.1 *Transmission and absorption*

Studies of transmission in whole human and mouse epidermis and human stratum corneum *in vitro* show that these tissues attenuate radiation in the solar UVR range. This attenuation, which is more pronounced for the UVB than for the UVA wavebands, affords some protection from solar UVR to dividing cells in the basal layer.

The different components of the human eye act as optical filters for the UVR range. Consequently, little or no UVR reaches the retina in the normal eye.

5.4.2 *Effects on the skin*

UVR produces erythema, melanin pigmentation and acute and chronic cellular and histological changes in humans. Generally consistent changes are seen in experimental species, including the hairless mouse.

The action spectra for erythema and tanning in humans and for oedema in hairless mice are similar. UVB is three to four times more effective than UVA in producing erythema. In humans, pigmentation protects against erythema and histopathological changes. People with a poor ability to tan, who burn easily and have light eye and hair colour are at a higher risk of developing melanoma, basal-cell and squamous-cell carcinomas (see section 5.2).

In humans, acquired pigmented naevi and solar keratoses, indicators of melanomas and squamous-cell carcinomas, respectively, are induced by exposure to the sun.

Xeroderma pigmentosum patients have a high frequency of pigmentary abnormalities and skin cancers on sun-exposed skin. These patients also have defective DNA repair.

5.4.3 *Effects on the immune response*

Relatively few investigations have been reported of the effects of UVR on immunity in humans, but changes do occur. There is evidence that contact allergy is suppressed by exposure to UVB and possibly to UVA radiation. The number of Langerhans' cells in the epidermis is decreased by exposure to UVR and sunlight, and the morphological loss of these cells is associated with changes in antigen-presenting cell function in the direction of suppression; this change may be due not only to simple loss of function but also to active migration of other antigen-presenting cells into the skin. A reduction in natural killer cell activity also occurs, which can be produced by UVA radiation. These changes are short-lived, and their functional significance is unknown. Pigmentation of the skin may not protect against some UVR-induced alterations of immune function.

Several immune responses are suppressed by UVR in mice and other rodents. Suppression of contact hypersensitivity has received most attention, and this response may be impaired locally, at the site of exposure to radiation, or systemically, at a distant, unexposed site. The two forms of suppression have different dose dependencies—systemic suppression requiring much higher doses—and their mechanisms appear to differ, but the efferent limb of each involves generation of hapten-specific T-suppressor cells that block induction but not elicitation of contact hypersensitivity. Systemic suppression of delayed hypersensitivity to injected antigens can also be produced by exposure to UVB radiation, and several observations suggest that the mechanism of this suppression differs from that of systemic suppression of contact hypersensitivity.

Alterations in immune function induced by exposure to UVR play a central role in photocarcinogenesis in mice. UVR-induced T-suppressor cells block a normal immunosurveillance system that prevents the growth of highly antigenic UVR-induced tumours. It is not known whether this mechanism operates in humans.

5.4.4 DNA photoproducts

Solar UVR induces a variety of photoproducts in DNA, including cyclobutane-type pyrimidine dimers, pyrimidine–pyrimidone (6-4) photoproducts, thymine glycols, cytosine damage, purine damage, DNA strand breaks and DNA–protein cross-links. Substantial information on biological consequences is available only for the first two classes. Both are potentially cytotoxic and can lead to mutations in cultured cells, and there is evidence that cyclobutane-type pyrimidine dimers may be precarcinogenic lesions. The relative and absolute levels of each type of lesion vary with wavelength. Substantial levels of thymidine glycols, strand breaks and DNA–protein cross-links are induced by solar UVA and UVB radiation, but not by UVC radiation. The ratio of strand breaks to cyclobutane-type dimer lesions increases as a function of increasing wavelength. In narrow band-width studies, the longest wavelength at which cyclobutane-type pyrimidine dimers have been observed is 365 nm, whereas the induction of strand breaks and DNA–protein cross-links has been observed at wavelengths in the UVB, UVA and visible ranges. Non-DNA chromophores such as porphyrins, which absorb solar UVR, appeared to be important in generating active intermediates that can lead to damage. Solar UVR also induces membrane damage.

5.4.5 Genetic and related effects

Measurable DNA damage is induced in human skin cells *in vivo* after exposures to UVA, UVB and UVC radiation, including doses in the range commonly experienced by humans. Most of the DNA damage after a single exposure is repaired within 24 h. The importance of these wavelength ranges depends on several factors. UVB is the most effective, UVC being somewhat less effective and UVA being much less effective, when compared on a per photon basis, probably owing to a combination of the biological effectiveness of the different wavebands and of their absorption in the outer layers of the skin.

Summary table of genetic and related effects of ultraviolet A radiation

Nonmammalian systems																Mammalian systems																				
Proka-ryotes		Lower eukaryotes				Plants			Insects							In vitro												In vivo							
																Animal cells							Human cells					Animals					Humans			
D	G	D	R	G	A	A	D	G	C	R	G	C	A			D	G	S	M	C	A	T	I	D	G	S	M	C	A	T	D	G	S	M	C	DL	A	D	S	M	C	A
+	+	+	+													+¹	+	+¹						+	+																+¹	

A, aneuploidy; C, chromosomal aberrations; D, DNA damage; DL, dominant lethal mutation; G, gene mutation; I, inhibition of intercellular communication; M, micronuclei; R, mitotic recombination and gene conversion; S, sister chromatid exchange; T, cell transformation

In completing the tables, the following symbols indicate the consensus of the Working Group with regard to the results for each endpoint:

+ considered to be positive for the specific endpoint and level of biological complexity
+¹ considered to be positive, but only one valid study was available to the Working Group; sperm abnormality, mouse
– considered to be negative
–¹ considered to be negative, but only one valid study was available to the Working Group
? considered to be equivocal or inconclusive (e.g., there were contradictory results from different laboratories; there were confounding exposures; the results were equivocal)

SUMMARY OF DATA REPORTED AND EVALUATION

Summary table of genetic and related effects of ultraviolet B radiation

Nonmammalian systems												Mammalian systems																						
Prokaryotes		Lower eukaryotes				Plants			Insects			In vitro																In vivo						
												Animal cells								Human cells								Animals						Humans
D	G	D	R	G	A	D	G	A	C	R	G	A	D	G	S	M	C	A	T	I	D	G	S	M	C	A	T	I	D	G	S	M	C	DL	A	D	S	M	C	A
+									+¹				+	+	+¹				+		+	+						+¹							+					+

A, aneuploidy; C, chromosomal aberrations; D, DNA damage; DL, dominant lethal mutation; G, gene mutation; I, inhibition of intercellular communication; M, micronuclei; R, mitotic recombination and gene conversion; S, sister chromatid exchange; T, cell transformation

In completing the tables, the following symbols indicate the consensus of the Working Group with regard to the results for each endpoint:

+ considered to be positive for the specific endpoint and level of biological complexity
+¹ considered to be positive, but only one valid study was available to the Working Group; sperm abnormality, mouse
– considered to be negative
–¹ considered to be negative, but only one valid study was available to the Working Group
? considered to be equivocal or inconclusive (e.g. there were contradictory results from different laboratories; there were confounding exposures; the results were equivocal)

Summary table of genetic and related effects of ultraviolet C radiation

Nonmammalian systems																	Mammalian systems																				
Prokaryotes			Lower eukaryotes				Plants				Insects						In vitro														In vivo					
																	Animal cells								Human cells								Animals						Humans				
D	G	R	D	R	G	A	D	G	C	R	G	C	A				D	G	S	M	C	A	T	I	D	G	S	M	C	A	T	I	D	G	S	M	C	DL	A	D	S	M	C	A
+	+	+	+	+¹	+	+¹	+¹	+¹									+¹	+	+		+				+	+	+	+¹	+¹	+		+	+¹							+¹				

A, aneuploidy; C, chromosomal aberrations; D, DNA damage; DL, dominant lethal mutation; G, gene mutation; I, inhibition of intercellular communication; M, micronuclei; R, mitotic recombination and gene conversion; S, sister chromatid exchange; T, cell transformation

In completing the tables, the following symbols indicate the consensus of the Working Group with regard to the results for each endpoint:

+   considered to be positive for the specific endpoint and level of biological complexity

+¹   considered to be positive, but only one valid study was available to the Working Group; sperm abnormality, mouse

−   considered to be negative

−¹   considered to be negative, but only one valid study was available to the Working Group

?   considered to be equivocal or inconclusive (e.g., there were contradictory results from different laboratories; there were confounding exposures; the results were equivocal)

Solar and 'solar-simulated' radiation and radiation from sunlamps (UVA and UVB) are mutagenic to prokaryotes and plants, induce DNA damage in fish and in amphibian cells *in vitro*, are mutagenic to and induce sister chromatid exchange in amphibian cells, induce micronucleus formation and transformation in mammalian cells *in vitro*, are mutagenic to and induce DNA damage and sister chromatid exchange in human cells *in vitro* and induce DNA damage in mammalian skin cells irradiated *in vivo*.

UVA radiation is mutagenic to prokaryotes and induces DNA damage in fungi. It is mutagenic to and induces DNA damage, chromosomal aberrations and sister chromatid exchange in mammalian cells and induces DNA damage and mutation in human cells *in vitro*.

UVB radiation is mutagenic to prokaryotes and induces chromosomal aberrations in plants. It is mutagenic to and induces DNA damage, sister chromatid exchange and transformation in mammalian cells, is mutagenic and induces DNA damage and transformation in human cells *in vitro* and induces DNA damage in mammalian skin cells irradiated *in vivo*.

UVC radiation induces DNA damage in and is mutagenic to prokaryotes, fungi and plants and induces DNA damage in insects and aneuploidy in yeast. It induces sister chromatid exchange in amphibian and avian cells *in vitro*; it is mutagenic to and induces DNA damage, chromosomal aberrations, sister chromatid exchange and transformation in mammalian and human cells *in vitro*; and it induces DNA damage in mammalian skin cells irradiated *in vivo*.

UVR in the three wavelength ranges can induce or enhance cellular and viral gene expression.

5.5 Evaluation[1]

There is *sufficient evidence* in humans for the carcinogenicity of solar radiation. Solar radiation causes cutaneous malignant melanoma and nonmelanocytic skin cancer.

There is *limited evidence* in humans for the carcinogenicity of exposure to ultraviolet radiation from sunlamps and sunbeds.

There is *inadequate evidence* in humans for the carcinogenicity of exposure to fluorescent lighting.

There is *inadequate evidence* in humans for the carcinogenicity of other sources of artificial ultraviolet radiation.

There is *sufficient evidence* for the carcinogenicity of solar radiation in experimental animals.

There is *sufficient evidence* for the carcinogenicity of broad-spectrum ultraviolet radiation in experimental animals.

There is *sufficient evidence* for the carcinogenicity of ultraviolet A radiation in experimental animals.

There is *sufficient evidence* for the carcinogenicity of ultraviolet B radiation in experimental animals.

---

[1]For definition of the italicized terms, see Preamble, pp. 32-35.

There is *sufficient evidence* for the carcinogenicity of ultraviolet C radiation in experimental animals.

Overall evaluation

Solar radiation *is carcinogenic to humans* (Group 1).

Ultraviolet A radiation *is probably carcinogenic to humans* (Group 2A).

Ultraviolet B radiation *is probably carcinogenic to humans* (Group 2A).

Ultraviolet C radiation *is probably carcinogenic to humans* (Group 2A).

Use of sunlamps and sunbeds *entails exposures that are probably carcinogenic to humans* (Group 2A).

Exposure to fluorescent lighting *is not classifiable as to its carcinogenicity to humans* (Group 3).

6. References

Aberer, W., Schuler, G., Stingl, G., Hönigsmann, H. & Wolff, K. (1981) Ultraviolet light depeletes surface markers of Langerhans cells. *J. invest. Dermatol.*, **76**, 202–210

Adam, S.A., Sheaves, J.K., Wright, N.H., Mosser, G., Harris, R.W. & Vessey, M.P. (1981) A case-control study of the possible association between oral contraceptives and malignant melanoma. *Br. J. Cancer*, **44**, 45–50

Agin, P.P., Rose, A.P., III, Lane, C.C., Akin, F.J. & Sayre, R.M. (1981a) Changes in epidermal forward scattering absorption after UVA or UVA–UVB irradiation. *J. invest. Dermatol.*, **76**, 174–177

Agin, P.P., Lane, C.C. & Sayre, R.M. (1981b) Ultraviolet irradiation induces optical and structural changes in the skin of hairless mice. *Photochem. Photobiophys.*, **3**, 185–194

Alcalay, J., Freeman, S.E., Goldberg, L.H. & Wolf, J.E. (1990) Excision repair of pyrimidine dimers induced by simulated solar radiation in the skin of patients with basal cell carcinoma. *J. invest. Dermatol.*, **95**, 506–509

Alter, B.J., Schendel, D.J., Bach, M.L., Bach, F.H., Klein, J. & Stimpfling, J.H. (1973) Cell-mediated lympholysis. Importance of serologically defined *H*-2 regions. *J. exp. Med.*, **137**, 1303–1309

Ambach, W. & Rehwald, W. (1983) Measurements of the annual variation of the erythema dose of global radiation. *Radiat. environ. Biophys.*, **21**, 295–303

American Conference of Governmental Industrial Hygienists (1991) *1991–1992 Threshold Limit Values for Chemical Substances and Physical Agents and Biological Exposure Indices*, Cincinnati, OH, pp. 123–128

Ananthaswamy, H.N. (1984a) Lethality and transformation of 10T1/2 mouse embryo fibroblast cell line by various wavelengths of ultraviolet radiation. *Photodermatology*, **1**, 265–176

Ananthaswamy, H.N. (1984b) Neoplastic transformation of neonatal mouse skin fibroblasts in culture after a single exposure to ultraviolet radiation *in vivo*. *Photodermatology*, **1**, 106–113

Ananthaswamy, H.N. & Fisher, M.S. (1981) Photoreactivation of ultraviolet radiation-induced pyrimidine dimers in neonatal BALB/c mouse skin. *Cancer Res.*, **41**, 1829–1833

Ananthaswamy, H.N. & Kripke, M.L. (1981) In vitro transformation of primary cultures of neonatal BALB/c mouse epidermal cells with ultraviolet-B radiation. *Cancer Res.*, **41**, 2882–2890

Ananthaswamy, H.N. & Pierceall, W.E. (1990) Molecular mechanisms of ultraviolet radiation carcinogenesis. *Photochem. Photobiol.*, **52**, 1119–1136

Ananthaswamy, H.N., Price, J.E., Goldberg, L.H. & Bales, E.S. (1988) Detection and identification of activated oncogenes in human skin cancers occurring on sun-exposed body sites. *Cancer Res.*, **48**, 3341–3346

Anderson, D.E. (1963) Effect of pigment on bovine ocular squamous carcinoma. *Ann. N.Y. Acad. Sci.*, **100**, 436–446

Anderson, R.R. & Parrish, J.A. (1981) The optics of human skin. *J. invest. Dermatol.*, **77**, 13–19

Anderson, R.R. & Parrish, J.A. (1982) Optical properties of human skin. In: Regan, J.D. & Parrish, J.A., eds, *The Science of Photo-medicine*, New York, Plenum, pp. 147–194

Anderson, L.M. & Rice, J.M. (1987) Tumorigenesis in athymic nude mouse skin by chemical carcinogenesis and ultraviolet light. *J. natl Cancer Inst.*, **78**, 125–134

Andley, U.P. (1987) Yearly review. Photodamage to the eye. *Photochem. Photobiol.*, **46**, 1057–1066

Anon. (1979) *Sun-lamps, Solariums, Sunbeds and Health Lamps*, London, Which? Consumer Association, pp. 268–272

Anon. (1987) *The Truth About Tanning*, London, Which? Consumer Association, pp. 214–216

Ansel, J.C., Luger, T.A. & Green, I. (1983) The effect of in vitro and in vivo UV irradiation on the production of ETAF [epidermal cell derived thymocyte-activating factor] activity by human and murine keratinocytes. *J. invest. Dermatol.*, **81**, 519–523

Applegate, L.A. & Ley, R.D. (1988) Ultraviolet radiation-induced lethality and repair of pyrimidine dimers in fish embryos. *Mutat. Res.*, **198**, 85–92

Applegate, L.A. & Ley, R.D. (1991) DNA damage is involved in the induction of opacification and neovascularization of the cornea by ultraviolet radiation. *Exp. Eye Res.*, **52**, 493–497

Applegate, L.A., Ley, R.D., Alcalay, J. & Kripke, M.L. (1989) Identification of the molecular target for the suppression of contact hypersensitivity by ultraviolet radiation. *J. exp. Med.*, **170**, 1117–1131

Arlett, C.F. & Harcourt, S.A. (1983) Variation in response to mutagens amongst normal and repair-defective human cells. In: Lawrence, C.W., ed., *Induced Mutagenesis: Molecular Mechanisms and their Implications for Environmental Protection*, New York, Plenum, pp. 249–267

Arlett, C.F., Harcourt, S.A., Cole, J., Green, M.H. & Anstey, A.V. (1992) A comparison of the response of unstimulated and stimulated T-lymphocytes and fibroblasts from normal, xeroderma pigmentosum and trichothiodystrophy donors to the lethal action of UV-C. *Mutat. Res.*, **273**, 127–135

Armstrong, B.K. (1984) Melanoma of the skin. *Br. med. Bull.*, **40**, 346–350

Armstrong, R.B. (1986) Solar urticaria. *Dermatol. Clin.*, **4**, 253–259

Armstrong, B.K. (1988) Epidemiology of malignant melanoma: intermittent or total accumulated exposure to the sun? *J. Dermatol. surg. Oncol.*, **14**, 835–849

Armstrong, B.K., Woodings, T.L., Stenhouse, N.S. & McCall, M.G. (1983) *Mortality from Cancer in Migrants to Australia 1962 to 1971*, Perth, University of Western Australia, pp. 21, 81–83

Armstrong, B.K., Heenan, P.J., Caruso, V., Glancy, R.J. & Holman, C.D.J. (1984) Seasonal variation in the junctional component of pigmented naevi (Letter to the Editor). *Int. J. Cancer*, **34**, 441–442

Armstrong, B.K., de Klerk, N.H. & Holman, C.D.J. (1986) Etiology of common acquired melanocytic nevi: constitutional variables, sun exposure, and diet. *J. natl Cancer Inst.*, **77**, 329–335

Athas, W.F., Hedayati, M.A., Matanoski, G.M., Farmer, E.R. & Grossman, L. (1991) Development and field-test validation of an assay for DNA repair in circulating human lymphocytes. *Cancer Res.*, **51**, 5786–5793

Atkin, M., Fenning, J., Heady, J.A., Kennaway, E.L. & Kennaway, N.M. (1949) The mortality from cancer of the skin and lip in certain occupations. *Br. J. Cancer*, **3**, 1–15

Aubry, F. & MacGibbon, B. (1985) Risk factors of squamous cell carcinoma of the skin. A case-control study in the Montreal region. *Cancer*, **55**, 907–911

Auerbach, H. (1961) Geographic variation in incidence of skin cancer in the United States. *Public Health Rep.*, **76**, 345–348

Augustsson, A., Stierner, U., Rosdahl, I. & Suurküla, M. (1990) Melanocytic naevi in sun-exposed and protected skin in melanoma patients and controls. *Acta derm. venereol.*, **71**, 512–517

Azizova, A.O., Islomov, A.I., Roshchupkin, D.I., Predvoditelev, D.A., Remizov, A.N. & Vladimirov, Y.A. (1980) Free radicals formed on ultraviolet irradiation of the lipids of biological membranes. *Biophysics*, **24**, 407–414

REFERENCES

Baadsgaard, O., Fox, D.A. & Cooper, K.D. (1988) Human epidermal cells from ultraviolet light-exposed skin preferentially activate autoreactive CD4+2H4+ suppressor-inducer lymphocytes and CD8+ suppressor/cytotoxic lymphocytes. *J. Immunol.*, **140**, 1738–1744

Bach, F.H., Grillot-Courvalin, C., Kuperman, O.J., Sollinger, H.W., Hayes, C., Sondel, P.M., Alter, B.J. & Bach, M.L. (1977) Antigenic requirements for triggering of cytotoxic T lymphocytes. *Immunol. Rev.*, **35**, 76–96

Bachem, A. (1956) Ophthalmic ultraviolet action spectra. *Am. J. Ophthalmol.*, **41**, 969–975

Baden, H.P. & Pathak, M.A. (1967) The metabolism and function of urocanic acid in skin. *J. invest. Dermatol.*, **48**, 11–17

Balshi, J.D., Francfort, J.W. & Perloff, L.J. (1985) The influence of ultraviolet irradiation on the blood transfusion effect. *Surgery*, **98**, 243–249

Banerjee, S.K., Christensen, R.B., Lawrence, C.W. & LeClerc, J.E. (1988) Frequency and spectrum of mutations produced by a single cis–syn thymine–thymine cyclobutane dimer in a single-stranded vector. *Proc. natl Acad. Sci. USA*, **85**, 8141–8145

Banerjee, S.K., Borden, A., Christensen, R.B., LeClerc, J.E. & Lawrence, C.W. (1990) SOS-Dependent replication past a single *trans-syn* T–T cyclobutane dimer gives a different mutation spectrum and increased error rate compared to replication past this lesion in uninduced cells. *J. Bacteriol.*, **172**, 2105–2112

Barrett, J.C., Tsutsui, T. & Ts'o, P.O.P. (1978) Neoplastic tranformation induced by a direct perturbation of DNA. *Nature*, **274**, 229–232

Barrett, S.F., Robbins, J.H., Tarone, R.E. & Kraemer, K.H. (1991) Evidence for defective repair of cyclobutane pyrimidine dimers with normal repair and other DNA photoproducts in a transcriptionally active gene transfected into Cockayne syndrome cells. *Mutat. Res.*, **255**, 281–291

Barth, C., Knuschke, P. & Barth, J. (1990) Examination of welders exposed to UV radiation (Ger.). *Z. ges. Hyg.*, **36**, 654–655

Baturay, N.Z., Targovnik, H.S., Reynolds, R.J. & Kennedy, A.R. (1985) Induction of in vitro transformation by near-uv light and its interaction with β-propiolactone. *Carcinogenesis*, **6**, 465–468

Beadle, P.C. & Burton, J.L. (1981) Absorption of ultraviolet radiation by skin surface lipid. *Br. J. Dermatol.*, **104**, 549–551

Beardmore, G.L. (1972) The epidemiology of malignant melanoma in Australia. In: McCarthy, W.H., ed., *Melanoma and Skin Cancer*, Sydney, V.C.N. Blight (Goverment Printer), pp. 39–64

Bech-Thomsen, N., Wulf, H.C., Poulsen, T. & Lundgren, K. (1988a) Pretreatment with long-wave ultraviolet light inhibits ultraviolet-induced skin tumor development in hairless mice. *Arch. Dermatol.*, **124**, 1215–1218

Bech-Thomsen, N., Wulf, H.C. & Lundgren, K. (1988b) Pre-treatment with UVA influences broad-spectrum UV photocarcinogenesis in hairless mice (Abstract). In: Rikles, E., ed., *Proceedings of the 10th International Congress on Photobiology, Jerusalem*, Jerusalem, International Association of Biology, p. 34

Beer, J.Z. & Zmudzka, B.Z. (1991) Activation of human immunodeficiency virus by radiation. In: Seymour, C.B. & Mothersill, C., eds, *New Developments in Fundamental and Applied Radiology*, London, Taylor & Francis, pp. 113–123

Beer, J.Z., Olvey, K.M., Miller, S.A. & Zmudzka, B.Z. (1991) Studies on the human immunodeficiency virus promoter induction by UVA and UVB radiation (Abstract). *Photochem. Photobiol.*, **53**, 25S

Beitner, H. (1986) The effect of high dose long-wave ultraviolet radiation (UVA) on epidermal melanocytes in human skin: a transmission electron microscopic study. *Photodermatology*, **3**, 133-139

Beitner, H. & Wennersten, G. (1983) The immediate action of long-wave ultraviolet radiation (UVA) on suprabasal melanocytes in human skin: a transmission electron microscopical study. *Acta dermatol. venereol.*, **63**, 328-334

Beitner, H., Norell, S.E., Ringborg, U., Wennersten, G. & Mattson, B. (1990) Malignant melanoma: aetiological importance of individual pigmentation and sun exposure. *Br. J. Dermatol.*, **122**, 43-51

Bender, M.A., Griggs, H.G. & Walker, P.L. (1973) Mechanisms of chromosomal aberration production. I. Aberration induction by ultraviolet light. *Mutat. Res.*, **20**, 387-402

Ben-Hur, E. & Ben-Ishai, R. (1968) *Trans-syn* thymine dimers in ultraviolet-irradiated denatured DNA: identification and photoreactivability. *Biochim. biophys. Acta*, **166**, 9-15

Beral, V. & Robinson, N. (1981) The relationship of malignant melanoma, basal and squamous skin cancers to indoor and outdoor work. *Br. J. Cancer*, **44**, 886-891

Beral, V., Evans, S., Shaw, H. & Milton, G. (1982) Malignant melanoma and exposure to fluorescent lighting at work. *Lancet*, **ii**, 290-293

Berg, M. (1987) Ultraviolet radiation in operation rooms. In: Passchier, W.F. & Bosnjakovic, B.F.M., eds, *Human Exposure to Ultraviolet Radiation: Risks and Regulations*, Amsterdam, Elsevier, pp. 359-363

Berger, D.S. (1976) The sunburning ultraviolet meter: design and performance. *Photochem. Photobiol.*, **24**, 587-593

Berger, H. & Kaase, H. (1983) Risk for carcinoma from sources of UVA radiation (Ger.). *Z. Hautkr.*, **58**, 63

Berger, D.S. & Urbach, F. (1982) A climatology of sunburning ultraviolet radiation. *Photochem. Photobiol.*, **35**, 187-192

Berger, H., Tsambaos, D. & Mahrle, G. (1980a) Experimental elastosis induced by chronic ultraviolet exposure. Light and electron-microscopic study. *Arch. dermatol. Res.*, **269**, 39-49

Berger, H., Tsambaos, D. & Kaase, H. (1980b) Experimental actinic elastosis by chronic exposure to filtered UVA radiation (Ger.). *Z. Hautkr.*, **55**, 1510-1527

Bergstresser, P.R. (1986) Ultraviolet B radiation induces 'local immunosuppression'. *Current Probl. Dermatol.*, **15**, 205-218

Bernhard, J.D., Pathak, M.A., Kochevar, I.E. & Parrish, J.A. (1987) Abnormal reactions to ultraviolet radiation. In: Fitzpatrick, T.B., Eisen, A.Z., Wolff, K., Freedberg, I.M. & Austen, K.F., eds, *Dermatology in General Medicine*, 3rd ed., New York, McGraw-Hill, pp. 1481-1507

Beukers, R. & Berends, W. (1960) Isolation and identification of the irradiation product of thymine. *Biochim. biophys. Acta*, **41**, 550-551

Beukers, R., Ylstra, J. & Berends, W. (1958) The effect of ultraviolet light on some components of the nucleic acids. II. In rapidly frozen solutions. *Recl. Trav. chim. Pays-Bas*, **77**, 729-732

Bielfeld, V., Weichenthal, M., Roser, M., Breitbart, E., Berger, J., Seemanova, E. & Rüdiger, H.W. (1989) Ultraviolet-induced chromosomal instability in cultured fibroblasts of heterozygote carriers for xeroderma pigmentosum. *Cancer Genet. Cytogenet.*, **43**, 219-226

Bissett, D.L., Hannon, D.P. & Orr, T.V. (1987) An animal model of solar-aged skin: histological, physical, and visible changes in UV-irradiated hairless mouse skin. *Photochem. Photobiol.*, **46**, 367-378

Bissett, D.L., Hannon, D.P. & Orr, T.V. (1989) Wavelength dependence of histological, physical, and visible changes in chronically UV-irradiated hairless mouse skin. *Photochem. Photobiol.*, **50**, 763–769

Blum, H.F. (1948) Sunlight as a causal factor in cancer of the skin of man. *J. natl Inst. Cancer*, **9**, 247–258

Blum, H.F. (1959) *Carcinogenesis by Ultraviolet Light*, Princeton, NJ, Princeton University Press

Blum, H.F. & Lippincott, S.W. (1942) Carcinogenic effectiveness of ultraviolet radiation of wavelength 2537 Å. *J. natl Cancer Inst.*, **3**, 211–216

Blum, H.F., Butler, E.G., Dailey, T.H., Daube, J.R., Mawe, R.C. & Soffen, G.A. (1959) Irradiation of mouse skin with single doses of ultraviolet light. *J. natl Cancer Inst.*, **22**, 979–993

Blumthaler, M., Canaval, H. & Ambach, W. (1983) Measurement of erythemally active doses of solar UVB radiation (Ger.). *Z. Med.-Meteorol.*, **2**, 14–17

Blumthaler, M., Ambach, W. & Daxecker, F. (1985a) Computation of radiation doses from different solar UVB radiations producing photoelectric keratosis (Ger.). *Klin. Mbl. Augenheilk.*, **186**, 275–278

Blumthaler, M., Ambach, W. & Canaval, H. (1985b) Seasonal variation of solar UV-radiation at a high mountain station. *Photochem. Photobiol.*, **42**, 147–152

Boettner, E.A. & Wolter, J.R. (1962) Transmission of the ocular media. *Invest. Ophthalmol.*, **1**, 776–783

Bootsma, D. & Hoeijmakers, J.H.J. (1991) The genetic basis of xeroderma pigmentosum. *Ann. Génét.*, **34**, 143–150

Borek, C., Ong, A. & Mason, H. (1989) Ozone and ultraviolet light act as additive cocarcinogens to induce in vitro neoplastic transformation. *Teratog. Carcinog. Mutag.*, **9**, 71–74

Bos, A.J.J. & de Haas, M.P. (1987) On the safe use of a high power ultraviolet laser. In: Passchier, W.F. & Bosnjakovic, B.F.M., eds, *Human Exposure to Ultraviolet Radiation: Risks and Regulations*, Amsterdam, Elsevier, pp. 377–382

Bose, B., Agarwal, S. & Chatterjee, S.N. (1989) UVA induced lipid peroxidation in liposomal membrane. *Radiat. environ. Biophys.*, **28**, 59–65

Bouissou, H., Pieraggi, M.-T., Julian, M. & Savit, T. (1988) The elastic tissue of the skin. A comparison of spontaneous and actinic (solar) aging. *Int. J. Dermatol.*, **27**, 327–335

Bowden, G.T., Trosko, J.E., Shapas, B.G. & Boutwell, R.K. (1975) Excision of pyrimidine dimers from epidermal DNA and nonsemiconservative epidermal DNA synthesis following ultraviolet irradiation of mouse skin. *Cancer Res.*, **35**, 3599–3607

Bowker, K.W. & Longford, A.R. (1987) Ultraviolet radiation hazards from the use of solaria. In: Passchier, W.F. & Bosnjakovic, B.F.M., eds, *Human Exposure to Ultraviolet Radiation: Risks and Regulations*, Amsterdam, Elsevier, pp. 365–369

Boyd, R.W. (1983) *Radiometry and the Detection of Optical Radiation*, New York, John Wiley & Sons

Boyle, J., MacKie, R.M., Briggs, J.D., Junor, B.J.R. & Aitchison, T.C. (1984) Cancer, warts and sunshine in renal transplant patients. A case–control study. *Lancet*, **i**, 702–705

Bradley, M.O. & Sharkey, N.A. (1977) Mutagenicity and toxicity of visible fluorescent light to cultured mammalian cells. *Nature*, **266**, 724–726

Bradley, M.O., Hsu, I.C. & Harris, C.C. (1979) Relationships between sister chromatid exchange and mutagenicity, toxicity and DNA damage. *Nature*, **282**, 318–320

Brash, D.E. & Haseltine, W.A. (1982) UV-Induced mutation hotspots occur at DNA damage hotspots. *Nature*, **298**, 189–192

Brash, D.E., Rudolph, J.A., Simon, J.A., Lin, A., McKenna, G.J., Baden, H.P., Halperin, A.J. & Pontén, J. (1991) A role for sunlight in skin cancer: UV-induced p53 mutations in squamous cell carcinoma. *Proc. natl Acad. Sci. USA*, **88**, 10124–10128

Bredberg, A., Kraemer, K.H. & Seidman, M.M. (1986) Restricted ultraviolet mutational spectrum in a shuttle vector propagated in xeroderma pigmentosum cells. *Proc. natl Acad. Sci. USA*, **83**, 8273–8277

Bridges, B.A. (1988) Mutagenic DNA repair in *Escherichia coli*. XVI. Mutagenesis by ultraviolet light plus delayed photoreversal in *rec*A strains. *Mutat. Res.*, **198**, 343–350

Bridges, B.A. (1990) Sunlight, DNA damage and skin cancer: a new perspective. *Jpn. J. Cancer Res.*, **81**, 105–107

Bridges, B.A. & Bates, H. (1990) Mutagenic DNA repair in *Escherichia coli*. VIII. Involvement of DNA polymerase III α-subunit (DnaE protein) in mutagenesis after exposure to UV light. *Mutagenesis*, **5**, 35–38

Bridges, B.A. & Brown, G.M. (1992) Mutagenic repair in *Escherichia coli*. XXI. A stable SOS-inducing signal persisting after excision repair of ultraviolet damage. *Mutat. Res.* (in press)

Bridges, B.A., Woodgate, R., Ruiz-Rubio, M., Sharif, F., Sedgwick, S.G. & Hübscher, U. (1987) Current understanding of UV-induced base pair substitution mutation in *E. coli* with particular reference to the DNA polymerase III complex. *Mutat. Res.*, **181**, 219–226

Broughton, B.C., Lehmann, A.R., Harcourt, S.A., Arlett, C.F., Sarasin, A., Kleijer, W.J., Beemer, F.A., Nairn, R. & Mitchell, D.L. (1990) Relationship between pyrimidine dimers, 6-4 photoproducts, repair synthesis and cell survival: studies using cells from patients with trichothiodystrophy. *Mutat. Res.*, **235**, 33–40

Brown, L.R. & Wüthrich, K. (1977) A spin label study of lipid oxidation catalyzed by heme proteins. *Biochim. biophys. Acta*, **464**, 356–369

Brown, J., Kopf, A.W., Rigel, D.S. & Fridman, R.J. (1984) Malignant melanoma in World War II veterans. *Int. J. Dermatol.*, **23**, 661–663

Bruggers, J.H.A., de Jong, W.E., Bosnjakovic, B.F.M. & Passchier, W.F. (1987) Use of artificial tanning equipment in the Netherlands. In: Passchier, W.F. & Bosnjakovic, B.F.M., eds, *Human Exposure to Ultraviolet Radiation: Risk and Regulations*, Amsterdam, Elsevier, pp. 235–239

Bruls, W.A.G., Slaper, H., van der Leun, J.C. & Berrens, L. (1984a) Transmission of human epidermis and stratum corneum as a function of thickness in the ultraviolet and visible wavelengths. *Photochem. Photobiol.*, **40**, 485–494

Bruls, W.A.G., van Weelden, H. & van der Leun, J.C. (1984b) Transmission of UV-radiation through human epidermal layers as a factor influencing the minimal erythema dose. *Photochem. Photobiol.*, **39**, 63–67

Bruyneel-Rapp, F., Dorsey, S.B. & Guin, J.D. (1988) The tanning salon: an area survey of equipment, procedures, and practices. *J. Am. Acad. Dermatol.*, **18**, 1030–1038

Burki, H.J. & Lam, C.K. (1978) Comparison of the lethal and mutagenic effects of gold and white fluorescent lights on cultured mammalian cells. *Mutat. Res.*, **54**, 373–377

Burnet, F.M. (1971) Implications of immunological surveillance for cancer therapy. *Isr. J. med. Sci.*, **7**, 9–16

Calkins, J., Selby, C. & Enoch, H.G. (1987) Comparison of UV action spectra for lethality and mutation in *Salmonella typhimurium* using a broad band source and monochromatic radiations. *Photochem. Photobiol.*, **45**, 631–636

Cameron, L.L., Vitasa, B.C., Lewis, P.G., Taylor, H.R. & Emmett, E.A. (1988) Visual assessment of facial elastosis using photographs as a measure of cumulative ultraviolet exposure. *Photodermatology*, **5**, 277–282

Caplan, R.M. (1967) Medical uses of the Wood's lamp. *J. Am. med. Assoc.*, **202**, 1035–1038

Cavard, C., Zider, A., Vernet, M., Bennoun, M., Saragosti, S., Grimber, B. & Briand, P. (1990) In vivo activation by ultraviolet rays of the human immunodeficiency virus type 1 long terminal repeat. *J. clin. Invest.*, **86**, 1369–1374

Cerutti, P.A. (1981) Measurement of thymine damage induced by oxygen radical species. In: Friedberg, E.C. & Hanawalt, P.C., eds, *DNA Repair. A Laboratory Manual of Research Procedures*, Vol. 1, Part A, New York, Marcel Dekker, pp. 57–68

Cerutti, P.A. & Netrawali, M. (1979) Formation and repair of DNA damage induced by indirect action of ultraviolet light in normal and xeroderma pigmentosum skin fibroblasts. *Radiat. Res.*, Suppl., 423–432

Cesarini, J.-P. & Muel, B. (1989) Erythema induced by quartz–halogen sources. *Photodermatology*, **6**, 222–227

Challoner, A.V.J., Corless, D., Davis, A., Deane, G.H.W., Diffey, B.L., Gupta, S.P. & Magnus, I.A. (1976) Personnel monitoring of exposure to ultraviolet radiation. *Clin. exp. Dermatol.*, **1**, 175–179

Challoner, A.V.J., Corbett, M.F., Davis, A., Diffey, B.L., Leach, J.F. & Magnus, I.A. (1978) Description and application of a personal ultraviolet dosimeter: a review of preliminary studies. *Natl Cancer Inst. Monogr.*, **50**, 97–100

Chamberlain, J. & Moss, S.H. (1987) Lipid peroxidation and other membrane damage produced in *Escherichia coli* K1060 by near-UV radiation and deuterium oxide. *Photochem. Photobiol.*, **45**, 625–630

Chan, G.L. & Little, J.B. (1976) Induction of oncogenic transformation *in vitro* by ultraviolet light. *Nature*, **264**, 442–444

Chan, G.L. & Little, J.B. (1979) Correlation of in vitro transformation with in vivo tumorigenicity in 10T½ mouse cells exposed to UV light. *Br. J. Cancer*, **39**, 590–593

Chan, G.L., Peak, M.J., Peak, J.G. & Haseltine, W.A. (1986) Action spectrum for the formation of endonuclease-sensitive sites and (6-4) photoproducts induced in a DNA fragment by ultraviolet radiation. *Int. J. Radiat. Biol.*, **50**, 641–648

Chao, C.C.-K. & Rosenstein, B.S. (1985) Induction of sister-chromatid exchanges in ICR 2A frog cells exposed to 254 nm and solar UV wavelengths. *Photochem. Photobiol.*, **41**, 625–627

Chao, C.C.-K. & Rosenstein, B.S. (1986) Use of metabolic inhibitors to investigate the excision repair of pyrimidine dimers and non-dimer DNA damages induced in human and ICR 2A frog cells by solar ultraviolet radiation. *Photochem. Photobiol.*, **43**, 165–170

Chu, E.H.Y. (1965a) Effects of ultraviolet radiation on mammalian cells. I. Induction of chromosome aberrations. *Mutat. Res.*, **2**, 75–94

Chu, E.H.Y. (1965b) Effects of ultraviolet radiation on mammalian cells. II. Differential UV and x-ray sensitivity of chromosomes to breakage in 5-aminouracil synchronized cell populations. *Genetics*, **52**, 1279–1294

Chung, H.-T., Burnham, D.K., Robertson, B., Roberts, L.K. & Daynes, R.A. (1986) Involvement of prostaglandins in the immune alterations caused by the exposure of mice to ultraviolet radiation. *J. Immunol.*, **137**, 2478–2484

Churchill, M.E., Peak, J.G. & Peak, M.J. (1991) Correlation between cell survival and DNA single-strand break repair proficiency in the Chinese hamster ovary cell lines AA8 and EM9 irradiated with 365-nm ultraviolet A radiation. *Photochem. Photobiol.*, **53**, 229–236

Cieminis, K.G.K., Rančelienė, V.M., Prijalgauskienė, A.J., Tiunaitienė, N.V., Rudzianskaité, A.M. & Jančys, Z.J. (1987) Chromosome and DNA damage and their repair in higher plants irradiated with short-wave ultraviolet light. *Mutat. Res.*, **181**, 9–16

Claas, F.H.J., Rothert, M., Havinga, I., Schothorst, A.A., Vermeer, B.-J. & van Rood, J.J. (1985) Influence of ultraviolet radiation treatment on the survival of heterotopic skin grafts in the mouse. *J. invest. Dermatol.*, **84**, 31–32

Clark, J.H. (1964) The effect of long ultraviolet radiation in the development of tumors induced by 20-methylcholanthrene. *Cancer Res.*, **24**, 207–211

Cleaver, J.E. (1968) Defective repair replication of DNA in xeroderma pigmentosum. *Nature*, **218**, 652–656

Cleaver, J.E. & Kraemer, K.H. (1989) Xeroderma pigmentosum. In: Scriver, C.R., Beaudet, A.L., Sly, W.S. & Valle, D., eds, *The Metabolic Basis of Inherited Disease*, New York, McGraw-Hill, pp. 2949–2971

Clemmesen, J. (1965) *Statistical Studies in the Aetiology of Malignant Neoplasms. I. Review and Results*, Copenhagen, Munksgaard, p. 409

Cole, C.A., Forbes, P.D., Davies, R.E. & Urbach, F. (1985) Effect of indoor lighting on normal skin. *Ann. N.Y. Acad. Sci.*, **453**, 305–316

Cole, C.A., Forbes, P.D. & Davies, R.E. (1986) An action spectrum for UV photocarcinogenesis. *Photochem. Photobiol.*, **43**, 275–284

Colella, C.M., Bogani, P., Agati, G. & Fusi, F. (1986) Genetic effects of UVB: mutagenicity of 308 nm light in Chinese hamster V79 cells. *Photochem. Photobiol.*, **43**, 437–442

Commission Internationale de l'Eclairage (International Commission on Illumination) (1987) *Vocabulaire International de l' Eclairage* [International Lighting Vocabulary] (CIE Publication No. 17.4), 4th ed., Geneva, Bureau Central de la Commission Electrotechnique Internationale, p. 3

Coohill, T.P., Peak, M.J. & Peak, J.G. (1987) The effects of the ultraviolet wavelengths of radiation present in sunlight on human cells *in vitro*. *Photochem. Photobiol.*, **46**, 1043–1050

Cooke, K.R. & Fraser, J. (1985) Migration and death from malignant melanoma. *Int. J. Cancer*, **36**, 175–178

Cooke, A. & Johnson, B.E. (1978) Dose response, wavelength dependence and rate of excision of ultraviolet radiation induced pyrimidine dimers in mouse skin DNA. *Biochim. biophys. Acta*, **517**, 24–30

Cooke, K.R., Skegg, D.C.G. & Fraser, J. (1984) Socio-economic status, indoor and outdoor work, and malignant melanoma. *Int. J. Cancer*, **34**, 57–62

Cooper, K.D., Fox, P., Neises, G. & Katz, S.I. (1985) Effects of ultraviolet radiation on human epidermal cell alloantigen presentation: initial depression of Langerhans cell-dependent function is followed by the appearance of $T6^-Dr^+$ cells that enhance epidermal alloantigen presentation. *J. Immunol.*, **134**, 129–137

Cooper, K.D., Neises, G.R. & Katz, S.I. (1986) Antigen-presenting $OKM5^+$ melanophages appear in human epidermis after ultraviolet radiation. *J. invest. Dermatol.*, **86**, 363–370

Corominas, M., Kamino, H., Leon, J. & Pellicer, A. (1989) Oncogene activation in human benign tumors of the skin (keratoacanthomas): Is H-*ras* involved in differentiation as well as proliferation? *Proc. natl Acad. Sci. USA*, **86**, 6372–6376

Cox, C.W.J. (1987) Ultraviolet irradiance levels in welding processes. In: Passchier, W.F. & Bosnjakovic, B.F.M., eds, *Human Exposure to Ultraviolet Radiation: Risks and Regulations*, Amsterdam, Elsevier, pp. 383–386

Cripps, D.J., Horowitz, S. & Hong, R. (1978) Spectrum of ultraviolet radiation on human B and T lymphocyte viability. *Clin. exp. Dermatol.*, **3**, 43–50

Cristofolini, M., Franceschi, S., Tasin, L., Zumiani, G., Piscioli, F., Talamini, R. & La Vecchia, C. (1987) Risk factors for cutaneous malignant melanoma in a northern Italian population. *Int. J. Cancer*, **39**, 150–154

Crombie, I.K. (1981) Distribution of malignant melanoma on the body surface. *Br. J. Cancer*, **43**, 842–849

Cullen, A.P. (1980) Additive effects of ultraviolet radiation. *Am. J. Optometry physiol. Optics*, **57**, 808–814

Cunningham, M.L., Johnson, J.S., Giovanazzi, S.M. & Peak, M.J. (1985) Photosensitized production of superoxide anion by monochromatic (290-405 nm) ultraviolet irradiation of NADH and NADPH coenzymes. *Photochem. Photobiol.*, **42**, 125–128

Czernielewski, J.M., Masouye, I., Pisani, A., Ferracin, J., Auvolat, D. & Ortonne, J.-P. (1988) Effect of chronic sun exposure on human Langerhans cell densities. *Photodermatology*, **5**, 116–120

Czochralska, B., Kawczynski, W., Bartosz, G. & Shugar, D. (1984) Oxidation of excited-state NADH and NAD dimer in aqueous medium—involvement of O_2^- as a mediator in the presence of oxygen. *Biochim. biophys. Acta*, **801**, 403–409

D'Ambrosio, S.M., Slazinski, L., Whetstone, J.W. & Lowney, E. (1981) Excision repair of UV-induced pyrimidine dimers in human skin *in vivo*. *J. invest. Dermatol.*, **77**, 311–313

D'Ambrosio, S.M., Bisaccia, E., Whetstone, J.W., Scarborough, D.A. & Lowney, E. (1983) DNA repair in skin of lupus erythematosus following in vivo exposure to ultraviolet radiation. *J. invest. Dermatol.*, **81**, 452–454

Danpure, H.J. & Tyrrell, R.M. (1976) Oxygen-dependence of near-UV (365 nm) lethality and the interaction of near-UV and x-rays in two mammalian cell lines. *Photochem. Photobiol.*, **23**, 171–177

Dardanoni, L., Gafá, L., Paterno, R. & Pavone, G. (1984) A case–control study on lip cancer risk factors in Ragusa (Sicily). *Int. J. Cancer*, **34**, 335–337

Davies, R.E. & Forbes, P.D. (1988) Retinoids and photocarcinogenesis: a review. *J. Toxicol. cutaneous ocul. Toxicol.*, **7**, 241–253

Davies, R.E., Dodge, H.A. & DeShields, L.H. (1972) Alteration of the carcinogenic activity of 7,12-dimethylbenz(a)anthracene (DMBA) by light (Abstract no. 54). *Proc. Am. Assoc. Cancer Res.*, **13**, 14

Davis, A. & Sims, D. (1983) *Weathering of Polymers*, London, Applied Science

Daynes, R.A., Spellman, C.W., Woodward, J.G. & Stewart, D.A. (1977) Studies into the transplantation biology of ultraviolet light-induced tumors. *Transplantation*, **23**, 343–348

Daynes, R.A., Spikes, J.D. & Krueger, G. (1983) *Experimental and Clinical Photoimmunology*, Vols I and II, Boca Raton, FL, CRC Press

Dearden, S.J., Hunter, T.F. & Philp, J. (1981) Bilayer microviscosity changes due to O_2 ($^1\Delta_g$) peroxidation in lipid vesicles. *Chem. Phys. Lett.*, **81**, 606–609

Dearden, S.J., Hunter, T.F. & Philp, J. (1985) Fatty acid analysis as a function of photo-oxidation in egg yolk lecithin vesicles. *Photochem. Photobiol.*, **41**, 213–215

De Fabo, E.C. & Kripke, M.L. (1980) Wavelength dependence and dose-rate independence of UV radiation-induced immunologic unresponsiveness of mice to a UV-induced fibrosarcoma. *Photochem. Photobiol.*, **32**, 183–188

De Fabo, E.C. & Noonan, F.P. (1983) Mechanism of immune suppression by ultraviolet irradiation *in vivo*. I. Evidence for the existence of a unique photoreceptor in skin and its role in photoimmunology. *J. exp. Med.*, **157**, 84–98

De Flora, S., Camoirano, A., Izzotti, A. & Bennicelli, C. (1990) Potent genotoxicity of halogen lamps, compared to fluorescent light and sunlight. *Carcinogenesis*, **11**, 2171-2177

Denkins, Y., Fidler, I.J. & Kripke, M.L. (1989) Exposure of mice to UVB radiation suppresses delayed hypersensitivity to *Candida albicans*. *Photochem. Photobiol.*, **49**, 615-619

Desai, I.D., Sawant, P.L. & Tappel, A.L. (1964) Peroxidative and radiation damage to isolated lysosomes. *Biochim. biophys. Acta*, **86**, 277-285

De Weerd-Kastelein, E.A., Keijzer, W., Rainaldi, G. & Bootsma, D. (1977) Induction of sister chromatid exchanges in xeroderma pigmentosum cells after exposure to ultraviolet light. *Mutat. Res.*, **45**, 253-261

Diffey, B.L. (1982) The consistency of studies of ultraviolet erythema in normal human skin. *Phys. med. Biol.*, **27**, 715-720

Diffey, B.L. (1986) Use of UVA sunbeds for cosmetic tanning. *Br. J. Dermatol.*, **115**, 67-76

Diffey, B. (1987a) A comparison of dosimeters used for solar ultraviolet radiometry. *Photochem. Photobiol.*, **46**, 55-60

Diffey, B.L. (1987b) Analysis of the risk of skin cancer from sunlight and solaria in subjects living in northern Europe. *Photodermatology*, **4**, 118-126

Diffey, B.L. (1987c) Cosmetic and medical applications of ultraviolet radiation: risk evaluation and protection techniques. In: Passchier, W.F. & Bosnjakovic, B.F.M., eds, *Human Exposure to Ultraviolet Radiation: Risks and Regulations*, Amsterdam, Elsevier, pp. 305-315

Diffey, B.L. (1988) The risk of skin cancer from occupational exposure to ultraviolet radiation in hospitals. *Phys. med. Biol.*, **33**, 1187-1193

Diffey, B.L. (1989a) Ultraviolet radiation and skin cancer. Are physiotherapists at risk? *Physiotherapy*, **75**, 615-616

Diffey, B.L. (1989b) Ultraviolet radiation dosimetry with polysulphone film. In: Diffey, B.L., ed., *Radiation Measurement in Photobiology*, London, Academic Press, pp. 135-159

Diffey, B.L. (1990a) Human exposure to ultraviolet radiation. *Semin. Dermatol.*, **9**, 2-10

Diffey, B.L. (1990b) Ultraviolet radiation lamps for the phototherapy and photochemotherapy of skin disease. In: Champion, R.H. & Pye, R.J., eds, *Recent Advances in Dermatology*, Vol. 8, Edinburgh, Churchill Livingstone, pp. 21-40

Diffey, B.L. (1991) Solar ultraviolet radiation effects on biological systems. *Phys. med. Biol.*, **36**, 299-328

Diffey, B.L. & Farr, P.M. (1987) An appraisal of ultraviolet lamps used for the phototherapy of psoriasis. *Br. J. Dermatol.*, **117**, 49-56

Diffey, B.L. & Farr, P.M. (1989) The normal range in diagnostic phototesting. *Br. J. Dermatol.*, **120**, 517-524

Diffey, B.L. & Farr, P.M. (1991a) Quantitative aspects of ultraviolet erythema. *Clin. phys. physiol. Meas.*, **12**, 311-325

Diffey, B.L. & Farr, P.M. (1991b) Tanning with UVB or UVA: an appraisal of risks. *J. Photochem. Photobiol. B. Biol.*, **8**, 219-223

Diffey, B.L. & Langley, F.C. (1986) *Evaluation of Ultraviolet Radiation Hazards in Hospitals* (Report No. 49), London, Institute of Physical Sciences in Medicine

Diffey, B.L. & McKinlay, A.F. (1983) The UVB content of 'UVA fluorescent lamps' and its erythemal effectiveness in human skin. *Phys. med. Biol.*, **28**, 351-358

Diffey, B.L., Kerwin, M. & Davis, A. (1977) The anatomical distribution of sunlight. *Br. J. Dermatol.*, **97**, 407-410

Diffey, B.L., Tate, T.J. & Davis, A. (1979) Solar dosimetry of the face: the relationship of natural ultraviolet radiation exposure to basal cell carcinoma localisation. *Phys. med. Biol.*, **24**, 931–939

Diffey, B.L., Challoner, A.V.J. & Key, P.J. (1980) A survey of the ultraviolet radiation emissions of photochemotherapy units. *Br. J. Dermatol.*, **102**, 301–306

Diffey, B.L., Larkö, O., Meding, B., Edeland, H.G. & Wester, U. (1986) Personal monitoring of exposure to ultraviolet radiation in the car manufacturing industry. *Ann. occup. Hyg.*, **30**, 163–170

Diffey, B.L., Farr, P.M. & Oakley, A.M. (1987) Quantitative studies on UVA-induced erythema in human skin. *Br. J. Dermatol.*, **117**, 57–66

DiPaolo, J.A. & Donovan, P.J. (1976) In vitro morphologic transformation of Syrian hamster cells by UV-irradiation is enhanced by X-irradiation and unaffected by chemical carcinogens. *Int. J. radiat. Biol.*, **30**, 41–53

Doda, D.D. & Green, A.E.S. (1980) Surface reflectance measurements in the UV from an airborne platform. Part 1. *Appl. Optics*, **19**, 2140–2145

Doll, R. (1991) Urban and rural factors in the aetiology of cancer. *Int. J. Cancer*, **47**, 803–810

Doll, R., Muir, C. & Waterhouse, J., eds (1970) *Cancer Incidence in Five Continents*, Vol. II, Berlin, Springer

Doniger, J., Jacobson, E.D., Krell, K. & DiPaolo, J.A. (1981) Ultraviolet light action spectra for neoplastic transformation and lethality of Syrian hamster embryo cells correlate with spectrum for pyrimidime dimer formation in cellular DNA. *Proc. natl Acad. Sci. USA*, **78**, 2378–2382

Dorado, G., Steingrimsdottir, H., Arlett, C.F. & Lehmann, A.R. (1991) Molecular analysis of ultraviolet-induced mutations in a xeroderma pigmentosum cell line. *J. mol. Biol.*, **217**, 217–222

Dorn, H.F. (1944a) Illness from cancer in the United States. *Public Health Rep.*, **59**, 33–48

Dorn, H.F. (1944b) Illness from cancer in the United States: IV. Illness from cancer of specific sites classed in broad groups. *Public Health Rep.*, **59**, 65–77

Dorn, H.F. (1944c) Illness from cancer in the United States: VI. Regional differences in illness from cancer. *Public Health Rep.*, **59**, 97–114

Dorn, C.R., Taylor, D.O.N. & Schneider, R. (1971) Sunlight exposure and risk of developing cutaneous and oral squamous cell carcinomas in white cats. *J. natl Cancer Inst.*, **46**, 1073–1078

Doudney, C.O. (1976) Mutation in ultraviolet light-damaged microorganisms. In: Wang, S.Y., ed., *Photochemistry and Photobiology of Nucleic Acids*, Vol. 2, New York, Academic Press, pp. 309–374

Dougherty, M.A., McDermott, R.J. & Hawkins, M.J. (1988) A profile of users of commercial tanning salons. *Health Values*, **12**, 21–28

Dreosti, I.E., Baghurst, P.A., Partick, E.J. & Turner, J. (1990) Induction of micronuclei in cultured murine splenocytes exposed to elevated levels of ferrous ions, hydrogen peroxide and ultraviolet irradiation. *Mutat. Res.*, **244**, 337–343

Drobetsky, E.A. & Glickman, B.W. (1990) The nature of ultraviolet light-induced mutations at the heterozygous *aprt* locus in Chinese hamster ovary cells. *Mutat. Res.*, **232**, 281–289

Dubin, N., Moseson, M. & Pasternack, B.S. (1986) Epidemiology of malignant melanoma: pigmentary traits, ultraviolet radiation, and the identification of high-risk populations. *Recent Results Cancer Res.*, **102**, 56–75

Dubin, N., Moseson, M. & Pasternack, B.S. (1989) Sun exposure and malignant melanoma among susceptible individuals. *Environ. Health Perspectives*, **81**, 139–151

Eaton, G.J., Custer, R.P. & Crane, A.R. (1978) Effects of ultraviolet light on nude mice. Cutaneous carcinogenesis and possible leukemogenesis. *Cancer*, **42**, 182–188

Ebbesen, P. (1981) Enhanced lymphoma incidence in BALB/c mice after ultraviolet light treatment. *J. natl Cancer Inst.*, **67**, 1077–1078

Eggset, G., Volden, G. & Krokan, H. (1983) UV-Induced DNA damage and its repair in human skin *in vivo* studied by sensitive immunohistochemical methods. *Carcinogenesis*, **4**, 745-750

Eggset, G., Krokan, H. & Volden, G. (1986) UV-Induced DNA damage and its repair in tanned and untanned human skin *in vivo*. *Photobiochem. Photobiophys.*, **10**, 181-187

Eisenstark, A. (1987) Mutagenic and lethal effects of near-ultraviolet radiation (290-400 nm) on bacteria and phage. *Environ. mol. Mutag.*, **10**, 317-337

Eklund, G. & Malec, E. (1978) Sunlight and incidence of cutaneous malignant melanoma. Effect of latitude and domicile in Sweden. *Scand. J. plast. reconstr. Surg.*, **12**, 231-241

Elkind, M.M. & Han, A. (1978) DNA single-strand lesions due to 'sunlight' and UV light: a comparison of their induction in Chinese hamster and human cells, and their fate in Chinese hamster cells. *Photochem. Photobiol.*, **27**, 717-724

Ellison, M.J. & Childs, J.D. (1981) Pyrimidine dimers induced in *Escherichia coli* DNA by ultraviolet radiation present in sunlight. *Photochem. Photobiol.*, **34**, 465-469

Elmets, C.A., Bergstresser, P.R., Tigelaar, R.E., Wood, P.J. & Streilein, J.W. (1983) Analysis of the mechanism of unresponsiveness produced by haptens painted on skin exposed to low dose ultraviolet radiation. *J. exp. Med.*, **158**, 781-794

Elmets, C.A., LeVine, M.J. & Bickers, D.R. (1985) Action spectrum studies for induction of immunologic unresponsiveness to dinitrofluorobenzene following in vivo low dose ultraviolet radiation. *Photochem. Photobiol.*, **42**, 391-397

Elmets, C.A., Larson, K., Urda, G.A. & Schacter, B. (1987) Inhibition of postbinding target cell lysis and of lymphokine-induced enhancement of human natural killer cell activity by in vitro exposure to ultraviolet B radiation. *Cell. Immunol.*, **104**, 47-58

Elwood, J.M. (1989a) Epidemiology and control of melanoma in white populations and in Japan. *J. invest. Dermatol.*, **92**, 214S-221S

Elwood, J.M. (1989b) Epidemiology of melanoma: its relationship to ultraviolet radiation and ozone depletion. In: Jones, R.R. & Wigley, T., eds, *Ozone Depletion. Health and Environmental Consequences*, New York, John Wiley & Sons, pp. 169-189

Elwood, J.M. & Gallagher, R.P. (1983) Site distribution of malignant melanoma. *Can. med. Assoc. J.*, **128**, 1400-1404

Elwood, J.M., Lee, J.A.H., Walter, S.D., Mo, T. & Green, A.E.S. (1974) Relationship of melanoma and other skin cancer mortality to latitude and ultraviolet radiation in the United States and Canada. *Int. J. Epidemiol.*, **3**, 325-332

Elwood, J.M., Gallagher, R.P., Hill, G.B., Spinelli, J.J., Pearson, J.C.G. & Threlfall, W. (1984) Pigmentation and skin reaction to sun as risk factors for cutaneous melanoma: Western Canada Melanoma Study. *Br. med. J.*, **288**, 99-102

Elwood, J.M., Gallagher, R.P., Davison, J. & Hill, G.B. (1985a) Sunburn, suntan and the risk of cutaneous malignant melanoma: the Western Canada Melanoma Study. *Br. J. Cancer*, **51**, 543-549

Elwood, J.M., Gallagher, R.P., Hill, G.B. & Pearson, J.C.G. (1985b) Cutaneous melanoma in relation to intermittent and constant sun exposure: the Western Canada Melanoma Study. *Int. J. Cancer*, **35**, 427-433

Elwood, J.M., Williamson, C. & Stapleton, P.J. (1986) Malignant melanoma in relation to moles, pigmentation, and exposure to fluorescent and other lighting sources. *Br. J. Cancer*, **53**, 65-74

Elwood, J.M., Gallagher, R.P., Worth, A.J., Wood, W.S. & Pearson, J.C.G. (1987) Etiological differences between subtypes of cutaneous malignant melanoma: Western Canada Melanoma Study. *J. natl Cancer Inst.*, **78**, 37-44

Elwood, J.M., Whitehead, S.M., Davison, J., Stewart, M. & Galt, M. (1990) Malignant melanoma in England: risk associated with naevi, freckles, social class, hair colour, and sunburn. *Int. J. Epidemiol.*, **19**, 801–810

Emmett, E.A. (1973) Ultraviolet radiation as a cause of skin tumors. *Crit. Rev. Toxicol.*, **2**, 211–255

Emmett, E.A. & Horstman, S.W. (1976) Factors influencing the output of ultraviolet radiation during welding. *J. occup. Med.*, **18**, 41–44

Engel, A., Johnson, M.-L. & Haynes, S.G. (1988) Health effects of sunlight exposure in the United States. Results from the first National Health and Nutrition Examination Survey, 1971–1974. *Arch. Dermatol.*, **124**, 72–79

English, D.R., Rouse, I.L., Xu, Z., Watt, J.D., Holman, C.D.J., Heenan, P.J. & Armstrong, B.K. (1985) Cutaneous malignant melanoma and fluorescent lighting. *J. natl Cancer Inst.*, **74**, 1191–1197

Enninga, I.C., Groenendijk, R.T.L., Filon, A.R., van Zeeland, A.A. & Simons, J.W.I.M. (1986) The wavelength dependence of UV-induced pyrimidine dimer formation, cell killing and mutation induction in human diploid skin fibroblasts. *Carcinogenesis*, **7**, 1829–1836

Epstein, J.H. (1965) Comparison of the carcinogenic and cocarcinogenic effects of ultraviolet light on hairless mice. *J. natl Cancer Inst.*, **34**, 741–745

Epstein, J.H. (1985) Animal models for studying photocarcinogenesis. In: Maibach, H. & Lowe, N., eds, *Models in Dermatology*, Vol.2, Basel, Karger, pp. 303–312

Epstein, J.H. (1988) Photocarcinogenesis promotion studies with benzoyl peroxide (BPO) and croton oil. *J. invest. Dermatol.*, **91**, 114–116

Epstein, J.H. (1990) Induction of melanomas in experimental animals. *Photodermatol. Photoimmunol. Photomed.*, **7**, 95–97

Epstein, J.H. & Epstein, W.L. (1962) Cocarcinogenic effect of ultraviolet light on DMBA tumor initiation in albino mice. *J. invest. Dermatol.*, **39**, 455–460

Epstein, J.H. & Epstein, W.L. (1963) A study of tumor types produced by ultraviolet light in hairless mice and hairy mice. *J. invest. Dermatol.*, **41**, 463–473

Epstein, J.H. & Roth, H.L. (1968) Experimental ultraviolet light carcinogenesis. A study of croton oil promoting effects. *J. invest. Dermatol.*, **50**, 387–389

Epstein, J.H., Epstein, W.L. & Nakai, T. (1967) Production of melanomas from DMBA-induced 'blue nevi' in hairless mice with ultraviolet light. *J. natl Cancer Inst.*, **38**, 19–30

Epstein, W.L., Fukuyama, K. & Epstein, J.H. (1969) Early effects of ultraviolet light on DNA synthesis in human skin *in vivo*. *Arch. Dermatol.*, **100**, 84–89

Epstein, J.H., Fukuyama, K., Reed, W.B. & Epstein, W.L. (1970) Defect in DNA synthesis in skin of patients with xeroderma pigmentosum demonstrated *in vivo*. *Science*, **168**, 1477–1478

Eriksen, P. (1987) Occupational applications of ultraviolet radiation: risk evaluation and protection techniques. In: Passchier, W.F. & Bosnjakovic, B.F.M., eds, *Human Exposure to Ultraviolet Radiation: Risks and Regulations*, Amsterdam, Elsevier, pp. 317–331

Eriksen, P., Moscato, P.M., Franks, J.K. & Sliney, D.H. (1987) Optical hazard evaluation of dental curing lights. *Community Dent. oral Epidemiol.*, **15**, 197–201

Everett, M.A., Yeargers, E., Sayre, R.M. & Olson, R.L. (1966) Penetration of epidermis by ultraviolet rays. *Photochem. Photobiol.*, **5**, 533–542

Everett, M.A., Nordquist, J. & Olson, R.L. (1970) Ultrastructure of human epidermis following chronic sun exposure. *Br. J. Dermatol.*, **84**, 248–257

Fan, J., Schoenfeld, R.J. & Hunter, R. (1959) A study of the epidermal clear cells with special reference to their relationship to the cells of Langerhans. *J. invest. Dermatol.*, **32**, 445–450

Farr, P.M. & Diffey, B.L. (1985) The erythemal response of human skin to ultraviolet radiation. *Br. J. Dermatol.*, **113**, 65–76

Fears, T.R., Scotto, J. & Schneiderman, M.A. (1976) Skin cancer, melanoma, and sunlight. *Am. J. public Health*, **66**, 461–464

Fears, T.R., Scotto, J. & Schneiderman, M.A. (1977) Mathematical models of age and ultraviolet effects on the incidence of skin cancer among whites in the United States. *Am. J. Epidemiol.*, **105**, 420–427

Fincham, S. & Hill, G.B. (1989) Risk factors for non-melanocytic skin cancer in Alberta males (Abstract). *Am. J. Epidemiol.*, **130**, 841

Findlay, G.M. (1928) Ultra-violet light and skin cancer. *Lancet*, **ii**, 1070–1073

Findlay, G.M. (1930) Cutaneous papillomata in the rat following exposure to ultra-violet light. *Lancet*, **i**, 1229–1231

Fisher, M.S. & Cifone, M.A. (1981) Enhanced metastatic potential of murine fibrosarcomas treated *in vitro* with ultraviolet radiation. *Cancer Res.*, **41**, 3018–3023

Fisher, G.J. & Johns, H.E. (1976) Pyrimidine photohydrates. In: Wang, S., ed., *Photochemistry and Photobiology of Nucleic Acids*, Vol. 1, New York, Academic Press, pp. 169–224

Fisher, M.S. & Kripke, M.L. (1977) Systemic alteration induced in mice by ultraviolet light irradiation and its relationship to ultraviolet carcinogenesis. *Proc. natl Acad. Sci. USA*, **74**, 1688–1692

Fisher, M.S. & Kripke, M.L. (1978) Further studies on the tumor-specific suppressor cells induced by ultraviolet radiation. *J. Immunol.*, **121**, 1139–1144

Fisher, M.S. & Kripke, M.L. (1982) Suppressor T lymphocytes control the development of primary skin cancers in ultraviolet-irradiated mice. *Science*, **216**, 1133–1134

Fix, D.F. (1986) Thermal resistance of UV-mutagenesis to photoreactivation in *E. coli* B/r *uvrA ung*: estimates of activation energy and further analysis. *Mol. gen. Genet.*, **204**, 452–456

Foley, P., Lanzer, D. & Marks, R. (1986) Are solar keratoses more common on the driver's side? *Br. med. J.*, **293**, 18

Forbes, P.D. & Urbach, F. (1975) Experimental modification of photocarcinogenesis. I. Fluorescent whitening agents and short-wave UVR. *Food Cosmet. Toxicol.*, **13**, 335–337

Forbes, P.D., Davies, R.E. & Urbach, F. (1978) Experimental ultraviolet photocarcinogenesis: wavelength interactions and time-dose relationships. *Natl Cancer Inst. Monogr.*, **50**, 31–38

Forbes, P.D., Blum, H.F. & Davies, R.E. (1981) Photocarcinogenesis in hairless mice: dose-response and the influence of dose-delivery. *Photochem. Photobiol.*, **34**, 361–365

Forbes, P.D., Davies, R.E., Urbach, F., Berger, D. & Cole, C. (1982) Simulated stratospheric ozone depletion and increased ultraviolet radiation: effects on photocarcinogenesis in hairless mice. *Cancer Res.*, **42**, 2796–2803

Forbes, P.D., Davies, R.E., Urbach, F. & Dunnick, J.K. (1990) Long-term toxicity of oral 8-methoxypsoralen plus ultraviolet radiation in mice. *J. Toxicol. cutaneous ocul. Toxicol.*, **9**, 237–250

Fornace, A.J., Jr, Schalch, H. & Alamo, I., Jr (1988) Coordinate induction of metallothioneins I and II in rodent cells by UV irradiation. *Mol. cell. Biol.*, **8**, 4716–4720

Fortner, G.W. & Kripke, M.L. (1977) In vitro reactivity of splenic lymphocytes from normal and UV-irradiated mice against syngeneic UV-induced tumors. *J. Immunol.*, **118**, 1483–1487

Fortner, G.W. & Lill, P.H. (1985) Immune response to ultraviolet-induced tumors. I. Transplantation immunity developing in syngeneic mice in response to progressor ultraviolet-induced tumors. *Transplantation*, **39**, 44–49

Foukal, P. (1990) *Solar Astrophysics*, New York, John Wiley & Sons, pp. 66–75

Fox, I.J., Sy, M.-S., Benacerraf, B. & Greene, M.I. (1981) Impairment of antigen-presenting cell function by ultraviolet radiation. II. Effect of in vitro ultraviolet irradiation on antigen-presenting cells. *Transplantation*, 31, 262-265

Francis, A.J. & Giannelli, F. (1991) Cooperation between human cells sensitive to UVA radiations: a clue to the mechanism of cellular hypersensitivity associated with different clinical conditions. *Exp. Cell Res.*, 195, 47-52

Franklin, W.A. & Haseltine, W.A. (1986) The role of the (6-4) photoproduct in ultraviolet light-induced transition mutations in *E. coli. Mutat. Res.*, 165, 1-7

Franklin, S.A., Lo, K.M. & Haseltine, W.A. (1982) Alkaline lability of fluorescent photoproducts produced in ultraviolet light-irradiated DNA. *J. biol. Chem.*, 257, 13535-13543

Frederick, J.E. (1990) Trends in atmospheric ozone and ultraviolet radiation: mechanisms and observations for the northern hemisphere. *Photochem. Photobiol.*, 51, 757-763

Freeman, R.G. (1975) Data on the action spectrum for ultraviolet carcinogenesis. *J. natl Cancer Inst.*, 55, 1119-1122

Freeman, R.G. (1978) Action spectrum for ultraviolet carcinogenesis. *Natl Cancer Inst. Monogr.*, 50, 27-29

Freeman, S.E. (1988) Variations in excision repair of UVB-induced pyrimidine dimers in DNA of human skin *in situ. J. invest. Dermatol.*, 90, 814-817

Freeman, R.G. & Knox, J.M. (1964) Ultraviolet-induced corneal tumors in different species and strains of animals. *J. invest. Dermatol.*, 43, 431-436

Freeman, S.E., Gange, R.W., Matzinger, E.A. & Sutherland, B.M. (1986) Higher pyrimidine dimer yields in skin of normal humans with higher UVB sensitivity. *J. invest. Dermatol.*, 86, 34-36

Freeman, S.E., Gange, R.W., Sutherland, J.C., Matzinger, E.A. & Sutherland, B.M. (1987) Production of pyrimidine dimers in DNA of human skin exposed *in situ* to UVA radiation. *J. invest. Dermatol.*, 88, 430-433

Freeman, S.E., Applegate, L.A. & Ley, R.D. (1988a) Excision repair of UVR-induced pyrimidine dimers in corneal DNA. *Photochem. Photobiol.*, 47, 159-163

Freeman, S.E., Ley, R.D. & Ley, K.D. (1988b) Sunscreen protection against UV-induced pyrimidine dimers in DNA of human skin *in situ. Photodermatology*, 5, 243-247

Freeman, S.E., Hacham, H., Gange, R.W., Maytum, D.J., Sutherland, J.C. & Sutherland, B.M. (1989) Wavelength dependence of pyrimidine dimer formation in DNA of human skin irradiated *in situ* with ultraviolet light. *Proc. natl Acad. Sci. USA*, 86, 5605-5609

Friedburg, E. (1984) *DNA Repair*, New York, W.H. Freeman & Co.

Frucht, D.M., Lamperth, L., Vicenzi, E., Belcher, J.H. & Martin, M.A. (1991) Ultraviolet radiation increases HIV-long terminal repeat-directed expression in transgenic mice. *AIDS Res. hum. Retroviruses*, 7, 729-733

Fujiwara, Y., Ichihashi, M., Kano, Y., Goto, K. & Shimizu, K. (1981) A new human photosensitive subject with a defect in the recovery of DNA synthesis after ultraviolet-light irradiation. *J. invest. Dermatol.*, 77, 256-263

Gafá, L., Filippazzo, M.G., Tumino, R., Dardanoni, G., Lanzarone, F. & Dardanoni, L. (1991) Risk factors of nonmelanoma skin cancer in Ragusa, Sicily: a case-control study. *Cancer Causes Control*, 2, 395-399

Gahring, L., Baltz, M., Pepys, M.B. & Daynes, R. (1984) Effect of ultraviolet radiation on production of epidermal cell thymocyte-activating factor/interleukin 1 *in vivo* and *in vitro. Immunology*, 81, 1198-1202

Gallagher, R.P. (1988) Ocular melanoma in farmers (Letter to the Editor). *Am. J. ind. Med.*, **13**, 523–525

Gallagher, P.E. & Duker, N.J. (1986) Detection of UV purine photoproducts in a defined sequence of human DNA. *Mol. cell. Biol.*, **6**, 707–709

Gallagher, P.E. & Duker, N.J. (1989) Formation of purine photoproducts in a defined human DNA sequence. *Photochem. Photobiol.*, **49**, 599–605

Gallagher, C.H., Path, F.R.C., Canfield, P.J., Greenoak, G.E. & Reeve, V.E. (1984a) Characterization and histogenesis of tumors in the hairless mouse produced by low-dosage incremental ultraviolet radiation. *J. invest. Dermatol.*, **83**, 169–174

Gallagher, C.H., Greenoak, G.E., Reeve, V.E., Canfield, P.J., Baker, R.S.U. & Bonin, A.M. (1984b) Ultraviolet carcinogenesis in the hairless mouse skin. Influence of the sunscreen 2-ethylhexyl-p-methoxycinnamate. *Aust. J. exp. Biol. med. Sci.*, **62**, 577–588

Gallagher, R.P., Threlfall, W.J., Jeffries, E., Band, P.R., Spinelli, J. & Coldman, A.J. (1984) Cancer and aplastic anemia in British Columbia farmers. *J. natl Cancer Inst.*, **72**, 1311–1315

Gallagher, R.P., Elwood, J.M., Rootman, J., Spinelli, J.J., Hill, G.B., Threlfall, W.J. & Birdsell, J.M. (1985) Risk factors for ocular melanoma: Western Canada Melanoma Study. *J. natl Cancer Inst.*, **74**, 775–778

Gallagher, R.P., Elwood, J.M. & Hill, G.B. (1986) Risk factors for cutaneous malignant melanoma: the Western Canada Melanoma Study. *Recent Results Cancer Res.*, **102**, 38–55

Gallagher, R.P., Elwood, J.M., Threlfall, W.J., Spinelli, J.J., Fincham, S. & Hill, G.B. (1987) Socio-economic status, sunlight exposure, and risk of malignant melanoma: the Western Canada Melanoma Study. *J. natl Cancer Inst.*, **79**, 647–652

Gallagher, P.E., Weiss, R.B., Brent, T.P. & Duker, N.J. (1989) Wavelength dependence of DNA incision by a human ultraviolet endonuclease. *Photochem. Photobiol.*, **49**, 363–367

Gallagher, R.P., McLean, D.I., Yang, C.P., Coldman, A.J., Silver, H.K.B., Spinelli, J.J. & Beagrie, M. (1990a) Anatomic distribution of acquired melanocytic nevi in white children. A comparison with melanoma: the Vancouver mole study. *Arch. Dermatol.*, **126**, 466–471

Gallagher, R.P., McLean, D.I., Yang, C.P., Coldman, A.J., Silver, H.K.B., Spinelli, J.J. & Beagrie, M. (1990b) Suntan, sunburn and pigmentation factors and the frequency of acquired melanocytic nevi in children. Similarities to melanoma: the Vancouver mole study. *Arch. Dermatol.*, **126**, 770–776

Gange, R.W. (1987) Acute effects of ultraviolet radiation in the skin. In: Fitzpatrick, T.B., Eisen, A.Z., Wolff, K., Freedberg, I.M. & Austen, K.F., eds, *Dermatology in General Medicine*, 3rd ed., New York, McGraw-Hill, pp. 1451–1457

Gange, R.W., Blackett, A.D., Matzinger, E.A., Sutherland, B.M. & Kochevar, I.E. (1985) Comparative protection efficiency of UVA- and UVB-induced tans against erythema and formation of endonuclease-sensitive sites in DNA by UVB in human skin. *J. invest. Dermatol.*, **85**, 362–364

Garbe, C., Krüger, S., Stadler, R., Guggenmoos-Holzmann, I. & Orfanos, C.E. (1989) Markers and relative risk in a German population for developing malignant melanoma. *Int. J. Dermatol.*, **28**, 517–523

Gardiner, B.G. & Kirsch, P.J. (1991) *European Intercomparison of Ultraviolet Spectrometers, Panorama, Greece, 3–12 July 1991* (Report to the Commission of the European Communities [STEP Project 76], Science and Technology for Environmental Protection), Luxembourg, Commission of the European Communities

Garland, F.C., White, M.R., Garland, C.F., Shaw, E. & Gorham, E.D. (1990) Occupational sunlight exposure and melanoma in the US Navy. *Arch. environ. Health*, **45**, 261–267

REFERENCES

Garrison, L.M., Murray, L.E., Doda, D.D. & Green, A.E.S. (1978) Diffuse-direct ultraviolet ratios with a compact double monochromator. *Appl. Optics*, **17**, 827–836

Gellin, G.A., Kopf, A.W. & Garfinkel, L. (1965) Basal cell epithelioma. A controlled study of associated factors. *Arch. Dermatol.*, **91**, 38–45

Gellin, G.A., Kopf, A.W. & Garfinkel, L. (1969) Malignant melanoma. A controlled study of possibly associated factors. *Arch. Dermatol.*, **99**, 43–48

Gensler, H.L. (1988a) Enhancement of chemical carcinogenesis in mice by systemic effects of ultraviolet radiation. *Cancer Res.*, **48**, 620–623

Gensler, H.L. (1988b) Prevention of chemically induced two-stage skin carcinogenesis in mice by systemic effects of ultraviolet irradiation. *Carcinogenesis*, **9**, 767–769

Gensler, H.L. & Bowden, G.T. (1987) UVB-Induced modulation of mouse skin tumor induction by benzo[*a*]pyrene (Abstract no. 546). *Proc. Am. Assoc. Cancer Res.*, **28**, 137

Gensler, H.L. & Welch, K. (1992) Prevalence of tumor prevention rather than tumor enhancement when repetitive UV radiation treatments precede initiation and promotion. *Carcinogenesis*, **13**, 9–13

Giam, Y.C. (1987) Cutaneous gerontology: solar degeneration of the skin. *Ann. Acad. Med. Singapore*, **16**, 94–97

Giannini, M.S.H. (1986) Suppression of pathogenesis in cutaneous leishmaniasis by UV irradiation. *Infect. Immun.*, **51**, 838–843

Gibbs, N.K., Young, A.R. & Magnus, I.A. (1985) Failure of UVR dose reciprocity for skin tumorigenesis in hairless mice treated with 8-methoxypsoralen. *Photochem. Photobiol.*, **42**, 39–42

Gibson, P. & Diffey, B.L. (1989) Techniques for spectroradiometry and broadband radiometry. In: Diffey, B.L., ed., *Radiation Measurement in Photobiology*, London, Academic Press, pp. 135–159

Gies, P., Roy, C. & Elliott, G. (1988) The anatomical distribution of solar UVR with emphasis on the eye. In: *Radiation Protection Practice*, Vol. 1, Sydney, Pergamon Press, pp. 341–344

Gilchrest, B.A., Soter, N.A., Stoff, J.S. & Mihm, M.C., Jr (1981) The human sunburn reaction: histologic and biochemical studies. *J. Am. Acad. Dermatol.*, **5**, 411–422

Gilchrest, B.A., Murphy, G.F. & Soter, N.A. (1982) Effect of chronologic aging and ultraviolet irradiation on Langerhans cells in human epidermis. *J. invest. Dermatol.*, **79**, 85–88

Giles, G.G., Marks, R. & Foley, P. (1988) Incidence of non-melanocytic skin cancer treated in Australia. *Br. med. J.*, **296**, 13–17

Glazier, A., Morison, W.L., Bucana, C., Hess, A.D. & Tutschka, P.J. (1984) Suppression of epidermal graft-*versus*-host disease with ultraviolet radiation. *Transplantation*, **37**, 211–213

Glickman, B.W., Schaaper, R.M., Haseltine, W.A., Dunn, R.L. & Brash, D.E. (1986) The C-C (6-4) UV photoproduct is mutagenic in *Escherichia coli*. *Proc. natl Acad. Sci. USA*, **83**, 6945–6949

Goldberg, L.H. & Altman, A. (1984) Benign skin changes associated with chronic sunlight exposure. *Cutis*, **34**, 33–39

Goñi, F.M., Ondarroa, M., Azpiazu, I. & Macarulla, J.M. (1985) Phospholipid oxidation catalysed by cytochrome *c* in liposomes. *Biochim. biophys. Acta*, **835**, 549–556

Goodman, G.J., Marks, R., Selwood, T.S., Ponsford, M.W. & Pakes, W. (1984) Non-melanocytic skin cancer and solar keratoses in Victoria—clinical studies II. *Aust. J. Derm.*, **25**, 103–106

Gordon, D. & Silverstone, H. (1976) Worldwide epidemiology of premalignant and malignant cutaneous lesions. In: Andrade, R., Gumport, S.L., Popkin, G.L. & Rees, R.D., eds, *Cancer of the Skin*, Vol. 1, London, W.B. Saunders, pp. 405–434

Gorlin, R.J. (1987) Nevoid basal-cell carcinoma syndrome. *Medicine*, **66**, 98–113

Grabner, G. & Brenner, W. (1981) Unscheduled DNA repair in the human cornea following solar simulating radiation. *Acta ophthalmol.*, **59**, 847–853

Grady, H.G., Blum, H.F. & Kirby-Smith, J.S. (1941) Pathology of tumors of the external ear in mice induced by ultraviolet radiation. *J. natl Cancer Inst.*, **2**, 269–276

Grady, H.G., Blum, H.F. & Kirby-Smith, J.S. (1943) Types of tumor induced by ultraviolet radiation and factors influencing their relative incidence. *J. natl Cancer Inst.*, **3**, 371–378

Graham, S., Marshall, J., Haughey, B., Stoll, H., Zielezny, M., Brasure, J. & West, D. (1985) An inquiry into the epidemiology of melanoma. *Am. J. Epidemiol.*, **122**, 606–619

Granstein, R.D. (1985) Epidermal I-J-bearing cells are responsible for transferable suppressor cell generation after immunization of mice with ultraviolet radiation-treated epidermal cells. *J. invest. Dermatol.*, **84**, 206–209

Granstein, R.D. & Sauder, D.N. (1987) Whole-body exposure to ultraviolet radiation results in increased serum interleukin-1 activity in humans. *Lymphokine Res.*, **6**, 187–193

Granstein, R.D., Lowy, A. & Greene, M.I. (1984) Epidermal antigen-presenting cells in activation of suppression: identification of a new functional type of ultraviolet radiation-resistant epidermal cell. *J. Immunol.*, **132**, 563–565

Granstein, R.D., Askari, M., Whitaker, D. & Murphy, G.F. (1987) Epidermal cells in activation of suppressor lymphocytes: further characterization. *J. Immunol.*, **138**, 4055–4062

Greaves, M.W., Hensby, C.N., Black, A.K., Plummer, N.A., Fincham, N., Warin, A.P. & Camp, R. (1978) Inflammatory reactions induced by ultraviolet irradiation. *Bull. Cancer*, **65**, 299–304

Green, A.C. (1984) Sun exposure and the risk of melanoma. *Aust. J. Dermatol.*, **25**, 99–102

Green, A.C. (1991) Premature ageing of the skin in a Queensland population. *Med. J. Aust.*, **155**, 473–478

Green, A.C. & Battistutta, D. (1990) Incidence and determinants of skin cancer in a high-risk Australian population. *Int. J. Cancer*, **46**, 356–361

Green, A. & O'Rourke, M.G.E. (1985) Cutaneous malignant melanoma in association with other skin cancers. *J. natl Cancer Inst.*, **74**, 977–980

Green, A. & Siskind, V. (1983) Geographical distribution of cutaneous melanoma in Queensland. *Med. J. Aust.*, **1**, 407–410

Green, A.E.S., Findley, G.B., Jr, Klenk, K.F., Wilson, W.M. & Mo, T. (1976) The ultraviolet dose dependence of non-melanoma skin cancer incidence. *Photochem. Photobiol.*, **24**, 353–362

Green, A.C., Siskind, V., Bain, C. & Alexander, J. (1985a) Sunburn and malignant melanoma. *Br. J. Cancer*, **51**, 393–397

Green, A.C., McLennan, R. & Siskind, V. (1985b) Common acquired naevi and the risk of malignant melanoma. *Int. J. Cancer*, **35**, 297–300

Green, A.C., Bain, C., McLennan, R. & Siskind, V. (1986) Risk factors for cutaneous melanoma in Queensland. *Recent Results Cancer Res.*, **102**, 76–97

Green, A.C., Beardmore, G., Hart, V., Leslie, D., Marks, R. & Staines, D. (1988a) Skin cancer in a Queensland population. *J. Am. Acad. Dermatol.*, **19**, 1045–1052

Green, A.C., Sorahan, T., Pope, D., Siskind, V., Hansen, M., Hanson, L., Leech, P., Ball, P.M. & Grimley, R.P. (1988b) Moles in Australian and British schoolchildren (Letter to the Editor). *Lancet*, **ii**, 1497

Greene, M.I., Sy, M.S., Kripke, M. & Benacerraf, B. (1979) Impairment of antigen-presenting cell function by ultraviolet radiation. *Proc. natl Acad. Sci. USA*, **76**, 6591–6595

Griggs, H.G. & Bender, M.A. (1973) Photoreactivation of ultraviolet-induced chromosomal aberrations. *Science*, **179**, 86–88

Grob, J.J., Gouvernet, J., Aymar, D., Mostaque, A., Romano, M.H., Collet, A.M., Noe, M.C., Diconstanzo, M.P. & Bonerandi, J.J. (1990) Count of benign melanocytic nevi as a major indicator of risk for nonfamilial nodular and superficial spreading melanoma. *Cancer*, **66**, 387–395

de Gruijl, F.R. & van der Leun, J.C. (1982a) Effect of chronic UV exposure on epidermal transmission in mice. *Photochem. Photobiol.*, **36**, 433–438

de Gruijl, F.R. & van der Leun, J.C. (1982b) Systemic influence of pre-irradiation of a limited skin area on UV-tumorigenesis. *Photochem. Photobiol.*, **35**, 379–383

de Gruijl, F.R. & van der Leun, J.C. (1983) Follow up on systemic influence of partial pre-irradiation on UV-tumorigenesis. *Photochem. Photobiol.*, **38**, 381–383

de Gruijl, F.R., van der Meer, J.B. & van der Leun, J.C. (1983) Dose–time dependency of tumor formation by chronic UV exposure. *Photochem. Photobiol.*, **37**, 53–62

Gupta, A.K. & Anderson, T.F. (1987) Psoralen photochemotherapy. *J. Am. Acad. Dermatol.*, **17**, 703–734

Gurish, M.F., Lynch, D.H. & Daynes, R.A. (1982) Changes in antigen-presenting cell function in the spleen and lymph nodes of ultraviolet-irradiated mice. *Transplantation*, **33**, 280–284

Gurish, M.F., Lynch, D.H., Yowell, R. & Daynes, R.A. (1983) Abrogation of epidermal antigen-presenting cell function by ultraviolet radiation administered *in vivo*. *Transplantation*, **36**, 304–309

Haenszel, W. (1963) Variations in skin cancer incidence within the United States. *Natl Cancer Inst. Monogr.*, **10**, 225–243

Halasz, C.L.G., Leach, E.E., Walther, R.R. & Poh-Fitzpatrick, M.B. (1983) Hydroa vacciniforme: induction of lesions with ultraviolet A. *J. Am. Acad. Dermatol.*, **8**, 171–176

Hall, J.D. & Mount, D.W. (1981) Mechanisms of DNA replication and mutagenesis in ultraviolet-irradiated bacteria and mammalian cells. *Prog. nucleic Acid Res. mol. Biol.*, **25**, 53–126

Halprin, K.M., Comerford, M., Presser, S.E. & Taylor, J.R. (1981) Ultraviolet light treatment delays contact sensitization to nitrogen mustard. *Br. J. Dermatol.*, **105**, 71–76

Haniszko, J. & Suskind, R.R. (1963) The effect of ultraviolet radiation on experimental cutaneous sensitization in guinea pigs. *J. invest. Dermatol.*, **40**, 183–191

Hannan, M.A., Paul, M., Amer, M.H. & Al-Watban, F.H. (1984) Study of ultraviolet radiation and genotoxic effects of natural sunlight in relation to skin cancer in Saudi Arabia. *Cancer Res.*, **44**, 2192–2197

Hanson, D.L. & DeLeo, V.A. (1989) Long wave ultraviolet radiation stimulates arachidonic acid release and cyclooxygenase activity in mammalian cells in culture. *Photochem. Photobiol.*, **49**, 423–430

Harber, L.C. & Bickers, D.R. (1981) *Photosensitivity Diseases. Principles of Diagnosis and Treatment*, Philadelphia, PA, W.B. Saunders

Hardie, I.R., Strong, R.W., Hartley, L.C.J., Woodruff, P.W.H. & Clunie, G.J.A. (1980) Skin cancer in Caucasian renal allograft recipients living in a subtropical climate. *Surgery*, **87**, 177–183

Hariharan, P.V. & Cerutti, P.A. (1976) Excision of ultraviolet and gamma ray products of the 5,6-dihydroxy-dihydrothymine type by nuclear preparations of xeroderma pigmentosum cells. *Biochim. biophys. Acta*, **447**, 375–378

Hariharan, P.V. & Cerutti, P.A. (1977) Formation of products of the 5,6-dihydroxydihydrothymine type by ultraviolet light in HeLa cells. *Biochemistry*, **16**, 2791–2795

Hartman, Z., Hartman, P.E. & McDermott, W.L. (1991) Mutagenicity of coolwhite fluorescent light for *Salmonella*. *Mutat. Res.*, **260**, 25–38

Hawk, J.L.M. & Parrish, J.A. (1982) Responses of normal skin to ultraviolet radiation. In: Reegan, J.D. & Parrish, J.A., eds, *The Science of Photomedicine*, New York, Plenum, pp. 219–260

Hayashi, Y. & Aurelian, L. (1986) Immunity to herpes simple virus type 2: viral antigen-presenting capacity of epidermal cells and its impairment by ultraviolet irradiation. *J. Immunol.*, **136**, 1087–1092

Health Council of the Netherlands (1986) *UV Radiation: Human Exposure to Ultraviolet Radiation* (Report 1986/93), The Hague

Hefferren, J.J., Cooley, R.O., Hall, J.B., Olsen, N.H. & Lyon, H.W. (1971) Use of ultraviolet illumination in oral diagnosis. *J. Am. dent. Assoc.*, **82**, 1353–1360

Henderson, S.T. (1970) *Daylight and Its Spectrum*, New York, American Elsevier

Henderson, L.M., Arlett, C.F., Harcourt, S.A., Lehmann, A.R. & Broughton, B.C. (1985) Cells from an immunodeficient patient (46BR) with a defect in DNA ligation are hypomutable but hypersensitive to the induction of sister chromatid exchanges. *Proc. natl Acad. Sci. USA*, **82**, 2044–2048

Henseler, T., Wolff, K., Hönigsmann, H. & Christophers, E. (1981) Oral 8- methoxypsoralen photochemotherapy of psoriasis. The European PUVA study: a cooperative study among 18 European centres. *Lancet*, **i**, 853–857

Herity, B., O'Loughlin, G., Moriarty, M.J. & Conroy, R. (1989) Risk factors for non-melanoma skin cancer. *Irish med. J.*, **82**, 151–152

Hersey, P., Bradley, M., Hasic, E., Haran, G., Edwards, A. & McCarthy, W.H. (1983a) Immunological effects of solarium exposure. *Lancet*, **i**, 545–548

Hersey, P., Haran, G., Hasic, E. & Edwards, A. (1983b) Alteration of T cell subsets and induction of suppressor T cell activity in normal subjects after exposure to sunlight. *J. Immunol.*, **31**, 171–174

Hersey, P., MacDonald, M., Burns, C., Schibeci, S., Matthews, H. & Wilkinson, F.J. (1987) Analysis of the effect of a sunscreen agent on the suppression of natural killer cell activity induced in human subjects by radiation from solarium lamps. *J. invest. Dermatol.*, **88**, 271–276

Hersey, P., MacDonald, M., Henderson, C., Schibeci, S., D'Alessandro, G., Pryor, M. & Wilkinson, F.J. (1988) Suppression of natural killer cell activity in humans by radiation from solarium lamps depleted of UVB. *J. invest. Dermatol.*, **90**, 305–310

Hietanen, M. (1990) *Spectroradiometric and Photometric Studies on Ocular Exposure to Ultraviolet and Visible Radiation from Light Sources*, PhD Thesis, Helsinki, University of Helsinki, Department of Physics

Higgins, S.A., Frenkel, K., Cummings, A. & Teebor, G.W. (1987) Definitive characterisation of human thymine glycol *N*-glycosylase activity. *Biochemistry*, **26**, 1683–1688

Hinds, M.W. & Kolonel, L.N. (1980) Malignant melanoma of the skin in Hawaii, 1960–1977. *Cancer*, **45**, 811–817

Hitchins, V.M., Beer, J.Z., Strickland, A.G. & Olvey, K.M. (1987) Lethal and mutagenic effects of UVA radiation in six strains of L5178Y mouse lymphoma cells. In: Passchier, W.F. & Bosnjakovic, B.F.M., eds, *Human Exposure to Ultraviolet Radiation: Risks and Regulations*, Amsterdam, Elsevier, pp. 65–69

Hoeijmakers, J.H.J. & Bootsma, D. (1990) Molecular genetics of eukaryotic DNA excision repair. *Cancer Cells*, **2**, 311–320

Hoffman, J.S. (1987) *Ultraviolet Radiation and Melanoma (with a Special Focus on Assessing the Risks of Stratospheric Ozone Depletion)* (EPA 400/1-87-001D), Washington DC, Office of Air and Radiation, US Environmental Protection Agency

Hogan, D.J., To, T., Gran, L., Wong, D. & Lane, P.R. (1989) Risk factors for basal cell carcinoma. *Int. J. Dermatol.*, **28**, 591–594

Holick, M.F., MacLaughlin, J.A., Clark, M.B., Holick, S.A., Potts, J.T., Jr, Anderson, R.R., Blank, I.H., Parrish, J.A. & Elias, P. (1980) Photosynthesis of previtamin D_3 in human skin and the physiologic consequences. *Science*, 210, 203–205

Hollaender, A. (1943) Effect of long ultraviolet and short visible radiation (3500 to 4900 Å) on *Escherichia coli. J. Bacteriol.*, 46, 531–541

Hollander, M.C. & Fornace, A.J., Jr (1989) Induction of *fos* RNA by DNA-damaging agents. *Cancer Res.*, 49, 1687–1692

Hollis, D.E. & Scheibner, A. (1988) Ultrastructural changes in epidermal Langerhans cells and melanocytes in response to ultraviolet irradiation, in Australians of aboriginal and Celtic descent. *Br. J. Dermatol.*, 119, 21–31

Holly, E.A., Kelly, J.W., Shpall, S.N. & Chiu, S.-H. (1987) Number of melanocytic nevi as a major risk factor for malignant melanoma. *J. Am. Acad. Dermatol.*, 17, 459–468

Holly, E.A., Aston, D.A., Char, D.H., Kristiansen, J.J. & Ahn, D.K. (1990) Uveal melanoma in relation to ultraviolet light exposure and host factors. *Cancer Res.*, 50, 5773–5777

Holly, E.A., Aston, D.A., Ahn, D.K., Kristiansen, J.J. & Char, D.H. (1991) No excess prior cancer in patients with uveal melanoma. *Ophthalmology*, 98, 608–611

Holman, C.D.J. & Armstrong, B.K. (1984a) Pigmentary traits, ethnic origin, benign nevi, and family history as risk factors for cutaneous malignant melanoma. *J. natl Cancer Inst.*, 72, 257–266

Holman, C.D.J. & Armstrong, B.K. (1984b) Cutaneous malignant melanoma and indicators of total accumulated exposure to the sun: an analysis separating histogenetic types. *J. natl Cancer Inst.*, 73, 75–82

Holman, C.D.J., Gibson, I.M., Stephenson, M. & Armstrong, B.K. (1983a) Ultraviolet irradiation of human body sites in relation to occupation and outdoor activity: field study using personal UVR dosimeters. *Clin. exp. Dermatol.*, 8, 269–277

Holman, C.D.J., Heenan, P.J., Caruso, V., Glancy, R.J. & Armstrong, B.K. (1983b) Seasonal variation in the junctional component of pigmented naevi. *Int. J. Cancer*, 31, 213–215

Holman, C.D.J., Armstrong, B.K., Evans, P.R., Lumsden, G.J., Dallimore, K.J., Meehan, C.J., Beagley, J. & Gibson, I.M. (1984a) Relationship of solar keratosis and history of skin cancer to objective measures of actinic skin damage. *Br. J. Dermatol.*, 110, 129–138

Holman, C.D., Evans, P.R., Lumsden, G.J. & Armstrong, B.K. (1984b) The determinants of actinic skin damage: problems of confounding among environmental and constitutional variables. *Am. J. Epidemiol.*, 120, 414–422

Holman, C.D.J., Armstrong, B.K. & Heenan, P.J. (1986a) Relationship of cutaneous malignant melanoma to individual sunlight-exposure habits. *J. natl Cancer Inst.*, 76, 403–414

Holman, C.D.J., Armstrong, B.K., Heenan, P.J., Blackwell, J.B., Cumming, F.J., English, D.R., Holland, S., Kelsall, G.R.H., Matz, L.R., Rouse, I.L., Singh, A., Ten Seldam, R.E.J., Watt, J.D. & Xu, Z. (1986b) The causes of malignant melanoma: results from the West Australian Lions Melanoma Research Project. *Recent Results Cancer Res.*, 102, 18–37

Holmberg, M. & Gumauskas, E. (1990) Chromosome-type exchange aberrations are induced by inhibiting repair of UVC-induced DNA lesions in quiescent human lymphocytes. *Mutat. Res.*, 232, 261–266

Hönigsmann, H., Brenner, W., Tanew, A. & Ortel, B. (1987) UV-Induced unscheduled DNA synthesis in human skin: dose response, correlation with erythema, time course and split dose exposure *in vivo. J. Photochem. Photobiol. B Biol.*, 1, 33–43

Hoover, T.L., Morison, W.L. & Kripke, M.L. (1987) Ultraviolet carcinogenesis in athymic nude mice. *Transplantation*, 44, 693–695

Howie, S.E.M., Norval, N., Maingay, J. & Ross, J.A. (1986a) Two phenotypically distant T cells (Ly1$^+$2$^-$ and Ly1$^-$2$^+$) are involved in ultraviolet-B light-induced suppression of the efferent DTH response to HSV-1 *in vivo. Immunology*, **58**, 653–658

Howie, S.E.M., Norval, M. & Maingay, J.P. (1986b) Alterations in epidermal handling of HSV-1 antigens *in vitro* induced by in vivo exposure to UVB light. *Immunology*, **57**, 225–230

Howie, S., Norval, M. & Maingay, J. (1986c) Exposure to low-dose ultraviolet radiation suppresses delayed-type hypersensitivity to herpes simplex virus in mice. *J. invest. Dermatol.*, **86**, 125–128

Hsia, H.C., Lebkowski, J.S., Leong, P.-M., Calos, M.P. & Miller, J.H. (1989) Comparison of ultraviolet irradiation-induced mutagenesis of the *lac*I gene in *Escherichia coli* and in human 293 cells. *J. mol. Biol.*, **205**, 103–113

Hsie, A.W., Li, A.P. & Machanoff, R. (1977) A fluence response study of lethality and mutagenicity of white, black, and blue fluorescent light, sunlamp, and sunlight irradiation in Chinese hamster ovary cells. *Mutat. Res.*, **45**, 333–342

Hsu, J., Forbes, P.D., Harber, L.C. & Lakow, E. (1975) Induction of skin tumors in hairless mice by a single exposure to UV radiation. *Photochem. Photobiol.*, **21**, 185–188

Hueper, W.C. (1942) Morphological aspects of experimental actinic and arsenic carcinomas in the skin of rats. *Cancer Res.*, **2**, 551–559

Huldschinsky, K. (1933) Eye sarcoma in rats caused by abnormally long UV (Ger.). *Dtsch. med. Wochenschr.*, **59**, 530–531

Hunter, D.J., Colditz, G.A., Stampfer, M.J., Rosner, B., Willett, W.C. & Speizer, F.E. (1990) Risk factors for basal cell carcinoma in a prospective cohort of women. *Ann. Epidemiol.*, **1**, 13–23

Husain, Z., Yang, Q. & Biswas, D.K. (1990) cHa-*ras* Proto oncogene. Amplification and overexpression in UVB-induced mouse skin papillomas and carcinomas. *Arch. Dermatol.*, **126**, 324–330

Husain, Z., Pathak, M.A., Flotte, T. & Wick, M.M. (1991) Role of ultraviolet radiation in the induction of melanocytic tumors in hairless mice following 7,12-dimethylbenz(*a*)anthracene application and ultraviolet irradiation. *Cancer Res.*, **51**, 4964–4970

Huselton, C.A. & Hill, H.Z. (1990) Melanin photosensitizes ultraviolet light (UVC) DNA damage in pigmented cells. *Environ. mol. Mutag.*, **16**, 37–43

Hutchinson, F. (1987) A review of some topics concerning mutagenesis by ultraviolet light. *Photochem. Photobiol.*, **45**, 897–903

Hutchinson, F. & Wood, R.D. (1986) Mechanisms of mutagenesis of *E. coli* by ultraviolet light. *Basic Life Sci.*, **38**, 377–383

Hutchinson, F., Yamamoto, K., Stein, J. & Wood, R.D. (1988) Effect of photoreactivation on mutagenesis of lambda phage by ultraviolet light. *J. mol. Biol.*, **202**, 593–601

IARC (1980) *IARC Monographs on the Evaluation of the Carcinogenic Risk of Chemicals to Humans*, Vol. 24, *Some Pharmaceutical Drugs*, Lyon, pp. 101–124

IARC (1985) *IARC Monographs on the Evaluation of the Carcinogenic Risk of Chemicals to Humans*, Vol. 37, *Tobacco Habits Other than Smoking; Betel Quid and Areca-nut Chewing; and Some Related Nitrosamines*, Lyon, pp. 37–136

IARC (1986a) *IARC Monographs on the Evaluation of the Carcinogenic Risk of Chemicals to Humans*, Vol. 40, *Some Naturally Occurring and Synthetic Food Compounds, Furocoumarins and Ultraviolet Radiation*, Lyon, pp. 317–371

IARC (1986b) *IARC Monographs on the Evaluation of the Carcinogenic Risk of Chemicals to Humans*, Vol. 38, *Tobacco Smoking*, Lyon

IARC (1987a) *IARC Monographs on the Evaluation of Carcinogenic Risks to Humans*, Suppl. 7, *Overall Evaluations of Carcinogenicity: An Updating of* IARC Monographs *Volumes 1 to 42*, Lyon, pp. 175–176

IARC (1987b) *IARC Monographs on the Evaluation of Carcinogenic Risks to Humans*, Suppl. 7, *Overall Evaluations of Carcinogenicity: An Updating of* IARC Monographs *Volumes 1 to 42*, Lyon, pp. 242–245

IARC (1988) *IARC Monographs on the Evaluation of Carcinogenic Risks to Humans*, Vol. 44, *Alcohol Drinking*, Lyon

IARC (1990) *IARC Monographs on the Evaluation of Carcinogenic Risks to Humans*, Vol. 49, *Chromium, Nickel and Welding*, Lyon, pp. 447–525

Ikushima, T. & Wolff, S. (1974) UV-Induced chromatid aberrations in cultured Chinese hamster cells after one, two, or three rounds of DNA replication. *Mutat. Res.*, 22, 193–201

International Electrotechnical Commission (1987) *Safety of Household and Similar Electrical Appliances*, Part 2, *Particular Requirements for Ultra-violet and Infra-red Radiation Skin Treatment Appliances for Household Use* (Publication 335-2-27: 1987), Geneva, Bureau Central de la Commission Electrotechnique Internationale

International Electrotechnical Commission (1989) *Amendment*, Geneva, Bureau Central de la Commission Electrotechnique Internationale

International Non-ionizing Radiation Committee of the International Radiation Protection Association (1985) Guidelines on limits of exposure to ultraviolet radiation of wavelengths between 180 nm and 400 nm (incoherent optical radiation). *Health Phys.*, 49, 331–340

International Non-ionizing Radiation Committee of the International Radiation Protection Association (1989) Proposed change to the IRPA 1985 guidelines on limits of exposure to ultraviolet radiation. *Health Phys.*, 56, 971–972

Iqbal, M. (1983) *An Introduction to Solar Radiation*, New York, Academic Press, pp. 335–373

Ito, M. (1966) The photodynamic effect of acridine orange and ultraviolet on the mouse skin carcinogenesis by 20-methylcholanthrene. *Nagoya med. J.*, 12, 247–259

Ito, A. & Ito, T. (1983) Possible involvement of membrane damage in the inactivation by broad-band near-UV radiation in *Saccharomyces cerevisiae* cells. *Photochem. Photobiol.*, 37, 395–401

Iversen, O.H. (1986) Carcinogenesis studies with benzoyl peroxide (Panoxyl gel 5%). *J. invest. Dermatol.*, 86, 442–448

Iversen, O.H. (1988) Skin tumorigenesis and carcinogenesis studies with 7,12-dimethylbenz[a]-anthracene, ultraviolet light, benzoyl peroxide (Panoxyl gel 5%) and ointment gel. *Carcinogenesis*, 9, 803–809

Jacobson, E.D., Krell, K., Dempsey, M.J., Lugo, M.H., Ellingson, O. & Hench, C.W., II (1978) Toxicity and mutagenicity of radiation from fluorescent lamps and a sunlamp in L5178Y mouse lymphoma cells. *Mutat. Res.*, 51, 61–75

Jacobson, E.D., Krell, K. & Dempsey, M.J. (1981) The wavelength dependence of ultraviolet light-induced cell killing and mutagenesis in L5178Y mouse lymphoma cells. *Photochem. Photobiol.*, 33, 257–260

Jagger, J. (1981) Near-UV radiation effects on microorganisms. *Photochem. Photobiol.*, 34, 761–768

Jagger, J. (1985) *Solar-UV Actions on Living Cells*, New York, Praeger

Jessup, J.M., Hanna, N., Palaszynski, E. & Kripke, M.L. (1978) Mechanisms of depressed reactivity to dinitrochlorobenzene and ultraviolet-induced tumors during ultraviolet carcinogenesis in BALB/c mice. *Cell. Immunol.*, 38, 105–115

Johns, H.E., Rapaport, S.A. & Delbrück, M. (1962) Photochemistry of thymine dimers. *J. mol. Biol.*, **4**, 104–114

Jones, C.A., Huberman, E., Cunningham, M.L. & Peak, M.J. (1987) Mutagenesis and cytotoxicity in human epithelial cells by far- and near-ultraviolet radiations: action spectra. *Radiat. Res.*, **110**, 244–254

Jones, S.K., Moseley, H. & Mackie, R. (1987) UVA-Induced melanocytic lesions. *Br. J. Dermatol.*, **117**, 111–115

Jose, J.G. & Pitts, D.G. (1985) Wavelength dependency of cataracts in albino mice following chronic exposure. *Exp. Eye Res.*, **41**, 545–563

Joshi, S.D., Shrikhande, S.S. & Ranade, S.S. (1984) Studies on UV-induced skin tumors in experimental animals. *Photobiochem. Photobiophys.*, **7**, 235–243

Joshi, S.D., Shrikhande, S.S., Naik, S.N. & Ranade, S.S. (1986) Multiplicity of tumours induced in the laboratory animals by ultraviolet (UVC) irradiation. *Indian J. anim. Sci.*, **56**, 736–743

Jun, B.-D., Roberts, L.K., Cho, B.-H., Robertson, B. & Daynes, R.A. (1988) Parallel recovery of epidermal antigen-presenting cell activity and contact hypersensitivity responses in mice exposed to ultraviolet irradiation: the role of a prostaglandin-dependent mechanism. *J. invest. Dermatol.*, **90**, 311–316

Kaase, H., Zidowitz, B. & Berger, H. (1984) Spectral radiation measurements of UV-irradiation apparatus and experiments to estimate the risk of photocarcinogenesis (Ger.). *Lichtforschung*, **1**, 43–47

Kaidbey, K.H. (1988) Wavelength dependence for DNA synthesis inhibition in hairless mouse epidermis. *Photodermatology*, **5**, 65–70

Kaidbey, K.H., Agin, P.P., Sayre, R.M. & Kligman, A.M. (1979) Photoprotection by melanin—a comparison of black and Caucasian skin. *J. Am. Acad. Dermatol.*, **1**, 249–260

Kalthoff, K. & Jäckle, H. (1982) Photoreactivation of pyrimidine dimers generated by a photosensitized reaction in RNA of insect embryos (*Smittia* spec.). In: Hélène, C., Charlier, M., Montenay-Garestier, T. & Laustriat, G., eds, *Trends in Photobiology*, New York, Plenum, pp. 173–188

Kantor, G.J. (1985) Effects of sunlight on mammalian cells. *Photochem. Photobiol.*, **41**, 741–746

Kawada, A. (1986) UVB-Induced erythema, delayed tanning, and UVA-induced immediate tanning in Japanese skin. *Photodermatology*, **3**, 327–333

Keijzer, W., Mulder, M.P., Langeveld, J.C.M., Smit, E.M.E., Bos, J.L., Bootsma, D. & Hoeijmakers, J.H.J. (1989) Establishment and characterization of a melanoma cell line from a xeroderma pigmentosum patient: activation of N-*ras* at a potential pyrimidine dimer site. *Cancer Res.*, **49**, 1229–1235

Kelfkens, G., de Gruijl, F.R. & van der Leun, J.C. (1991a) Tumorigenesis by short-wave ultraviolet A: papillomas *versus* squamous cell carcinomas. *Carcinogenesis*, **12**, 1377–1382

Kelfkens, G., van Weelden, H., de Gruijl, F.R. & van der Leun, J.C. (1991b) Influence of dose rate in ultraviolet tumorigenesis. *J. Photochem. Photobiol. B Biol.*, **10**, 41–50

Keller, A.Z. (1970) Cellular types, survival, race, nativity, occupations, habits and associated diseases in the pathogenesis of lip cancers. *Am. J. Epidemiol.*, **91**, 486–499

Kelly, G.E., Meikle, W.D. & Sheil, A.G.R. (1987) Effects of immunosuppressive therapy on the induction of skin tumors by ultraviolet irradiation in hairless mice. *Transplantation*, **44**, 429–434

Kelly, G.E., Meikle, W.D. & Moore, D.E. (1989) Enhancement of UV-induced skin carcinogenesis by azathioprine: role of photochemical sensitisation. *Photochem. Photobiol.*, **49**, 59–65

Kennedy, A.R., Ritter, M.A. & Little, J.B. (1980) Fluorescent light induces malignant transformation in mouse embryo cell cultures. *Science*, **207**, 1209–1211

Keyse, S.M. & Tyrrell, R.M. (1989) Heme oxygenase is the major 32-kDa stress protein induced in human skin fibroblasts by UVA radiation, hydrogen peroxide and sodium arsenite. *Proc. natl Acad. Sci. USA*, **86**, 99–103

Keyse, S.M., Moss, S.H. & Davies, D.J.G. (1983) Action spectra for inactivation of normal and xeroderma pigmentosum human skin fibroblasts by ultraviolet radiations. *Photochem. Photobiol.*, **37**, 307–312

Keyse, S.M., Applegate, L.A., Tromvoukis, Y. & Tyrrell, R.M. (1990) Oxidant stress leads to transcriptional activation of the human heme oxygenase gene in cultured skin fibroblasts. *Mol. cell. Biol.*, **10**, 4967–4969

Khlat, M., Vail, A., Parkin, M. & Green, A. (1992) Mortality from melanoma in migrants to Australia: variation by age at arrival and duration of stay. *Am. J. Epidemiol.*, **135**, 1103–1113

Kim, T.-Y., Kripke, M.L. & Ullrich, S.E. (1990) Immunosuppression by factors released from UV-irradiated epidermal cells: selective effects on the generation of contact and delayed hypersensitivity after exposure to UVA or UVB radiation. *J. invest. Dermatol.*, **94**, 26–32

Kinlen, L.J., Sheil, A.G.R., Peto, J. & Doll, R. (1979) Collaborative United Kingdom–Australasian study of cancer in patients treated with immunosuppressive drugs. *Br. med. J.*, **ii**, 1461–1466

Kinsey, V.E. (1948) Spectral transmission of the eye to ultraviolet radiations. *Arch. Opthalmol.*, **39**, 508–513

Kirby-Smith, J.S. & Craig, D.L. (1957) The induction of chromosome aberrations in *Tradescantia* by ultraviolet radiation. *Genetics*, **42**, 176–187

Kirkpatrick, C.S., Lee, J.A.H. & White, E. (1990) Melanoma risk by age and socio-economic status. *Int. J. Cancer*, **46**, 1–4

Kjeldstad, B. & Johnsson, A. (1986) An action spectrum for blue and near ultraviolet inactivation of *Propionibacterium acnes*; with emphasis on a possible porphyrin photosensitization. *Photochem. Photobiol.*, **43**, 67–70

Klamen, D.L. & Tuveson, R.W. (1982) The effect of membrane fatty acid composition on the near-UV (300–400 nm) sensitivity of *Escherichia coli* K1060. *Photochem. Photobiol.*, **35**, 167–173

Klein, K.L. & Linnemann, C.C., Jr (1986) Induction of recurrent genital herpes simplex virus type 2 infection by ultraviolet light (Letter to the Editor). *Lancet*, **i**, 796–797

Klepp, O. & Magnus, K. (1979) Some environmental and bodily characteristics of melanoma patients. A case–control study. *Int. J. Cancer*, **23**, 482–486

Kligman, A.M. (1969) Early destructive effect of sunlight on human skin. *J. Am. med. Assoc.*, **210**, 2377–2380

Kligman, L.H. (1988) UVA enhances low dose UVB tumorigenesis (Abstract). *Photochem. Photobiol.*, **47**, 8S

Kligman, L.H. (1989) The ultraviolet-irradiated hairless mouse: a model for photoaging. *J. Am. Acad. Dermatol.*, **21**, 623–631

Kligman, L.H. & Kligman, A.M. (1981) Histogenesis and progression of ultraviolet light-induced tumors in hairless mice. *J. natl Cancer Inst.*, **67**, 1289–1297

Kligman, L.H. & Sayre, R.M. (1991) An action spectrum for ultraviolet induced elastosis in hairless mice: quantification of elastosis by image analysis. *Photochem. Photobiol.*, **53**, 237–242

Kligman, L.H., Akin, F.J. & Kligman, A.M. (1985) The contributions of UVA and UVB to connective tissue damage in hairless mice. *J. invest. Dermatol.*, **84**, 272–276

Kligman, L.H., Kaidbey, K.H., Hitchins, V.M. & Miller, S.A. (1987) Long wavelength (> 340 nm) ultraviolet-A induced skin damage in hairless mice is dose dependent. In: Passchier, W.F. & Bosnjakovic, B.F.M., eds, *Human Exposure to Ultraviolet Radiation: Risks and Regulations*, Amsterdam, Elsevier, pp. 77–81

Kligman, L.H., Crosby, M.J., Miller, S.A., Hitchins, V.M. & Beer, J.Z. (1990) Skin cancer induction in hairless mice by long-wavelength UVA radiation: a progress report (Abstract). *Photochem. Photobiol.*, **51**, 18S–19S

Kligman, L.H., Crosby, M.J., Miller, S.A., Hitchins, V.M. & Beer, J.Z. (1992) Carcinogenesis by long wavelength UVA and failure of reciprocity. In: Urbach, F., ed., *Biological Responses to Ultraviolet A Radiation*, Overland Park, KS, Valdenmar, pp. 301–308

Knees-Matzen, S., Roser, M., Reimers, U., Ehlert, U., Weichenthal, M., Breitbart, E.W. & Rüdiger, H.W. (1991) Increased UV-induced sister-chromatid exchange in cultured fibroblasts of first-degree relatives of melanoma patients. *Cancer Genet. Cytogenet.*, **53**, 265–270

Kocsard, E. & Ofner, F. (1964) Contact eczematous sensitization and sensitivity of the solar elastotic skin. *Aust. J. Dermatol.*, **7**, 203–205

Kolari, P.J., Lauharanta, J. & Hoikkala, M. (1986) Midsummer solar UV-radiation in Finland compared with the UV-radiation from phototherapeutic devices measured by different techniques. *Photodermatology*, **3**, 340–345

Kopecky, K.E., Pugh, G.W., Jr, Hughes, D.E., Booth, G.D. & Cheville, N.F. (1979) Biological effect of ultraviolet radiation on cattle: bovine ocular squamous cell carcinoma. *Am. J. vet. Res.*, **40**, 1783–1788

Kopf, A.W., Lindsay, A.C., Rogers, G.S., Friedman, R.J., Rigel, D.S. & Levenstein, M. (1985) Relationship of nevocytic nevi to sun exposure in dysplastic nevus syndrome. *J. Am. Acad. Dermatol.*, **12**, 656–662

Kostkowski, H.J., Saunders, R., Ward, J., Popenoe, C. & Green, A. (1982) *Measurement of Solar Terrestrial Spectral Irradiance in the Ozone Cut-off Region, Self-study Manual on Optical Radiation Measurements*, Part III, *Applications* (NBS Technical Note 910-5; PB 83-184382), Washington DC, National Bureau of Standards

Koval, T.M. (1987) Photoreactivation of UV damage in cultured *Drosophila* cells. *Experientia*, **43**, 445–446

Kraemer, K.H. (1980) Xeroderma pigmentosum. A prototype disease of environmental–genetic interaction. *Arch. Dermatol.*, **116**, 541–542

Kraemer, K.H. & Greene, M.H. (1985) Dysplastic nevus syndrome. Familial and sporadic precursors of cutaneous melanoma. *Dermatol. Clin.*, **3**, 225–237

Kraemer, K.H. & Slor, H. (1984) Xeroderma pigmentosum. *Clin. Dermatol.*, **2**, 33–69

Kraemer, K.H. & Weinstein, G.D. (1977) Decreased thymidine incorporation in circulating leukocytes after treatment of psoriasis with psoralen and long-wave ultraviolet light. *J. invest. Dermatol.*, **69**, 211–214

Kraemer, K.H., Lee, M.M. & Scotto, J. (1987) Xeroderma pigmentosum. Cutaneous, ocular, and neurologic abnormalities in 830 published cases. *Arch. Dermatol.*, **123**, 241–250

Kraemer, K.H., Seetharam, S., Protić-Sabljić, M., Brash, D.E., Bredberg, A. & Seidman, M.M. (1988) Defective DNA repair and mutagenesis by dimer and non-dimer photoproducts in xeroderma pigmentosum measured with plasmid vectors. In: Friedberg, E.C. & Hanawalt, P.C., eds, *Mechanisms and Consequences of DNA Damage Processing*, New York, Alan R. Liss, pp. 325–335

Kramer, G.F. & Ames, B.N. (1987) Oxidative mechanisms of toxicity of low-intensity near-UV light in *Salmonella typhimurium*. *J. Bacteriol.*, **169**, 2259–2266

Krepinsky, A.B., Rainbow, A.J. & Heddle, J.A. (1980) Studies on the ultraviolet light sensitivity of Bloom's syndrome fibroblasts. *Mutat. Res.*, **69**, 357–368

Kricker, A., English, D.R., Randell, P.L., Heenan, P.J., Clay, C.D., Delaney, T.A. & Armstrong, B.K. (1990) Skin cancer in Geraldton, Western Australia: a survey of incidence and prevalence. *Med. J. Aust.*, **152**, 399–407

Kricker, A., Armstrong, B.K., English, D.R. & Heenan, P.J. (1991a) Pigmentary and cutaneous risk factors for non-melanocytic skin cancer—a case–control study. *Int. J. Cancer*, **48**, 650–662

Kricker, A., Armstrong, B.K., English, D., Heenan, P.J. & Randell, P.L. (1991b) A case–control study of non-melanocytic skin cancer and sun exposure in Western Australia (Abstract No. III, P2). *Cancer Res. clin. Oncol.*, **117** (Suppl. II), S75

Kripke, M.L. (1974) Antigenicity of murine skin tumors induced by ultraviolet light. *J. natl Cancer Inst.*, **53**, 1333–1336

Kripke, M.L. (1977) Latency, histology, and antigenicity of tumors induced by ultraviolet light in three inbred mouse strains. *Cancer Res.*, **37**, 1395–1400

Kripke, M.L. & Fisher, M.S. (1976) Immunologic parameters of ultraviolet carcinogenesis. *J. natl Cancer Inst.*, **57**, 211–215

Kripke, M.L. & McClendon, E. (1986) Studies on the role of antigen-presenting cells in the systemic suppression of contact hypersensitivity by UVB radiation. *J. Immunol.*, **137**, 443–447

Kripke, M.L. & Morison, W.L. (1985) Modulation of immune function by UV radiation. *J. invest. Dermatol.*, **85**, 625–665

Kripke, M.L. & Morison, W.L. (1986a) Studies on the mechanism of systemic suppression of contact hypersensitivity by ultraviolet B radiation. *Photodermatology*, **3**, 4–14

Kripke, M.L. & Morison, W.L. (1986b) Studies on the mechanism of systemic suppression of contact hypersensitivity by UVB radiation. II. Differences in the suppression of delayed and contact hypersensitivity in mice. *J. invest. Dermatol.*, **86**, 543–549

Kripke, M.L. & Sass, E.R., eds (1978) International conference on ultraviolet carcinogenesis. *Natl Cancer Inst. Monogr.*, **50**

Kripke, M.L., Lofgreen, J.S., Beard, J., Jessup, J.M. & Fisher, M.S. (1977) In vivo immune responses of mice during carcinogenesis by ultraviolet irradiation. *J. natl Cancer Inst.*, **59**, 1227–1230

Kromberg, J.G.R., Castle, D., Zwane, E.M. & Jenkins, T. (1989) Albinism and skin cancer in southern Africa. *Clin. Genet.*, **36**, 43–52

Krutmann, J. & Elmets, C.A. (1988) Recent studies on mechanisms in photoimmunology. *Photochem. Photobiol.*, **48**, 787–798

Kubitschek, H.E. (1967) Mutagenesis of near-visible light. *Science*, **155**, 1545–1546

Kumar, S., Joshi, P.C., Sharma, N.D., Bose, S.N., Davies, R.J.H., Takeda, N. & McCloskey, J.A. (1991) Adenine photodimerization in deoxyadenylate sequences: elucidation of the mechanism through structural studies of a major d(ApA) photoproduct. *Nucleic Acids Res.*, **19**, 2841–2847

Kunkel, T.A. (1984) Mutational specificity of depurination. *Proc. natl Acad. Sci. USA*, **81**, 1494–1498

Kupper, T.S., Chua, A.O., Flood, P., McGuire, J. & Gubler, U. (1987) Interleukin 1 gene expression in cultured human keratinocytes is augmented by ultraviolet irradiation. *J. clin. Invest.*, **80**, 430–436

Kurihara, T., Tatsumi, K., Takahashi, H. & Inoue, M. (1987) Sister-chromatid exchanges induced by ultraviolet light in Bloom's syndrome fibroblasts. *Mutat. Res.*, **183**, 197–202

Lai, L.-W. & Rosenstein, B.S. (1990) Induction of DNA strand breaks and DNA–protein cross-links in normal human skin fibroblasts following exposure to 254 nm UV radiation. *J. Photochem. Photobiol. B Biol.*, **6**, 395–404

Lancaster, H.O. (1956) Some geographical aspects of the mortality from melanoma in Europeans. *Med. J. Aust.*, **1**, 1082–1087

Lancaster, H.O. & Nelson, J. (1957) Sunlight as a cause of melanoma: a clinical survey. *Med. J. Aust.*, **1**, 452–456

Larcom, L.L., Morris, T.E. & Smith, M.E. (1991) Tanning salon exposure suppression of DNA repair capacity and mitogen-induced DNA synthesis. *Photochem. Photobiol.*, **53**, 511–516

Larkö, O. & Diffey, B.L. (1983) Natural UV-B radiation received by people with outdoor, indoor and mixed occupations and UV-B treatment of psoriasis. *Clin. exp. Dermatol.*, **8**, 279–285

Larkö, O. & Diffey, B.L. (1986) Occupational exposure to ultraviolet radiation in dermatology departments. *Br. J. Dermatol.*, **114**, 479–484

Lau, H., Reemtsma, K. & Hardy, M.A. (1983) Pancreatic islet allograft prolongation by donor-specific blood transfusions treated with ultraviolet irradiation. *Science*, **221**, 754–756

Lau, H., Reemtsma, K. & Hardy, M.A. (1984) Prolongation of rat islet allograft survival by direct ultraviolet irradiation of the graft. *Science*, **223**, 607–609

Leach, J.F., McLeod, V.E., Pingstone, A.R., Davis, A. & Deane, G.H.W. (1978) Measurement of the ultraviolet doses received by office workers. *Clin. exp. Dermatol.*, **3**, 77–79

LeClerc, J.E., Borden, A. & Lawrence C.W. (1991) The thymine–thymine pyrimidine–pyrimidone (6-4) ultraviolet light photoproduct is highly mutagenic and specifically induces 3′-thymine-to-cytosine transitions in *Escherichia coli*. *Proc. natl Acad. Sci. USA*, **88**, 9685–9689

Lee, J.A.H. (1982) Melanoma and exposure to sunlight. *Epidemiol. Rev.*, **4**, 110–136

Lee, J.A.H. & Storer, B.E. (1980) Excess of malignant melanomas in women in the British Isles. *Lancet*, **ii**, 1337–1339

Lee, J.A.H. & Strickland, D. (1980) Malignant melanoma: social status and outdoor work. *Br. J. Cancer*, **41**, 757–763

Lee, G.S.-F., Savage, E.A., Ritzel, R.G. & von Borstel, R.C. (1988) The base-alteration spectrum of spontaneous and ultraviolet radiation-induced forward mutations in the *URA3* locus of *Saccharomyces cerevisiae*. *Mol. gen. Genet.*, **214**, 396–404

Lehmann, A.R., Arlett, C.F., Broughton, B.C., Harcourt, S.A., Steingrimsdottir, H., Stefanini, M., Taylor, A.M.R., Natarajan, A.T., Green, S., King, M.D., MacKie, R.M., Stephenson, J.B.P. & Tolmie, J.L. (1988) Trichothiodystrophy, a human DNA repair disorder with heterogeneity in the cellular response to ultraviolet light. *Cancer Res.*, **48**, 6090–6096

Lerman, S. (1980) *Radiant Energy and the Eye*, New York, MacMillan

Lerman, S. (1988) Ocular phototoxicity. *New Engl. J. Med.*, **319**, 1475–1477

Letvin, N.L., Greene, M.I., Benacerraf, B. & Germain, R.N. (1980a) Immunologic effects of whole-body ultraviolet irradiation: selective defect in splenic adherent cell function *in vitro*. *Proc. natl Acad. Sci. USA*, **77**, 2881–2885

Letvin, N.L., Nepom, J.T., Greene, M.I., Benacerraf, B. & Germain, R.N. (1980b) Loss of Ia-bearing splenic adherent cells after whole body ultraviolet irradiation. *J. Immunol.*, **125**, 2550–2554

van der Leun, J.C. (1984) Yearly review: UV-carcinogenesis. *Photochem. Photobiol.*, **39**, 861–868

van der Leun, J.C. (1987a) Principles of risk reduction and protection. In: Passchier, W.F. & Bosnjakovic, B.F.M., eds, *Human Exposure to Ultraviolet Radiation: Risks and Regulations*, Amsterdam, Elsevier, pp. 293–303

van der Leun, J.C. (1987b) Interactions of different wavelengths in effects of UV radiation on skin. *Photodermatology*, **4**, 257–264 van der Leun, J.C. (1992) Interactions of UVA and UVB in photodermatology: what was photoaugmentation? In: Urbach, F., ed., *The Biological Responses to Ultraviolet A Radiation*, Overland Park, KS, Valdenmar, pp. 309-319

van der Leun, J.C. & van Weelden, H. (1986) UVB phototherapy: principles, radiation sources, regimens. *Curr. Probl. Dermatol.*, **15**, 39-51

Levi, F., La Vecchia, C., Te, V.-C. & Mezzanotte, G. (1988) Descriptive epidemiology of skin cancer in the Swiss canton of Vaud. *Int. J. Cancer*, **42**, 811-816

Lew, R.A., Sober, A.J., Cook, N., Marvell, R. & Fitzpatrick, T.B. (1983) Sun exposure habits in patients with cutaneous melanoma: a case control study. *J. Dermatol. surg. Oncol.*, **9**, 981-986

Ley, R.D. (1985) Photoreactivation of UV-Induced pyrimidine dimers and erythema in the marsupial *Monodelphis domestica*. *Proc. natl Acad. Sci. USA*, **82**, 2409-2411

Ley, R.D. & Applegate, L.A. (1989) Photodermatological studies with the marsupial *Monodelphis domestica*. In: Maibach, H.I. & Lowe, N.J., eds, *Models in Dermatology*, Vol. 4, Basel, Karger, pp. 265-275

Ley, R.D., Sedita, B.A. & Boye, E. (1978) DNA polymerase I-mediated repair of 365 nm-induced single-strand breaks in the DNA of *Escherichia coli*. *Photochem. Photobiol.*, **27**, 323-327

Ley, R.D., Peak, M.J. & Lyon, L.L. (1983) Induction of pyrimidine dimers in epidermal DNA of hairless mice by UVB: an action spectrum. *J. invest. Dermatol.*, **80**, 188-191

Ley, R.D., Applegate, L.A., Stuart, T.D. & Fry, R.J.M. (1987) UV radiation-induced skin tumors in *Monodelphis domestica*. *Photodermatology*, **4**, 144-147

Ley, R.D., Applegate, L.A. & Freeman, S.E. (1988) Photorepair of ultraviolet radiation-induced pyrimidine dimers in corneal DNA. *Mutat. Res.*, **194**, 49-55

Ley, R.D., Applegate, L.A., Padilla, R.S. & Stuart, T.D. (1989) Ultraviolet radiation-induced malignant melanoma in *Monodelphis domestica*. *Photochem. Photobiol.*, **50**, 1-5

Ley, R.D., Applegate, L.A., Fry, R.J.M. & Sanchez, A.B. (1991) Photoreactivation of ultraviolet radiation-induced skin and eye tumors of *Monodelphis domestica*. *Cancer Res.*, **51**, 6539-6542

Lightfoote, M.M., Zmundzka, B.Z., Olvey, K.M., Miller, S.A. & Beer, J.Z. (1992) Effects of UVB radiation on human immunodeficiency virus. In: Houck, M. & Kligman, A., eds, *Biological Effects of Light* (in press)

Lill, P.H. (1983) Latent period and antigenicity of murine tumors induced in C3H mice by shortwavelength ultraviolet radiation. *J. invest. Dermatol.*, **81**, 342-346

Lin, C.S., Goldthwait, D.A. & Samols, D. (1990) Induction of transcription from the long terminal repeat of Moloney murine sarcoma provirus by UV-irradiation, x-irradiation, and phorbol ester. *Proc. natl Acad. Sci. USA*, **87**, 36-40

Lindahl-Kiessling, K. & Säfwenberg, J. (1971) Inability of UV-irradiated lymphocytes to stimulate allogenic cells in mixed lymphocyte cultures. *Int. Arch. Allergy*, **41**, 670-678

Lindqvist, C. (1979) Risk factors in lip cancer: a questionnaire survey. *Am. J. Epidemiol.*, **109**, 521-530

Lippincott, S.W. & Blum, H.F. (1943) Neoplasms and other lesions of the eye induced by ultraviolet radiation in strain A mice. *J. natl Cancer Inst.*, **3**, 545-554

Lippke, J.A., Gordon, L.K., Brash, D.E. & Haseltine, W.A. (1981) Distribution of UV light-induced damage in a defined sequence of human DNA: detection of alkaline-sensitive lesions at pyrimidine nucleoside-cytidine sequences. *Proc. natl Acad. Sci. USA*, **78**, 3388-3392

Lischko, A.M., Seddon, J.M., Gragoudas, E.S., Egan, K.M. & Glynn, R.J. (1989) Evaluation of prior primary malignancy as a determinant of uveal melanoma. A case-control study. *Ophthalmology*, **96**, 1716-1721

Livingston, W. (1983) Landscape as viewed in the 320-nm ultraviolet. *J. opt. Soc. Am.*, **73**, 1653-1657

Logani, M.K., Sambuco, C.P., Forbes, P.D. & Davies, R.E. (1984) Skin-tumour promoting activity of methyl ethyl ketone peroxide—a potent lipid-peroxidizing agent. *Food chem. Toxicol.*, **22**, 879–882

Luande, J., Henschke, C.I. & Mohammed, N. (1985) The Tanzanian human albino skin. *Cancer*, **55**, 1823–1828

van der Lubbe, J.L.M., Rosdorff, H.J.M., Bos, J.L. & van der Eb, A.J. (1988) Activation of N-ras induced by ultraviolet irradiation *in vitro*. *Oncogene Res.*, **3**, 9–20

Lücke-Huhle, C., Mai, S. & Herrlich, P. (1989) UV-Induced early-domain binding factor as the limiting component of simian virus 40 DNA amplification in rodent cells. *Mol. cell. Biol.*, **9**, 4812–4818

Lundgren, K. & Wulf, H.C. (1988) Cytotoxicity and genotoxicity of UVA irradiation in Chinese hamster ovary cells measured by specific locus mutations, sister chromatid exchanges and chromosome aberrations. *Photochem. Photobiol.*, **47**, 559–563

Lunnon, R.J. (1984) Ultraviolet photography. In: Williams, R., ed., *Medical Photography Study Guide*, Lancaster, MTP Press, pp. 225–233

Lynch, D.H., Gurish, M.F. & Haynes, R.A. (1983) The effects of high-dose UV exposure on murine Langerhans cell function at exposed and unexposed sites as assessed using in vivo and in vitro assays. *J. invest. Dermatol.*, **81**, 336–341

Lynge, E. & Thygesen, L. (1990) Occupational cancer in Denmark. Cancer incidence in the 1970 census population. *Scand. J. Work Environ. Health*, **16** (Suppl. 2), 1–35

Machta, L., Cotton, G., Hass, W. & Komhyr, W. (1975) *CIAP Measurements of Solar Ultraviolet Radiation* (Final Report, Interagency Agreement DOT-A5-20082), Washington DC, US Department of Transportation

Mack, T.M. & Floderus, B. (1991) Malignant melanoma risk by nativity, place of residence at diagnosis, and age at migration. *Cancer Causes Control*, **2**, 401–411

MacKie, R.M. & Aitchison, T. (1982) Severe sunburn and subsequent risk of primary cutaneous malignant melanoma in Scotland. *Br. J. Cancer*, **40**, 955–960

MacKie, R.M., Freudenberger, T. & Aitchison, T.C. (1989) Personal risk-factor chart for cutaneous melanoma. *Lancet*, **ii**, 487–490

MacLaughlin, J. & Holick, M.F. (1985) Aging decreases the capacity of human skin to produce vitamin D_3. *J. clin. Invest.*, **76**, 1536–1538

Madewell, B.R., Conroy, J.D. & Hodgkins, E.M. (1981) Sunlight-skin cancer association in the dog: a report of three cases. *J. cutaneous Pathol.*, **8**, 434–443

Magee, M.J., Kripke, M.L. & Ullrich, S.E. (1989a) Suppression of the elicitation of the immune response to alloantigen by ultraviolet radiation. *Transplantation*, **47**, 1008–1013

Magee, M.J., Kripke, M.L. & Ullrich, S.E. (1989b) Inhibition of the immune response to alloantigen in the rat by exposure to ultraviolet radiation. *Photochem. Photobiol.*, **50**, 193–199

Magnus, K. (1973) Incidence of malignant melanoma of the skin in Norway, 1955–1970. Variations in time and space and solar radiation. *Cancer*, **32**, 1275–1286

Magnus, K. (1991) The Nordic profile of skin cancer incidence. A comparative epidemiological study of the three main types of skin cancer. *Int. J. Cancer*, **47**, 12–19

Maher, V.M., Dorney, D.J., Mendrala, A.L., Konze-Thomas, B. & McCormick, J.J. (1979) DNA excision-repair processes in human cells can eliminate the cytotoxic and mutagenic consequences of ultraviolet irradiation. *Mutat. Res.*, **62**, 311–323

Maher, V.M., Rowan, L.A., Silinskas, K.C., Kateley, S.A. & McCormick, J.J. (1982) Frequency of UV-induced neoplastic transformation of diploid human fibroblasts is higher in xeroderma pigmentosum cells than in normal cells. *Proc. natl Acad. Sci. USA*, **79**, 2613–2617

Maltzman, W. & Czyzyk, L. (1984) UV-Irradiation stimulates levels of p53 cellular tumor antigen in nontransformed mouse cells. *Mol. cell. Biol.*, **4**, 1689–1694

Mandal, T.K. & Chatterjee, S.N. (1980) Ultraviolet- and sunlight-induced lipid peroxidation in liposomal membrane. *Radiat. Res.*, **83**, 290–302

Mariutti, G. & Matzeu, M. (1987) Measurement of ultraviolet radiation in welding processes and hazard evaluation. In: Passchier, W.F. & Bosnjakovic, B.F.M., eds, *Human Exposure to Ultraviolet Radiation: Risks and Regulations*, Amsterdam, Elsevier, pp. 387–390

Marks, R. & Selwood, T.S. (1985) Solar keratoses. The association with erythemal ultraviolet radiation in Australia. *Cancer*, **56**, 2332–2336

Marks, R., Ponsford, M.W., Selwood, T.S., Goodman, G. & Mason, G. (1983) Non-melanocytic skin cancer and solar keratoses in Victoria. *Med. J. Aust.*, **2**, 619–622

Marks, R., Rennie, G. & Selwood, T.S. (1988) Malignant transformation of solar keratoses to squamous cell carcinoma. *Lancet*, **i**, 795–797

Marks, R., Jolley, D., Dorevitch, A.P. & Selwood, T.S. (1989) The incidence of non-melanocytic skin cancers in an Australian population: results of a five-year prospective study. *Med. J. Aust.*, **150**, 475–478

Marshall, R.R. & Scott, D. (1976) The relationship between chromosome damage and cell killing in UV-irradiated normal and xeroderma pigmentosum cells. *Mutat. Res.*, **36**, 397–400

Marshall, R.R., Arlett, C.F., Harcourt, S.A. & Broughton, B.A. (1980) Increased sensitivity of cell strains from Cockayne's syndrome to sister-chromatid exchange induction and cell killing by UV light. *Mutat. Res.*, **69**, 107–112

Matsumoto, K., Sugiyama, M. & Ogura, R. (1991) Non-dimer DNA damage in Chinese hamster V-79 cells exposed to ultraviolet-B light. *Photochem. Photobiol.*, **54**, 389–392

Matsunaga, T., Hieda, K. & Nikaido, O. (1991) Wavelength dependent formation of thymine dimers and (6-4)photoproducts in DNA by monochromatic ultraviolet light ranging from 150 to 365 nm. *Photochem. Photobiol.*, **54**, 403–410

Matsuoka, L.Y., Ide, L., Wortsman, J., MacLaughlin, J.A. & Holick, M.F. (1987) Sunscreens suppress cutaneous vitamin D_3 synthesis. *J. clin. Endocrinol. Metab.*, **64**, 1165–1168

Mayne, L.V., Mullenders, L.H.F. & van Zeeland, A.A. (1988) Cockayne's syndrome: a UV sensitive disorder with a defect in the repair of transcribing DNA but normal overall excision repair. In: Friedberg, E.C. & Hanawalt, P.C., eds, *Mechanisms and Consequences of DNA Damage Processing*, New York, Alan R. Liss, pp. 349–353

McCormick, J.P., Fisher, J.R., Pachlatko, J.P. & Eisenstark, A. (1976) Characterisation of a cell-lethal product from the photooxidation of tryptophan: hydrogen peroxide. *Science*, **191**, 468–669

McCredie, M. & Coates, M.S. (1989) *Cancer Incidence in Migrants to New South Wales. 1972 to 1984*, New South Wales Central Cancer Registry, Woolloomooloo, NSW, New South Wales Cancer Council, pp. 22-23, 62–63

McGovern, V.J., Shaw, H.M., Milton, G.W. & Farago, G.A. (1980) Is malignant melanoma arising in a Hutchinson's melanotic freckle a separate disease entity? *Histopathology*, **4**, 235–242

McGrath, H., Wilson, W.A. & Scopelitis, E. (1986) Acute effects of low-fluence ultraviolet light on human T-lymphocyte subsets. *Photochem. Photobiol.*, **43**, 627–631

McGregor, W.G., Chen, R.-H., Lukash, L., Maher, V.M. & McCormick, J.J. (1991) Cell cycle-dependent strand bias for UV-induced mutations in the transcribed strand of excision repair-proficient human fibroblasts but not in repair-deficient cells. *Mol. cell. Biol.*, **11**, 1927–1934

McKinlay, A.F. & Diffey, B.L. (1987) A reference action spectrum for ultraviolet induced erythema in human skin. *CIE (Commission Internationale de l'Eclairage) J.*, **6**, 17–22

McKinlay, A.F. & Whillock, M.J. (1987) Measurement of ultra-violet radiation from fluorescent lamps used for general lighting and other purposes in the UK. In: Passchier, W.F. & Bosnjakovic, B.F.M., eds, *Human Exposure to Ultraviolet Radiation: Risks and Regulations*, Amsterdam, Elsevier, pp. 253–258

McKinlay, A.F., Whillock, M.J. & Meulemans, C.C.E. (1989) *Ultraviolet Radiation and Blue-light Emissions from Spotlights Incorporating Tungsten Halogen Lamps* (Report NRPB-R228), Didcot, National Radiological Protection Board

McLennan, A.G. (1987) The repair of ultraviolet light-induced DNA damage in plant cells. *Mutat. Res.*, **181**, 1–7

Mecherikunnel, A.T. & Richmond, J.C. (1980) *Spectral Distribution of Solar Radiation* (NASA Technical Memorandum 82021), Greenbelt, MD, Goddard Space Flight Center

Melski, J.W., Tanenbaum, L., Parrish, J.A., Fitzpatrick, T.B. & Bleich, H.L. (1977) Oral methoxsalen photochemotherapy for the treatment of psoriasis: a cooperative clinical trial. *J. invest. Dermatol.*, **68**, 328–335

Menzies, S.W., Greenoak, G.E., Reeve, V.E. & Gallagher, C.H. (1991) Ultraviolet radiation-induced murine tumors produced in the absence of ultraviolet radiation-induced systemic tumor immunosuppression. *Cancer Res.*, **51**, 2773–2779

Miguel, A.G. & Tyrrell, R.M. (1983) Induction of oxygen-dependent lethal damage by monochromatic UVB (313 nm) radiation: strand breakage, repair and cell death. *Carcinogenesis*, **4**, 375–380

Milham, S., Jr (1983) *Occupational Mortality in Washington State 1950–1979* (DHSS (NIOSH) Publ. No. 83-116), Cincinnati, OH, National Institute for Occupational Safety and Health

Miller, J.H. (1985) Mutagenic specificity of ultraviolet light. *J. mol. Biol.*, **182**, 45–68

Miskin, R. & Ben-Ishai, R. (1981) Induction of plasminogen activator by UV light in normal and xeroderma pigmentosum fibroblasts. *Proc. natl Acad. Sci. USA*, **78**, 6236–6240

Mitchell, D.L. (1988) The relative cytotoxicity of (6-4)photoproducts and cyclobutane dimers in mammalian cells. *Photochem. Photobiol.*, **48**, 51–57

Mitchell, D.L., Haipek, C.A. & Clarkson, J.M. (1985) (6-4)Photoproducts are removed from the DNA of UV-irradiated mammalian cells more efficiently than cyclobutane pyrimidine dimers. *Mutat. Res.*, **143**, 109–112

Mitchell, D.L., Jen, J. & Cleaver, J.E. (1991) Relative induction of cyclobutane dimers and cytosine photohydrates in DNA irradiated *in vitro* and *in vivo* with ultraviolet-C and ultraviolet-B light. *Photochem. Photobiol.*, **54**, 741–746

Moan, J., Dahlback, A., Henriksen, T. & Magnus, K. (1989) Biological amplification factor for sunlight-induced nonmelanoma skin cancer at high latitudes. *Cancer Res.*, **49**, 5207–5212

Molina, M.J. & Rowland, F.S. (1974) Stratospheric sink for chlorofluoromethanes: chlorine atom-catalyzed destruction of ozone. *Nature*, **249**, 810–812

Mondal, S. & Heidelberger, C. (1976) Transformation of C3H/10T1/2 CL8 mouse embryo fibroblasts by ultraviolet irradiation and a phorbol skin. *Nature*, **260**, 710–711

Morgun, V.V., Logvinenko, V.F., Dvernyakov, V.S. & Timoshenko, V.M. (1988) Mutagenic activity of pulsed concentrated sunlight. *Sov. Genet.*, **24**, 189–193

Morison, W.L. (1983a) *Phototherapy and Photochemotherapy of Skin Disease*, New York, Praeger

Morison, W.L. (1983b) Solar urticaria. In: Parrish, J.A., Kripke, M.L. & Morison, W.L., eds, *Photoimmunology*, New York, Plenum, pp. 215-226

Morison, W.L. (1983c) Autoimmune disease. In: Parrish, J.A., Kripke, M.L. & Morison, W.L., eds, *Photoimmunology*, New York, Plenum, pp. 255-266

Morison, W.L. (1986) The effects of UVA radiation on immune function. In: Urbach, F. & Grange, R.W., eds, *The Biological Effects of UVA Radiation*, New York, Praeger, pp. 202-209

Morison, W.L. (1989) Effects of ultraviolet radiation on the immune system in humans. *Photochem. Photobiol.*, 50, 515-524

Morison, W.L. & Kelley, S.P. (1985) Sunlight suppressing rejection of 280 to 320 nm UV-radiation-induced skin tumors in mice. *J. natl Cancer Inst.*, 74, 525-527

Morison, W.L. & Kochevar, I.E. (1983) Photoallergy. In: Parrish, J.A., Kripke, M.L. & Morison, W.L., eds, *Photoimmunology*, New York, Plenum, pp. 227-253

Morison, W.L. & Kripke, M.L. (1984) Systemic suppression of contact hypersensitivity by ultraviolet B radiation or methoxsalen/ultraviolet A radiation in the guinea pig. *Cell. Immunol.*, 85, 270-277

Morison, W.L., Parrish, J.A., Bloch, K.J. & Krugler, J.I. (1979a) In vivo effect of UVB on lymphocyte function. *Br. J. Dermatol.*, 101, 513-519

Morison, W.L., Parrish, J.A., Anderson, R.R. & Bloch, K.J. (1979b) Sensitivity of mononuclear cells to UV radiation. *Photochem. Photobiol.*, 29, 1045-1047

Morison, W.L., Bucana, C. & Kripke, M.L. (1984) Systemic suppression of contact hypersensitivity by UVB radiation is unrelated to the UVB-induced alterations in the morphology and number of Langerhans cells. *Immunology*, 52, 299-306

Morison, W.L., Pike, R.A. & Kripke, M.L. (1985) Effect of sunlight and its component wavebands on contact hypersensitivity in mice and guinea pigs. *Photodermatology*, 2, 195-204

Morison, W.L., Jerdan, M.S., Hoover, T.L. & Farmer, E.R. (1986) UV radiation-induced tumors in haired mice: identification as squamous cell carcinomas. *J. natl Cancer Inst.*, 77, 1155-1162

Mørk, N.-J. & Austad, J. (1982) Short-wave ultraviolet light (UVB) treatment of allergic contact dermatitis of the hands. *Acta dermatovenereol.*, 63, 87-89

Morrison, H. (1985) Photochemistry and photobiology of urocanic acid. *Photodermatology*, 2, 158-165

Moseley, H. (1988) *Non-ionising Radiation: Microwaves, Ultraviolet and Laser Radiation* (Medical Physics Handbooks 18), Bristol, Adam Hilger, pp. 110-154

Moss, S.H. & Smith, K.C. (1981) Membrane damage can be a significant factor in the inactivation of *Escherichia coli* by near-ultraviolet radiation. *Photochem. Photobiol.*, 33, 203-210

Muir, C., Waterhouse, J., Mack, T., Powell, J. & Whelan, S., eds (1987) *Cancer Incidence in Five Continents, Vol. V* (IARC Scientific Publications No. 88), Lyon, IARC

Munakata, N. (1989) Genotoxic action of sunlight upon *Bacillus subtilis* spores: monitoring studies at Tokyo, Japan. *J. Radiat. Res.*, 30, 338-351

Murthy, P.B., Rahiman, M.A. & Tulpule, P.G. (1982) UV-Induced chromosome aberrations, sister-chromatid exchanges and cell survival in cultured lymphocytes from malnourished children. *Pediatr. Res.*, 16, 663-664

Musk, P., Campbell, R., Staples, J., Moss, D.J. & Parsons, P.G. (1989) Solar and UVC-induced mutation in human cells and inhibition by deoxynucleosides. *Mutat. Res.*, 227, 25-30

Mutzhas, M.F. (1986) UVA-Emitting light sources. In: Urbach, F. & Grange, R.W., eds, *The Biological Effects of UVA Radiation*, New York, Praeger, pp. 10-23

Myhr, B.C., Turnbull, D. & DiPaolo, J.A. (1979) Ultraviolet mutagenesis of normal and xeroderma pigmentosum variant human fibroblasts. *Mutat. Res.*, 62, 341-353

Natarajan, A.T., van Zeeland, A.A., Verdegaal-Immerzeel, E.A.M. & Filon, A.R. (1980) Studies on the influence of photoreactivation on the frequencies of UV-induced chromosomal aberrations, sister-chromatid exchanges and pyrimidine dimers in chicken embryonic fibroblasts. *Mutat. Res.*, **69**, 307-317

Nathanson, R.B., Forbes, P.D. & Urbach, F. (1976) Modification of photocarcinogenesis by two immunosuppressive agents. *Cancer Lett.*, **1**, 243-247

Nelson, E.W., Eichwald, E.J. & Shelby, J. (1987) Increased ultraviolet radiation-induced skin cancers in cyclosporine-treated mice. *Transplant. Proc.*, **19**, 526-527

Nikula, K.J., Benjamin, S.A., Angleton, G.M., Saunders, W.J. & Lee, A.C. (1992) Ultraviolet radiation, solar dermatosis, and cutaneous neoplasia in beagle dogs. *Radiat. Res.*, **129**, 11-18

Nishi, Y., Hasegawa, M.M., Taketomi, M., Ohkawa, Y. & Inui, N. (1984) Comparison of 6-thioguanine-resistant mutation and sister chromatid exchanges in Chinese hamster V79 cells with forty chemical and physical agents. *Cancer Res.*, **44**, 3270-3279

Noonan, F.P. & De Fabo, E.C. (1985) Immune suppression by ultraviolet radiation and its role in ultraviolet radiation induced carcinogenesis in mice. *Aust. J. Dermatol.*, **26**, 4-8

Noonan, F.P., De Fabo, E.C. & Kripke, M.L. (1981a) Suppression of contact hypersensitivity by UV radiation and its relationship to UV-induced suppression of tumor immunity. *Photochem. Photobiol.*, **34**, 683-689

Noonan, F.P., Kripke, M.L., Pedersen, G.M. & Greene, M.I. (1981b) Suppression of contact hypersensitivity in mice by ultraviolet irradiation is associated with defective antigen presentation. *Immunology*, **43**, 527-533

Noonan, F.P., Bucana, C., Sauder, D.N. & De Fabo, E.C. (1984) Mechanism of systemic immune suppression by UV irradiation *in vivo*. II. The UV effects on number and morphology of epidermal Langerhans cells and the UV-induced suppression of contact hypersensitivity have different wavelength dependencies. *J. Immunol.*, **132**, 2408-2416

Noonan, F.P., De Fabo, E.C. & Morrison, H. (1988) *cis*-Urocanic acid, a product formed by ultraviolet B irradiation of the skin, initiates an antigen presentation defect in splenic dendritic cells *in vivo*. *J. invest. Dermatol.*, **90**, 92-99

Norbury, K.C. & Kripke, M.L. (1978) Ultraviolet carcinogenesis in T-cell-depleted mice. *J. natl Cancer Inst.*, **61**, 917-921

Norimura, T., Maher, V.M. & McCormick, J.J. (1990) Cytotoxic and mutagenic effect of UV, ethylnitrosourea and ($\pm$)-7β,8α-dihydroxy-9α,10α-epoxy-7,8,9,10-tetrahydrobenzo[*a*]pyrene in diploid human T lymphocytes in culture: comparison with fibroblasts. *Mutagenesis*, **5**, 447-451

Norval, M., McIntyre, C.R., Simpson, T.J., Howie, S.E.M. & Bardshiri, E. (1988) Quantification of urocanic acid isomers in murine skin during development and after irradiation with UVB light. *Photodermatology*, **5**, 179-186

Nusbaum, B.P., Edwards, E.K., Jr, Horwitz, S.N. & Frost, P. (1983) Psoriasis therapy. The effect of UV radiation on sensitization to mechlorethamine. *Arch. Dermatol.*, **119**, 117-121

O'Beirn, S.F., Judge, P., Urbach, F., MacCon, C.F. & Martin, F. (1970) Skin cancer in County Galway, Ireland. *Proc. natl Cancer Conf.*, **6**, 489-500

O'Dell, B.L., Jessen, R.T., Becker, L.E., Jackson, R.T. & Smith, E.B. (1980) Diminished immune response in sun-damaged skin. *Arch. Dermatol.*, **116**, 559-561

Oettlè, A.G. (1963) Skin cancer in Africa. *Natl Cancer Inst. Monogr.*, **10**, 197-214

Office of Population Censuses and Surveys (1986) *Occupational Mortality: the Registrar General's Decennial Supplement for Great Britain 1979-80, 1982-83* (Series DS No. 6), London, Her Majesty's Stationery Office

Okamoto, H. & Kripke, M.L. (1987) Effector and suppressor circuits of the immune response are activated *in vivo* by different mechanisms. *Proc. natl Acad. Sci. USA*, **84**, 3841–3845

O'Loughlin, C., Moriarty, M.J., Herity, B. & Daly, L. (1985) A re-appraisal of risk factors for skin carcinoma in Ireland: a case control study. *Irish J. med. Sci.*, **154**, 61–65

Olsen, J.H. & Jensen, O.M. (1987) Occupation and risk of cancer in Denmark. An analysis of 93 810 cancer cases, 1970–1979. *Scand. J. Work Environ. Health*, **13** (Suppl. 1), 1–91

Olsen, W.M., Huitfeldt, H.S. & Eggset, G. (1989) UVB-Induced (6-4) photoproducts in hairless mouse epidermis studied by quantitative immunohistochemistry. *Carcinogenesis*, **10**, 1669–1673

Onuigbo, W.I.B. (1978) Lip lesions in Nigerian Igbos. *Int. J. oral Surg.*, **7**, 73–75

Østerlind, A. (1987) Trends in incidence of ocular malignant melanoma in Denmark 1943–1982. *Int. J. Cancer*, **40**, 161–164

Østerlind, A. (1990) Malignant melanoma in Denmark. Occurrence and risk factors. *Acta oncol.*, **29**, 1–22

Østerlind, A., Olsen, J.H., Lynge, E. & Ewertz, M. (1985) Second cancer following cutaneous melanoma and cancers of the brain, thyroid, connective tissue, bone, and eye in Denmark, 1943–80. *Natl Cancer Inst. Monogr.*, **68**, 361–388

Østerlind, A., Hou-Jensen, K. & Jensen, O.M. (1988a) Incidence of cutaneous malignant melanoma in Denmark 1978–1982. Anatomic site distribution, histologic types, and comparison with nonmelanoma skin cancer. *Br. J. Cancer*, **58**, 385–391

Østerlind, A., Tucker, M.A., Stone, B.J. & Jensen, O.M. (1988b) The Danish case–control study of cutaneous malignant melanoma. II. Importance of UV-light exposure. *Int. J. Cancer*, **42**, 319–324

Østerlind, A., Tucker, M.A., Hou-Jensen, K., Stone, B.J., Engholm, G. & Jensen, O.M. (1988c) The Danish case–control study of cutaneous malignant melanoma. I. Importance of host factors. *Int. J. Cancer*, **42**, 200–206

Otani, T. & Mori, R. (1987) The effects of ultraviolet irradiation of the skin on herpes simplex virus infection: alterations in immune function mediated by epidermal cells and in the course of infection. *Arch. Virol.*, **96**, 1–15

Oxholm, A., Oxholm, P., Staberg, B. & Bendtzen, K. (1988) Immunohistological detection of interleukin I-like molecules and tumour necrosis factor in human epidermis before and after UVB-irradiation *in vivo*. *Br. J. Dermatol.*, **118**, 369–376

Paffenbarger, R.S., Jr, Wing, A.L. & Hyde, R.T. (1978) Characteristics in youth predictive of adult-onset malignant lymphomas, melanomas, and leukemias: brief communication. *J. natl Cancer Inst.*, **60**, 89–92

Parrington, J.M. (1972) Ultraviolet-induced chromosome aberrations and mitotic delay in human fibroblast cells. *Cytogenetics*, **11**, 117–131

Parrington, J.M., Delhanty, J.D.A. & Baden, H.P. (1971) Unscheduled DNA synthesis, UV-induced chromosome aberrations and SV40 transformation in cultured cells from xeroderma pigmentosum. *Ann. hum. Genet.*, **35**, 149–160

Parrish, J.A. (1983) *The Effect of Ultraviolet Radiation on the Immune System*, New Brunswick, NJ, Johnson & Johnson

Parrish, J.A. & Jaenicke, K.F. (1981) Action spectrum for phototherapy of psoriasis. *J. invest. Dermatol.*, **76**, 359–362

Parrish, J.A., Fitzpatrick, T.B., Tanenbaum, L. & Pathak, M.A. (1974) Photochemotherapy of psoriasis with oral methoxsalen and longwave ultraviolet light. *New Engl. J. Med.*, **291**, 1207–1211

Parrish, J.A., Zaynoun, S. & Anderson, R.R. (1981) Cumulative effects of repeated subthreshold doses of ultraviolet radiation. *J. invest. Dermatol.*, **76**, 356–358

Parrish, J.A., Kripke, M.L. & Morison, W.L., eds (1983) *Photoimmunology*, New York, Plenum

Parry, J.M., Sharp, D., Tippins, R.S. & Parry, E.M. (1979) Radiation-induced mitotic and meiotic aneuploidy in the yeast *Saccharomyces cerevisiae*. *Mutat. Res.*, **61**, 37–55

Parshad, R., Taylor, W.G., Sanford, K.K., Camalier, R.F., Gantt, R. & Tarone, R.E. (1980a) Fluorescent light-induced chromosome damage in human IMR-90 fibroblasts. Role of hydrogen peroxide and related free radicals. *Mutat. Res.*, **73**, 115–124

Parshad, R., Sanford, K.K., Jones, G.M., Tarone, R.E., Hoffman, H.A. & Grier, A.H. (1980b) Susceptibility to fluorescent light-induced chromatid breaks associated with DNA repair deficiency and malignant transformation in culture. *Cancer Res.*, **40**, 4415–4419

Pathak, M.A., Fitzpatrick, T.B., Greiter, F. & Kraus, E.W. (1987) Preventive treatment of sunburn, dermatoheliosis, and skin cancer with sun-protective agents. In: Fitzpatrick, T.B., Eisen, A.Z., Wolff, K., Freedberg, I.M. & Austen, K.F., eds, *Dermatology in General Medicine*, 3rd ed., New York, McGraw-Hill, pp. 1507–1522

Patrick, M.H. (1977) Studies on thymine-derived UV photoproducts in DNA—I. Formation and biological role of pyrimidine adducts in DNA. *Photochem. Photobiol.*, **25**, 357–372

Patton, J.D., Rowan, L.A., Mendrala, A.L., Howell, J.N., Maher, V.M. & McCormick, J.J. (1984) Xeroderma pigmentosum fibroblasts including cells from XP variants are abnormally sensitive to the mutagenic and cytotoxic action of broad spectrum simulated sunlight. *Photochem. Photobiol.*, **39**, 37–42

Pauw, H. & Meulemans, C.C.E. (1987) Occupational exposure to artificial UV-A radiation within the Philips factories in the Netherlands. In: Passchier, W.F. & Bosnjakovic, B.F.M., eds, *Human Exposure to Ultraviolet Radiation: Risks and Regulations*, Amsterdam, Elsevier, pp. 265–268

Peak, M.J. & Peak, J.G. (1982) Single-strand breaks induced in Bacillus *subtilis* DNA by ultraviolet light: action spectrum and properties. *Photochem. Photobiol.*, **35**, 675–680

Peak, J.G. & Peak, M.J. (1990) Ultraviolet light induces double-strand breaks in DNA of cultured human P3 cells as measured by neutral filter elution. *Photochem. Photobiol.*, **52**, 387–393

Peak, J.G. & Peak, M.J. (1991) Comparison of initial yields of DNA-to-protein crosslinks and single-strand breaks induced in cultured human cells by far- and near-ultraviolet light, blue light, and x-rays. *Mutat. Res.*, **246**, 187–191

Peak, M.J., Peak, J.G., Moehring, M.P. & Webb, R.B. (1984) Ultraviolet action spectra for DNA dimer induction, lethality, and mutagenesis in *Escherichia coli* with emphasis on the UVB region. *Photochem. Photobiol.*, **40**, 613–620

Peak, J.G., Peak, M.J., Sikorski, R.S. & Jones, C.A. (1985) Induction of DNA–protein crosslinks in human cells by ultraviolet and visible radiations: action spectrum. *Photochem. Photobiol.*, **41**, 295–302

Peak, M.J., Peak, J.G. & Carnes, B.A. (1987) Induction of direct and indirect single-strand breaks in human cell DNA by far- and near-ultraviolet radiations: action spectrum and mechanisms. *Photochem. Photobiol.*, **45**, 381–387

Peak, J.G., Woloschak, G.E. & Peak, M.J. (1991a) Enhanced expression of protein kinase C gene caused by solar radiation. *Photochem. Photobiol.*, **53**, 395–397

Peak, J.G., Pilas, B., Dudek, E.J. & Peak, M.J. (1991b) DNA breaks caused by monochromatic 365 nm ultraviolet-A radiation or hydrogen peroxide and their repair in human epithelioid and xeroderma pigmentosum cells. *Photochem. Photobiol.*, **54**, 197–203

Pearse, A.D., Gaskell, S.A. & Marks, R. (1987) Epidermal changes in human skin following irradiation with either UVB or UVA. *J. invest. Dermatol.*, **88**, 83–87

Penn, I. (1980) Immunosuppression and skin cancer. *Clin. plast. Surg.*, **7**, 361–368

Pepino, P., Hardy, M.A., Chabot, J.A., Berger, C., Marboe, C. & Wasfie, T. (1989) UVB irradiated allogeneic bone marrow transplantation in rats: prevention of graft *versus* host disease without immunosuppression. *Transpl. Proc.*, **21**, 2995–2996

Perna, J.J., Mannix, M.L., Rooney, J.F., Notkins, A.L. & Straus, S.E. (1987) Reactivation of latent herpes simplex virus infection by ultraviolet light: a human model. *J. Am. Acad. Dermatol.*, **17**, 473–478

Perry, L.L. & Greene, M.I. (1982) Antigen presentation by epidermal Langerhans cells: loss of function following ultraviolet (UV) irradiation *in vivo*. *Clin. Immunol. Immunopathol.*, **24**, 204–219

Perticone, P., Hellgren, D. & Lambert, B. (1986) Effects of UV-irradiation on the SCE frequency in human lymphocyte cultures. *Mutat. Res.*, **175**, 83–89

Peto, R., Pike, M.C., Day, N.E., Gray, R.G., Lee, P.N., Parish, S., Peto, J., Richards, S. & Wahrendorf, J. (1980) Guidelines for simple sensitive significance tests for carcinogenic effects in long-term animal experiments. In: *IARC Monographs on the Evaluation of the Carcinogenic Risk of Chemicals to Humans*, Suppl. 2, *Long-term and Short-term Screening Assays for Carcinogens: A Critical Appraisal*, Lyon, pp. 311–426

Phelps, R.G., Bernstein, L.E., Harpaz, N., Gordon, R.E., Cruickshank, F.A. & Schwartz, E. (1989) Characterization of a dermal derived malignant mesenchymal tumor arising in ultraviolet irradiated mice. *Am. J. Pathol.*, **135**, 149–159

Phillips, R. (1983) *Sources and Applications of Ultraviolet Radiation*, London, Academic Press

Pierceall, W.E. & Ananthaswamy, H.N. (1991) Transformation of NIH 3T3 cells by transfection with UV-irradiated human c-Ha-*ras*-1 proto-oncogene DNA. *Oncogene*, **6**, 2085–2091

Pierceall, W.E., Mukhopadhyay, T., Goldberg, L.H. & Ananthaswamy, H.N. (1991) Mutations in the p53 tumor suppressor gene in human cutaneous squamous cell carcinomas. *Mol. Carcinog.*, **4**, 445–449

Pitcher, H.M. & Longstreth, J.D. (1991) Melanoma mortality and exposure to ultraviolet radiation: an empirical relationship. *Environ. int.*, **17**, 7–21

Pitts, D.G., Cullen, A.P. & Hacker, P.D. (1977) Ocular effects of ultraviolet radiation from 295 to 365 nm. *Invest. Ophthalmol. visual Sci.*, **16**, 932–939

Plummer, N.A., Greaves, M.W., Hensby, C.N. & Black, A.K. (1977) Inflammation in human skin induced by ultraviolet radiation. *Postgrad. med. J.*, **53**, 656–657

Poulsen, J.T., Staberg, B., Wulf, H.C. & Brodthagen, H. (1984) Dermal elastosis in hairless mice after UV-B and UV-A applied simultaneously, separately or sequentially. *Br. J. Dermatol.*, **110**, 531–538

Pound, A.W. (1970) Induced cell proliferation and the initiation of skin tumour formation in mice by ultraviolet light. *Pathology*, **2**, 269–275

Putschar, W. & Holtz, F. (1930) Production of skin tumours in rats by long duration UV irradiation (Ger.). *Z. Krebsforsch.*, **33**, 219–260

Putvinsky, A.V., Sokolov, A.I., Roshchupkin, D.I. & Vladimirov, Y.A. (1979) Electric breakdown of bilayer phospholipid membranes under ultraviolet irradiation-induced lipid peroxidation. *FEBS Lett.*, **106**, 53–55

Räsänen, L., Reunala, T., Lehto, M., Jansén, C., Rantala, I. & Leinikki, P. (1989) Immediate decrease in antigen-presenting function and delayed enhancement of interleukin-I production in human epidermal cells after in vivo UVB irradiation. *Br. J. Dermatol.*, **120**, 589–596

Rasmussen, R.E., Hammer-Wilson, M. & Berns, M.W. (1989) Mutation and sister chromatid exchange induction in Chinese hamster ovary (CHO) cells by pulsed excimer laser radiation at 193 nm and 308 nm and continuous UV radiation at 254 nm. *Photochem. Photobiol.*, **49**, 413–418

Ray-Keil, L. & Chandler, J.W. (1986) Reduction in the incidence of rejection of heterotopic murine corneal transplants by pretreatment with ultraviolet radiation. *Transplantation*, **42**, 403–406

Reddy, A.L., Fialkow, P.J. & Salo, A. (1978) Ultraviolet radiation-induced chromosomal abnormalities in fetal fibroblasts from New Zealand black mice. *Science*, **201**, 920–922

Reeve, V.E., Greenoak, G.E., Gallagher, C.H., Canfield, P.J. & Wilkinson, F.J. (1985) Effect of immunosuppressive agents and sunscreens on UV carcinogenesis in the hairless mouse. *Aust. J. exp. Biol. med. Sci.*, **63**, 655–665

Resnick, M.A., Westmoreland, J., Amaya, E. & Bloom, K. (1987) UV-Induced damage and repair in centromere DNA of yeast. *Mol. gen. Genet.*, **210**, 16–22

Rigel, D.S., Friedman, R.J., Levenstein, M.J. & Greenwald, D.I. (1983) Relationship of fluorescent lights to malignant melanoma: another view. *J. Dermatol. surg. Oncol.*, **9**, 836–838

Ringvold, A. (1980) Cornea and ultraviolet radiation. *Acta ophthalmol.*, **58**, 63–68

Rivers, J.K., Norris, P.G., Murphy, G.M., Chu, A.C., Midgley, G., Morris, J., Morris, R.W., Young, A.R. & Hawk, J.L.M. (1989) UVA sunbeds: tanning, photoprotection, acute adverse effects and immunological changes. *Br. J. Dermatol.*, **120**, 767–777

Roberts, L.K. & Daynes, R.A. (1980) Modification of the immunogenic properties of chemically induced tumors arising in hosts treated concomitantly with ultraviolet light. *J. Immunol.*, **125**, 438–447

Roberts, L.K., Samlowski, W.E. & Daynes, R.A. (1986) Immunological consequences of ultraviolet radiation exposure. *Photodermatology*, **3**, 284–298

Robertson, D.F. (1972) *Solar Ultraviolet Radiation in Relation to Human Sunburn and Skin Cancer*, PhD Thesis, University of Queensland, Australia

Robertson, B., Gahring, L., Newton, R. & Daynes, R. (1987) In vivo administration of interleukin 1 to normal mice depresses their capacity to elicit contact hypersensitivity responses: prostaglandins are involved in this modification of immune function. *J. invest. Dermatol.*, **88**, 380–387

Robinson, J.K. (1987) Risk of developing another basal cell carcinoma. A 5-year prospective study. *Cancer*, **60**, 118–120

Roffo, A.H. (1934) Cancer and the sun: carcinomas and sarcomas caused by the action of the sun in toto (Fr.). *Bull. Assoc. fr. Étude Cancer*, **23**, 590–616

Roffo, A.H. (1939) Physico-chemical etiology of cancer (with special emphasis on the association with solar radiation) (Ger.). *Strahlentherapie*, **66**, 328–350

Ronai, Z.A. & Weinstein, I.B. (1988) Identification of a UV-induced *trans*-acting protein that stimulates polyomavirus DNA replication. *J. Virol.*, **62**, 1057–1060

Ronai, Z.A., Lambert, M.E., Johnson, M.D., Okin, E. & Weinstein, I.B. (1987) Induction of asynchronous replication of polyoma DNA in rat cells by ultraviolet irradiation and the effects of various inhibitors. *Cancer Res.*, **47**, 4565–4570

Ronai, Z.A., Lambert, M.E. & Weinstein, I.B. (1990) Inducible cellular responses to ultraviolet light irradiation and other mediators of DNA damage in mammalian cells. *Cell Biol. Toxicol.*, **6**, 105–126

Rooney, J.F., Bryson, Y., Mannix, M.L., Dillon, M., Wohlenberg, C.R., Banks, S., Wallington, C.J., Notkins, A.L. & Straus, S.E. (1991) Prevention of ultraviolet-light-induced herpes labialis by sunscreen. *Lancet*, **338**, 1419–1422

Rosario, R., Mark, G.J., Parrish, J.A. & Mihm, M.C., Jr (1979) Histological changes produced in skin by equally erythemogenic doses of UVA, UVB, UVC with psoralens. *Br. J. Dermatol.*, **101**, 299–308

Rosdahl, I.K. (1979) Local and systemic effects on the epidermal melanocyte population in UV-irradiated mouse skin. *J. invest. Dermatol.*, **73**, 306–309

Rosenstein, B.S. (1988) The induction of DNA strand breaks in normal human skin fibroblasts exposed to solar ultraviolet radiation. *Radiat. Res.*, **116**, 313–319

Rosenstein, B.S. & Ducore, J.M. (1983) Induction of DNA strand breaks in normal human fibroblasts exposed to monochromatic ultraviolet and visible wavelengths in the 240–546 nm range. *Photochem. Photobiol.*, **38**, 51–55

Rosenstein, B.S. & Mitchell, D.L. (1987) Action spectra for the induction of pyrimidine (6-4)pyrimidone photoproducts and cyclobutane pyrimidine dimers in normal human skin fibroblasts. *Photochem. Photobiol.*, **45**, 775–780

Rosenstein, B.S. & Mitchell, D.L. (1991) The repair of DNA damages induced in normal human skin fibroblasts exposed to simulated sunlight. *Radiat. Res.*, **126**, 338–342

Rosenstein, B.S. & Rosenstein, R.B. (1985) Induction of chromosome aberrations in ICR 2A frog cells exposed to 265–313 nm monochromatic ultraviolet wavelengths and photoreactivating light. *Photochem. Photobiol.*, **41**, 57–61

Rosenstein, B.S., Murphy, J.T. & Ducore, J.M. (1985) Use of a highly sensitive assay to analyze the excision repair of dimer and non-dimer DNA damages induced in human skin fibroblasts by 254 nm and solar ultraviolet radiation. *Cancer Res.*, **45**, 5526–5531

Rosenstein, B.S., Lai, L.-W., Ducore, J.M. & Rosenstein, R.B. (1989) DNA–protein crosslinking with normal and solar UV-sensitive ICR 2A frog cell lines exposed to solar UV-radiation. *Mutat. Res.*, **217**, 219–226

Rosenthal, F.S., Phoon, C., Bakalian, A.E. & Taylor, H.R. (1988) The ocular dose of ultraviolet radiation to outdoor workers. *Invest. Ophthalmol. visual Sci.*, **29**, 649–656

Rosenthal, F.S., Lew, R.A., Rouleau, L.J. & Thomson, M. (1990) Ultraviolet exposure to children from sunlight: a study using personal dosimetry. *Photodermatol. Photoimmunol. Photomed.*, **7**, 77–81

Rosenthal, F.S., West, S.K., Munoz, B., Emmett, E.A., Strickland, P.T. & Taylor, H.R. (1991) Ocular and facial skin exposure to ultraviolet radiation in sunlight: a personal exposure model with application to a worker population. *Health Phys.*, **61**, 77–86

Roser, M., Böhm, A., Oldigs, M., Weichenthal, M., Reimers, U., Schmidt-Preuss, U., Breitbart, E.W. & Rüdiger, H.W. (1989) Ultraviolet-induced formation of micronuclei and sister chromatid exchange in cultured fibroblasts of patients with cutaneous malignant melanoma. *Cancer Genet. Cytogenet.*, **41**, 129–137

Roshchupkin, D.I., Pelenitsyn, A.B., Potapenko, A.Y., Talitsky, V.V. & Vladimirov, Y.A. (1975) Study of the effects of ultraviolet light on biomembranes—IV. The effect of oxygen on UV-induced hemolysis and lipid photoperoxidation in rat erythrocytes and liposomes. *Photochem. Photobiol.*, **21**, 63–69

Ross, J.A., Howie, S.E.M., Norval, M., Maingay, J. & Simpson, T.J. (1986) Ultraviolet-irradiated urocanic acid suppresses delayed-type hypersensitivity to herpes simplex virus in mice. *J. invest. Dermatol.*, **87**, 630–633

Ross, J.A., Howie, S.E.M., Norval, M. & Maingay, J. (1987) Two phenotypically distinct T cells are involved in ultraviolet-irradiated urocanic acid-induced suppression of the efferent delayed-type hypersensitivity response to herpes simplex virus, type 1 in vivo. *J. invest. Dermatol.*, **89**, 230–233

Roth, D.E., Hodge, S.J. & Callen, J.P. (1989) Possible ultraviolet A-induced lentigines: a side effect of chronic tanning salon usage. *J. Am. Acad. Dermatol.*, **20**, 950–954

Roza, L., van der Wulp, K.J.M., MacFarlane, S.J., Lohman, P.H.M. & Baan, R.A. (1988) Detection of cyclobutane thymine dimers in DNA of human cells with monoclonal antibodies raised against a thymine dimer-containing tetranucleotide. *Photochem. Photobiol.*, **48**, 627–633

Ruiz-Rubio, M., Woodgate, R., Bridges, B.A., Herrera, G. & Blanco, M. (1986) New role for photoreversible pyrimidine dimers in induction of prototrophic mutations in excision-deficient *Escherichia coli* by ultraviolet light. *J. Bacteriol.*, **166**, 1141–1143

Rusch, H.P., Kline, B.E. & Baumann, C.A. (1941) Carcinogenesis by ultraviolet rays with reference to wavelength and energy. *Arch. Pathol.*, **31**, 135–146

Russell, W.O., Wynne, E.S., Loquvam, G.S. & Mehl, D.A. (1956) Studies on bovine ocular squamous carcinoma ('cancer eye'). I. Pathological anatomy and historical review. *Cancer*, **9**, 1–52

Saftlas, A.F., Blair, A., Cantor, K.P., Hanrahan, L. & Anderson, H.A. (1987) Cancer and other causes of death among Wisconsin farmers. *Am. J. ind. Med.*, **11**, 119–129

Sagher, D. & Strauss, B. (1983) Insertion of nucleotides opposite apurinic/apyrimidinic sites in deoxyribonucleic acid during in vitro synthesis: uniqueness of adenine nucleotides. *Biochemistry*, **22**, 4518–4526

Sammartano, L.J. & Tuveson, R.W. (1985) Hydrogen peroxide induced resistance to broad-spectrum near-ultraviolet light (300–400 nm) inactivation in *Escherichia coli*. *Photochem. Photobiol.*, **41**, 367–370

Sammartano, L.J. & Tuveson, R.W. (1987) Escherichia coli strains carrying the cloned cytochrome d terminal oxidase complex are sensitive to near-UV inactivation. *J. Bacteriol.*, **169**, 5304–5307

Sancar, A. & Rupp, W.D. (1983) A novel repair enzyme: UVRABC excision nuclease of Escherichia coli cuts a DNA strand on both sides of the damaged region. *Cell*, **33**, 249–260

Sanderson, B.J.S., Dempsey, J.L. & Morley, A.A. (1984) Mutations in human lymphocytes: effects of x- and UV-irradiation. *Mutat. Res.*, **140**, 223–227

Sanford, K.K., Parshad, R., Jones, G., Handleman, S., Garrison, C. & Price, F. (1979) Role of photosensitization and oxygen in chromosome stability and 'spontaneous' malignant transformation in culture. *J. natl Cancer Inst.*, **63**, 1245–1255

Santamaria, L., Giordono, G.G., Alfisi, M. & Cascione, F. (1966) Effects of light on 3,4-benzpyrene carcinogenesis. *Nature*, **210**, 824–825

Santamaria, L., Bianchi, A., Arnaboldi, A., Daffara, P. & Andreoni, L. (1985) Photocarcinogenesis by methoxypsoralen, neutral red, proflavine, and long UV radiation. *Cancer Detect. Prev.*, **8**, 447–454

Sauder, D.N., Tamaki, K., Moshell, A.N., Fujiwara, H. & Katz, S.I. (1981) Induction of tolerance to topically applied TNCB using TNP-conjugated ultraviolet light-irradiated epidermal cells. *J. Immunol.*, **127**, 261–265

Sauder, D.N., Noonan, F.P., DeFabo, E.C. & Katz, S.I. (1983) Ultraviolet radiation inhibits alloantigen presentation by epidermal cells: partial reversal by the soluble epidermal cell product, epidermal cell-derived thymocyte-activating factor (ETAF). *J. invest. Dermatol.*, **80**, 485–489

Sayre, R.M. & Kligman, L.H. (1992) Discrepancies in the measurement of spectral sources. *Photochem. Photobiol.*, **55**, 141–143

Schaaper, R.M., Kunkel, T.A. & Loeb, L.A. (1983) Infidelity of DNA synthesis associated with bypass of apurinic sites. *Proc. natl Acad. Sci. USA*, **80**, 487–491

Schaaper, R.M., Dunn, R.L. & Glickman, B.W. (1987) Mechanisms of ultraviolet-induced mutation. Mutational spectra in the *Escherichia coli* lacI gene for a wild-type and an excision-repair deficient strain. *J. mol. Biol.*, **198**, 187–202

Schacter, B., Lederman, M.M., LeVine, M.J. & Ellner, J.J. (1983) Ultraviolet radiation inhibits human natural killer activity and lymphocyte proliferation. *J. Immunol.*, **130**, 2484–2487

Scheibner, A., McCarthy, W.H., Milton, G.W. & Nordlund, J.J. (1983) Langerhans cell and melanocyte distribution in 'normal' human epidermis. *Aust. J. Dermatol.*, **24**, 9–16

Scheibner, A., Hollis, D.E., McCarthy, W.H. & Milton, G.W. (1986a) Effects of exposure to ultraviolet light in a commercial solarium on Langerhans cells and melanocytes in human epidermis. *Aust. J. Dermatol.*, **27**, 35–41

Scheibner, A., Hollis, D.E., McCarthy, W.H. & Milton, G.W. (1986b) Effects of sunlight exposure on Langerhans cells and melanocytes in human epidermis. *Photodermatology*, **3**, 15–25

Schothorst, A.A., Slaper, H., Telgt, D., Alhadi, B. & Suurmond, D. (1987a) Amounts of ultraviolet B (UVB) received from sunlight or artificial UV sources by various population groups in the Netherlands. In: Passchier, W.F. & Bosnjakovic, B.F.M., eds, *Human Exposure to Ultraviolet Radiation: Risks and Regulations*, Amsterdam, Elsevier, pp. 269–273

Schothorst, A.A., Enninga, E.C. & Simons, J.W.I.M. (1987b) Mutagenic effects per erythemal dose of artificial and natural sources of ultraviolet light. In: Passchier, W.F. & Bosnjakovic, B.F.M., eds, *Human Exposure to Ultraviolet Radiation: Risks and Regulations*, Amsterdam, Elsevier, pp. 103–107

Schothorst, A.A., Evers, L.M., Noz, K.C., Filon, R. & van Zeeland, A.A. (1991) Pyrimidine dimer induction and repair in cultured human skin keratinocytes or melanocytes after irradiation with monochromatic ultraviolet radiation. *J. invest. Dermatol.*, **96**, 916–920

van der Schroeff, J.G., Evers, L.M., Boot, A.J.M. & Bos, J.L. (1990) *ras* Oncogene mutations in basal cell carcinomas and squamous cell carcinomas of human skin. *J. invest. Dermatol.*, **94**, 423–425

Schulze, R. (1962) On the climate of the Earth (Ger.). *Strahlentherapie*, **119**, 321–348

Schulze, R. (1970) Global radiation climate (Ger.). *Wiss. Forschungsber.*, **72**, 1–220

Schulze, R. & Gräfe, K. (1969) Consideration of sky ultraviolet radiation in the measurement of solar ultraviolet radiation. In: Urbach, F., ed., *The Biological Effects of Ultraviolet Radiation* (Proceedings of the First International Conference), Oxford, Pergamon Press, pp. 359–373

Schwartz, S.M. & Weiss, N.S. (1988) Place of birth and incidence of ocular melanoma in the United States. *Int. J. Cancer*, **41**, 174–177

Schwarz, T., Urbanska, A., Gschnait, F. & Luger, T.A. (1986) Inhibition of the induction of contact hypersensitivity by a UV-mediated epidermal cytokine. *J. invest. Dermatol.*, **87**, 289–291

Schwarz, T., Urbanski, A., Kirnbauer, R., Köck, A., Gschnait, F. & Luger, T.A. (1988) Detection of a specific inhibitor of interleukin 1 in sera of UVB-treated mice. *J. invest. Dermatol.*, **91**, 536–540

Scotto, J. & Fears, T.R. (1987) The association of solar ultraviolet and skin melanoma incidence among Caucasians in the United States. *Cancer Invest.*, **5**, 275–283

Scotto, J., Kopf, A.W. & Urbach, F. (1974) Non-melanoma skin cancer among Caucasians in four areas of the United States. *Cancer*, **34**, 1333–1338

Scotto, J., Fears, T. & Gori, G. (1976) *Measurements of Ultraviolet Radiations in the United States and Comparisons with Skin Cancer Data* (NIH Publ. No. 76-1029), Washington DC, US Department of Health, Education, and Welfare

Scotto, J., Fears, T.R. & Fraumeni, J.F., Jr (1982) Solar radiation. In: Schottenfeld, D. & Fraumeni, J.F., Jr, eds, *Cancer Epidemiology and Prevention*, Philadelphia, W.B. Saunders, pp. 254–276

Scotto, J., Fears, T.R. & Fraumeni, J.F., Jr (1983) *Incidence of Nonmelanoma Skin Cancer in the United States* (NIH Publ. No. 83-2433), Bethesda, MD, National Cancer Institute

Seddon, J.M., Gragoudas, E.S., Glynn, R.J., Egan, K.M., Albert, D.M. & Blitzer, P.H. (1990) Host factors, UV radiation and risk of uveal melanoma. A case–control study. *Arch. Ophthalmol.*, **108**, 1274–1280

Seetharam, S., Protić-Sabljić, M., Seidman, M.M. & Kraemer, K.H. (1987) Abnormal ultraviolet mutagenic spectrum in plasmid DNA replicated in cultured fibroblasts from a patient with the skin cancer-prone disease, xeroderma pigmentosum. *J. clin. Invest.*, **80**, 1613–1617

Seguin, L.R., Tarone, R.E., Liao, K. & Robbins, J.H. (1988) Ultraviolet light-induced chromosomal aberrations in cultured cells from Cockayne syndrome and complementation group C xeroderma pigmentosum patients: lack of correlation with cancer susceptibility. *Am. J. hum. Genet.*, **42**, 468–475

Sekiya, T., Fushimi, M., Hori, H., Hirohashi, S., Nishimura, S. & Sugimura, T. (1984) Molecular cloning and the total nucleotide sequence of the human c-Ha-*ras*-1 gene activated in a melanoma from a Japanese patient. *Proc. natl Acad. Sci. USA*, **81**, 4771–4775

Servilla, K.S., Burnham, D.K. & Daynes, R.A. (1987) Ability of cyclosporine to promote the growth of transplanted ultraviolet radiation-induced tumors in mice. *Transplantation*, **44**, 291–295

Setlow, R.B. (1968) Photoproducts in DNA irradiated *in vivo*. *Photochem. Photobiol.*, **7**, 643–649

Setlow, R.B. (1975) The relevance of photobiological repair. *An. Acad. Bras. Cienc.*, **45** (Suppl. 215), 215–220

Setlow, R.B. & Carrier, W.L. (1966) Pyrimidine dimers in ultraviolet-irradiated DNA's. *J. mol. Biol.*, **17**, 237–254

Setlow, R.B., Swenson, P.A. & Carrier, W.L. (1963) Thymine dimers and inhibition of DNA synthesis by ultraviolet irradiation of cells. *Science*, **142**, 1464–1466

Setlow, R.B., Woodhead, A.D. & Grist, E. (1989) Animal model for ultraviolet radiation-induced melanoma: platyfish–swordtail hybrid. *Proc. natl Acad. Sci. USA*, **86**, 8922–8926

Shabad, L.M. & Litvinova, S.N. (1972) The combined effect of UV irradiation and 7,12-DMBA on the skin of hairless mice (Russ.). *Biull. eksp. Biol. Med.*, **73**, 84–85

Shanmugaratnam, K., Lee, H.P. & Day, N.E., eds (1983) *Cancer Incidence in Singapore 1968–1977* (IARC Scientific Publications No. 47), Lyon, IARC, p. 73

Shetlar, M.D. (1980) Cross-linking of proteins to nucleic acids by ultraviolet light. *Photochem. Photobiol. Rev.*, **5**, 105–197

Shukla, V.K., Hughes, D.C., Hughes, L.E., McCormick, F. & Padua, R.A. (1989) *ras* Mutations in human melanotic lesions: K-*ras* activation is a frequent and early event in melanoma development. *Oncogene Res.*, **5**, 121–127

Siede, W. & Eckardt, F. (1986) Analysis of mutagenic DNA repair in a thermoconditional mutant of *Saccharomyces cerevisiae*. III. Dose–response pattern of mutation induction in UV-irradiated *rev2*^{ts} cells. *Mol. gen. Genet.*, **202**, 68–74

Siemiatycki, J. (1991) *Risk Factors for Cancer in the Workplace*, Boca Raton, FL, CRC Press

Silverstone, H. & Gordon, D. (1966) Regional studies in skin cancer. Second report: wet tropical and subtropical coasts of Queensland. *Med. J. Aust.*, **2**, 733–740

Silverstone, H. & Searle, J.H.A. (1970) The epidemiology of skin cancer in Queensland: the influence of phenotype and environment. *Br. J. Cancer*, **24**, 235–252

Singh, B. & Gupta, R.S. (1982) Mutagenic response to ultraviolet light and X-rays at five independent genetic loci in Chinese hamster ovary cells. *Environ. Mutag.*, **4**, 543–551

Sisson, T.R.C. & Vogl, T.P. (1982) Phototherapy of hyperbilirubinemia. In: Regan, J.D. & Parrish, J.A., eds, *The Science of Photomedicine*, New York, Plenum, pp. 477–509

Sjövall, P. & Christensen, O.B. (1986) Local and systemic effect of ultraviolet irradiation (UVB and UVA) on human allergic contact dermatitis. *Acta dermatol. venereol.*, **66**, 290–294

Slaga, T.J., Klein-Szanto, A.J.P., Triplett, L.L., Yotti, L.P. & Trosko, J.E. (1981) Skin tumor-promoting activity of benzoyl peroxide, a widely used free radical-generating compound. *Science*, **213**, 1023–1025

Slaper, H. (1987) *Skin Cancer and UV Exposure: Investigations on the Estimation of Risks*, PhD Thesis, University of Utrecht

Sliney, D.H. (1986) Physical factors in cataractogenesis: ambient ultraviolet radiation and temperature. *Invest. Ophthalmol. visual Sci.*, **27**, 781–790

Sliney, D.H. (1990) The IRPA/INIRC guidelines or limits of exposure to laser radiation. In: Grandolfo, M., Rindi, S. & Sliney, D.H., eds, *Light, Lasers, and Synchrotron Radiation. A Health Risk Assessment*, New York, Plenum, pp. 341–346

Sliney, D.H. & Wolbarsht, M. (1980) *Safety with Lasers and Other Optical Sources: A Comprehensive Handbook*, New York, Plenum

Sliney, D., Wood, R.L., Jr, Moscato, P.M., Marshall, W.J. & Eriksen, P. (1990) Ultraviolet exposure in the outdoor environment: measurement of ambient ultraviolet exposure levels at large zenith angles. In: Grandolfo, M., Rindi, S. & Sliney, D.H., eds, *Light, Lasers, and Synchrotron Radiation. A Health Risk Assessment*, New York, Plenum, pp. 169–180

Smith, K.C. (1964) The photochemistry of thymine and bromouracil *in vivo*. *Photochem. Photobiol.*, **3**, 1–10

Smith, K.C. (1976) The radiation-induced addition of proteins and other molecules to nucleic acids. In: Wang, S.Y., ed., *Photochemistry and Photobiology of Nucleic Acids*, Vol. 2, New York, Academic Press, pp. 187–218

Smith, K.C., ed. (1989) *The Science of Photobiology*, 2nd ed., New York, Plenum, pp. 47–53

Sorahan, T. & Grimley, R.P. (1985) The aetiological significance of sunlight and fluorescent lighting in malignant melanoma: a case–control study. *Br. J. Cancer*, **52**, 765–769

Spangrude, G.J., Bernhard, E.J., Ajioka, R.S. & Daynes, R.A. (1983) Alterations in lymphocyte homing patterns within mice exposed to ultraviolet radiation. *J. Immunol.*, **130**, 2974–2981

Spellman, C.W. & Daynes, R.A. (1978) Properties of ultraviolet light-induced suppressor lymphocytes within a syngeneic tumor system. *Cell. Immunol.*, **36**, 383–387

Spellman, C.W., Woodward, J.G. & Daynes, R.A. (1977) Modification of immunological potential by ultraviolet radiation. I. Immune status of short-term UV-irradiated mice. *Transplantation*, **24**, 112–119

Spikes, J.D., Kripke, M.L., Connor, R.J. & Eichwald, E.J. (1977) Time of appearance and histology of tumors induced in the dorsal skin of C3Hf mice by ultraviolet radiation from a mercury arc lamp. *J. natl Cancer Inst.*, **59**, 1637–1643

Spitzer, W.O., Hill, G.B., Chambers, L.W., Helliwell, B.E. & Murphy H.B. (1975) The occupation of fishing as a risk factor in cancer of the lip. *New Engl. J. Med.*, **293**, 419–424

Spruance, S.L. (1985) Pathogenesis of herpes simplex labialis: experimental induction of lesions with UV light. *J. clin. Microbiol.*, **22**, 366–368

Staberg, B., Wulf, H.C., Poulsen, T., Klemp, P. & Brodthagen, H. (1983a) Carcinogenic effect of sequential artificial sunlight and UVA irradiation in hairless mice. Consequences for solarium 'therapy'. *Arch. Dermatol.*, **119**, 641–643

Staberg, B., Wulf, H.C., Klemp, P., Poulsen, T. & Brodthagen, H. (1983b) The carcinogenic effect of UVA radiation. *J. invest. Dermatol.*, **81**, 517–519

Stanley, S.K., Folks, T.M. & Fauci, A.S. (1989) Induction of expression of human immunodeficiency virus in a chronically infected promonocytic cell line by ultraviolet irradiation. *AIDS Res. hum. Retroviruses*, **5**, 375–384

Stefanini, M., Lagomarsini, P., Arlett, C.F., Marinoni, S., Borrone, C., Crovato, F., Trevisan, G., Cordone, G. & Nuzzo, F. (1986) Xeroderma pigmentosum (complementation group D) mutation is present in patients affected by trichothiodystrophy with photosensitivity. *Hum. Genet.*, **74**, 107–112

Stein, B., Rahmsdorf, H.J., Steffen, A., Litfin, M. & Herrlich, P. (1989) UV-Induced DNA damage is an intermediate step in UV-induced expression of human immunodeficiency virus type 1, collagenase, c-*fos*, and metallothionein. *Mol. cell. Biol.*, **9**, 5169–5181

Steinitz, R., Parkin, D.M., Young, J.L., Bieber, C.A. & Katz, L., eds (1989) *Cancer Incidence in Jewish Migrants to Israel, 1961–1981* (IARC Scientific Publications No. 98), Lyon, IARC, pp. 114–115, 134–135

Stenbäck, F. (1975a) Species-specific neoplastic progression by ultraviolet light on the skin of rats, guinea pigs, hamsters and mice. *Oncology*, **31**, 209–225

Stenbäck, F. (1975b) Studies on the modifying effect of ultraviolet radiation on chemical skin carcinogenesis. *J. invest. Dermatol.*, **64**, 253–257

Stenbäck, F. (1975c) Ultraviolet light irradiation as initiating agent in skin tumor formation by the two-stage method. *Eur. J. Cancer*, **11**, 241–246

Stenbäck, F. (1978) Life history and histopathology of ultraviolet light-induced skin tumors. *Natl Cancer Inst. Monogr.*, **50**, 57–70

Stenbäck, F. & Shubik, P. (1973) Carcinogen-induced skin tumorigenesis in mice: enhancement and inhibition by ultraviolet light. *Z. Krebsforsch.*, **79**, 234–240

Sterenborg, H.J.C.M. & van der Leun, J.C. (1987) Action spectra for tumorigenesis by ultraviolet radiation. In: Passchier, W.F. & Bosnjakovic, B.F.M., eds, *Human Exposure to Ultraviolet Radiation: Risks and Regulations*, Amsterdam, Elsevier, pp. 173–191

Sterenborg, H.J.C.M. & van der Leun, J.C. (1988) Change in epidermal transmission due to UV-induced hyperplasia in hairless mice: a first approximation of the action spectrum. *Photodermatology*, **5**, 71–82

Sterenborg, H.J.C.M. & van der Leun, J.C. (1990) Tumorigenesis by a long wavelength UVA source. *Photochem. Photobiol.*, **51**, 325–330

Sterenborg, H.J.C.M., de Gruijl, F.R. & van der Leun, J.C. (1986) UV-Induced epidermal hyperplasia in hairless mice. *Photodermatology*, **3**, 206–214

Sterenborg, H.J.C.M., van der Putte, S.C.J. & van der Leun, J.C. (1988) The dose–response relationship of tumorigenesis by ultraviolet radiation of 254 nm. *Photochem. Photobiol.*, **47**, 245–253

Sterenborg, H.J.C.M., de Gruijl, F.R., Kelfkens, G. & van der Leun, J.C. (1991) Evaluation of skin cancer risk resulting from long term occupational exposure to radiation from ultraviolet lasers in the range from 190 to 400 nm. *Photochem. Photobiol.*, **54**, 775–780

Stierner, U., Rosdahl, I., Augustsson, A. & Kågedal, B. (1989) UVB irradiation induces melanocyte increase in both exposed and shielded human skin. *J. invest. Dermatol.*, **92**, 561–564

Stingl, L.A., Sauder, D.N., Iijima, M., Wolff, K., Pehamberger, H. & Stingl, G. (1983) Mechanism of UVB induced impairment of the antigen-presenting capacity of murine epidermal cells. *J. Immunol.*, **130**, 1586–1591

Streeter, P.R. & Fortner, G.W. (1988a) Immune response to ultraviolet-induced tumors. II. Effector cells in tumor immunity. *Transplantation*, **46**, 250–255

Streeter, P.R. & Fortner, G.W. (1988b) Immune response to ultraviolet induced tumors. III. Analysis of cloned lymphocyte populations exhibiting antitumor activity. *Transplantation*, **46**, 256–260

Streilein, J.W. & Bergstresser, P.R. (1981) Langerhans cell function dictates induction of contact hypersensitivity or unresponsiveness to DNFB in Syrian hamsters. *J. invest. Dermatol.*, **77**, 272–277

Streilein, J.W., Toews, G.T., Gilliam, J.N. & Bergstresser, P.R. (1980) Tolerance or hypersensitivity to 2,4-dinitro-1-fluorobenzene: the role of Langerhans cell density within epidermis. *J. invest. Dermatol.*, **74**, 319–322

Strickland, P.T. (1982) Tumor induction in Sencar mice in response to ultraviolet radiation. *Carcinogenesis*, **3**, 1487–1489

Strickland, P.T. (1986) Photocarcinogenesis by near-ultraviolet (UVA) radiation in Sencar mice. *J. invest. Dermatol.*, **87**, 272–275

Strickland, P.T., Burns, F.J. & Albert, R.E. (1979) Induction of skin tumors in the rat by single exposure to ultraviolet radiation. *Photochem. Photobiol.*, **30**, 683–688

Strickland, P.T., Creasia, D. & Kripke, M.L. (1985) Enhancement of two-stage skin carcinogenesis by exposure of distant skin to UV radiation. *J. natl Cancer Inst.*, **74**, 1129–1134

Strickland, P.T., Bruze, M. & Creasey, J. (1988) Cyclobuta-dithymidine induction by solar-simulating UV radiation in human skin. I. Protection by constitutive pigmentation. *Photodermatology*, **5**, 166–169

Sundararaman, N., St John, D.E. & Venkateswaran, S.V. (1975) *Solar Ultraviolet Radiation Received at the Earth's Surface Under Clear and Cloudless Conditions*, Springfield, VA, National Technical Information Service

Sutherland, B.M., Harber, L.C. & Kochevar, I.E. (1980) Pyrimidine dimer formation and repair in human skin. *Cancer Res.*, **40**, 3181–3185

Sutherland, B.M., Delihas, N.C., Oliver, R.P. & Sutherland, J.C. (1981) Action spectra for ultraviolet light-induced transformation of human cells to anchorage-independent growth. *Cancer Res.*, **41**, 2211–2214

Sutherland, B.M., Freeman, A.G. & Bennett, P.V. (1988) Human cell transformation in the study of sunlight-induced cancers in the skin of man. *Mutat. Res.*, **199**, 425–436

Suzuki, H., Fukuyama, K. & Epstein, W.L. (1977) Changes in nuclear DNA and RNA during epidermal keratinization. *Cell Tissue Res.*, **184**, 155–167

Suzuki, F., Han, A., Lankas, G.R., Utsumi, H. & Elkind, M.M. (1981) Spectral dependencies of killing, mutation, and transformation in mammalian cells and their relevance to hazards caused by solar ultraviolet radiation. *Cancer Res.*, **41**, 4916–4924

Swerdlow, A.J. (1979) Incidence of malignant melanoma of the skin in England and Wales and its relationship to sunshine. *Br. med. J.*, **ii**, 1324–1327

Swerdlow, A.J. (1983a) Epidemiology of melanoma of the eye in the Oxford region, 1952–78. *Br. J. Cancer*, **47**, 311–313

Swerdlow, A.J. (1983b) Epidemiology of eye cancer in adults in England and Wales, 1962–77. *Am. J. Epidemiol.*, **118**, 294–300

Swerdlow, A.J. (1985) Seasonality of presentation of cutaneous melanoma, squamous cell cancer and basal cell cancer in the Oxford region. *Br. J. Cancer*, **52**, 893–900

Swerdlow, A.J., English, J.S.C., MacKie, R.M., O'Doherty, C.J., Hunter, J.A.A., Clark, J. & Hole, D.J. (1988) Fluorescent lights, ultraviolet lamps, and risk of cutaneous melanoma. *Br. med. J.*, **297**, 647–650

Sylvania (undated) *Germicidal and Short-wave Ultraviolet Radiation*, Danvers, MA, GTE Products Corporation

Szabó, G., Gerald, A.B., Pathak, M.A. & Fitzpatrick, T.B. (1972) The ultrastructure of racial color differences in man. In: Riley, V., ed., *Pigmentation. Its Genesis and Biologic Control*, New York, Appleton-Century-Crofts, pp. 23–41

Szebeni, J. & Tollin, G. (1988) Some relationships between ultraviolet light and heme-protein-induced peroxidative lipid breakdown in liposomes, as reflected by fluorescence changes: the effect of negative surface charge. *Photochem. Photobiol.*, 47, 475–479

Takebe, H., Nishigori, C. & Satoh, Y. (1987) Genetics and skin cancer of xeroderma pigmentosum in Japan. *Jpn. J. Cancer Res. (Gann)*, 78, 1135–1143

Talve, L., Stenbäck, F. & Jansén, C.T. (1990) UVA irradiation increases the incidence of epithelial tumors in UVB-irradiated hairless mice. *Photodermatol. Photoimmunol. Photomed.*, 7, 109–115

Tamaki, K. & Iijima, M. (1989) The effect of ultraviolet B irradiation on delayed-type hypersensitivity, cytotoxic T lymphocyte activity, and skin graft rejection. *Transplantation*, 47, 372–376

Tan, E.M. & Stoughton, R.B. (1969) Ultraviolet light induced damage to deoxyribonucleic acid in human skin. *J. invest. Dermatol.*, 52, 537–542

Taylor, H.R., West, S.K., Rosenthal, F.S., Muñoz, B., Newland, H.S., Abbey, H. & Emmett, E.A. (1988) Effect of ultraviolet radiation on cataract formation. *New Engl. J. Med.*, 319, 1429–1433

Taylor, H.R., West, S.K., Rosenthal, F.S., Muñoz, B., Newland, H.S. & Emmett, E.A. (1989) Corneal changes associated with chronic UV irradiation. *Arch. Ophthalmol.*, 107, 1481–1484

Teppo, L., Pakkanen, M. & Hakulinen, T. (1978) Sunlight as a risk factor of malignant melanoma of the skin. *Cancer*, 41, 2018–2027

Teppo, L., Pukkala, E., Hakama, M., Hakulinen, T., Herva, A. & Saxén, E. (1980) Way of life and cancer incidence in Finland. A municipality-based ecological analysis. *Scand. J. soc. Med.*, **Suppl. 19**, 5–84

Tessman, I. & Kennedy, M.A. (1991) The two-step model of UV mutagenesis reassessed: deamination of cytosine in cyclobutane dimers as the likely source of the mutations associated with photoreactivation. *Mol. gen. Genet.*, 227, 144–148

Thiers, B.H., Maize, J.C., Spicer, S.S. & Cantor, A.B. (1984) The effect of aging and chronic sun exposure on human Langerhans cell population. *J. invest. Dermatol.*, 82, 223–226

Tjernlund, U. & Juhlin, L. (1982) Effect of UV radiation on immunological and histochemical markers of Langerhans cells in normal appearing skin of psoriatic patients. *Acta dermatol.*, 272, 171–176

Tlsty, T.D., Brown, P.C. & Schimke, R.T. (1984) UV radiation facilitates methotrexate resistance and amplification of the dihydrofolate reductase gene in cultured 3T6 mouse cells. *Mol. cell. Biol.*, 4, 1050–1056

Toda, K.-I., Miyachi, Y., Nesumi, N., Konishi, J. & Imamura, S. (1986) UVB/PUVA-induced suppression of human natural killer activity is reduced by superoxide dismutase and/or interleukin 2 *in vitro*. *J. invest. Dermatol.*, 86, 519–522

Toews, G.B., Bergstresser, P.R. & Streilein, J.W. (1980) Epidermal Langerhans cell density determines whether contact hypersensitivity or unresponsiveness follows skin painting with DNFB. *J. Immunol.*, 124, 445–453

Tominaga, A., Lefort, S., Mizel, S.B., Dambrauskas, J.T., Granstein, R., Lowy, A., Benacerraf, B. & Greene, M.I. (1983) Molecular signals in antigen presentation. I. Effects of interleukin 1 and 2 on radiation-treated antigen-presenting cells *in vivo* and *in vitro*. *Clin. Immunol. Immunopathol.*, 29, 282–293

Trosko, J.E. & Brewen, J.G. (1967) Inhibition of ultraviolet-induced chromosome breaks by cysteamine in 5-bromouracil-substituted mammalian cells. *Radiat. Res.*, **32**, 200–213

Tucker, M.A., Boice, J.D., Jr & Hoffman, D.A. (1985a) Second cancer following cutaneous melanoma and cancers of the brain, thyroid, connective tissue, bone and eye in Connecticut, 1935–82. *Natl Cancer Inst. Monogr.*, **68**, 161–189

Tucker, M.A., Shields, J.A., Hartge, P., Augsberger, J., Hoover, R.N. & Fraumeni, J.F., Jr (1985b) Sunlight exposure as risk factor for intraocular malignant melanoma. *New Engl. J. Med.*, **313**, 789–792

Turner, B.J., Siatkowski, R.M., Augsburger, J.J., Shields, J.A., Lustbader, E. & Mastrangelo, M.J. (1989) Other cancers in uveal melanoma patients and their families. *Am. J. Ophthalmol.*, **107**, 601–608

Tuveson, R.W. & Sammartano, L.J. (1986) Sensitivity of *Hem*A mutant *Escherichia coli* cells to inactivation by near-UV light depends on the level of supplementation with δ-aminolevulinic acid. *Photochem. Photobiol.*, **43**, 621–626

Tyrrell, R.M. (1973) Induction of pyrimidine dimers in bacterial DNA by 365 nm radiation. *Photochem. Photobiol.*, **17**, 69–73

Tyrrell, R.M. (1982) Cell inactivation and mutagenesis by solar ultraviolet radiation. In: Hélène, C., Charlier, M., Montenay-Garestier, T. & Laustriat, G., eds, *Trends in Photobiology*, New York, Plenum, pp. 155–172

Tyrrell, R.M. (1984) Mutagenic action of monochromatic UV radiation in the solar range on human cells. *Mutat. Res.*, **129**, 103–110

Tyrrell, R.M. (1985) A common pathway for protection of bacteria against damage by solar UVA (334 nm, 365 nm) and an oxidising agent (H_2O_2). *Mutat. Res.*, **145**, 129–136

Tyrrell, R.M. (1991) UVA (320–280 nm) radiation as an oxidative stress. In: Sir, H., ed., *Oxidative Stress: Oxidants and Antioxidants*, London, Academic Press, pp. 57–83

Tyrrell, R.M. & Pidoux, M. (1986) Endogenous glutathione protects human skin fibroblasts against the cytotoxic action of UVB, UVA and near-visible radiations. *Photochem. Photobiol.*, **44**, 561–564

Tyrrell, R.M. & Pidoux, M. (1988) Correlation between endogenous glutathione content and sensitivity of cultured human skin cells to radiation at defined wavelengths in the solar ultraviolet range. *Photochem. Photobiol.*, **47**, 405–412

Tyrrell, R.M. & Pidoux, M. (1989) Singlet oxygen involvement in the inactivation of cultured human fibroblasts by UVA (334 nm, 365 nm) and near visible (405 nm) radiations. *Photochem. Photobiol.*, **49**, 407–412

Tyrrell, R.M., Ley, R.D. & Webb, R.B. (1974) Induction of single-strand breaks (alkali-labile bonds) in bacterial and phage DNA by near UV (365 nm) radiation. *Photochem. Photobiol.*, **20**, 395–398

Ullrich, S.E. (1985) Suppression of lymphoproliferation by hapten-specific suppressor T lymphocytes from mice exposed to ultraviolet radiation. *Immunology*, **54**, 343–352

Ullrich, S.E. (1986) Suppression of the immune response to allogeneic histocompatibility antigens by a single exposure to ultraviolet radiation. *Transplantation*, **42**, 287–291

Ullrich, S.E. (1987) The effect of ultraviolet radiation-induced suppressor cells on T-cell activity. *Immunology*, **60**, 353–360

Ullrich, S.E. & Kripke, M.L. (1984) Mechanisms in the suppression of tumor rejection produced in mice by repeated UV radiation. *J. Immunol.*, **133**, 2786–2790

Ullrich, S.E., Yee, G.K. & Kripke, M.L. (1986a) Suppressor lymphocytes induced by epicutaneous sensitization of UV-irradiated mice control multiple immunological pathways. *Immunology*, **58**, 185-190

Ullrich, S.E., Azizi, E. & Kripke, M.L. (1986b) Suppression of the induction of delayed-type hypersensitivity reactions in mice by a single exposure to ultraviolet radiation. *Photochem. Photobiol.*, **43**, 633-638

Ultra-Violet Products, Inc. (1977) *Laboratory Applications for Ultraviolet Equipment*, San Gabriel, CA

United Kingdom Stratospheric Ozone Review Group (1991) *Stratospheric Ozone 1991*, London, Her Majesty's Stationery Office

United Nations Environment Programme (1989) *Environmental Effects Panel Report*, Geneva

Unrau, P., Wheatcroft, R., Cox, B. & Olive, T. (1973) The formation of pyrimidine dimers in the DNA of fungi and bacteria. *Biochim. biophys. Acta*, **312**, 626-632

Urbach, F., Davies, R.E. & Forbes, P.D. (1966) Ultraviolet radiation and skin cancer in man. In: Montagna, W. & Dobson, R.L., eds, *Advances in Biology of Skin*, Vol. VII, *Carcinogenesis*, Oxford, Pergamon Press, pp. 195-214

Urbach, F., Epstein, J.H. & Forbes, P.D. (1974) Ultraviolet carcinogenesis: experimental, global and genetic aspects. In: Pathak, M.A., Harber, L.C., Seiji, M. & Kukita, A., eds, *Sunlight and Man. Normal and Abnormal Photobiologic Responses*, Tokyo, University of Tokyo Press, pp. 259-283

Vågerö, D., Ringbäck, G. & Kiviranta, H. (1986) Melanoma and other tumours of the skin among office, other indoor and outdoor workers in Sweden 1961-1979. *Br. J. Cancer*, **53**, 507-512

Vågerö, D., Swerdlow, A.J. & Beral, V. (1990) Occupation and malignant melanoma: a study based on cancer registration data in England and Wales and in Sweden. *Br. J. ind. Med.*, **47**, 317-324

Valerie, K. & Rosenberg, M. (1990) Chromatin structure implicated in activation of HIV-1 gene expression by ultraviolet light. *New Biol.*, **2**, 712-718

Valerie, K., Delers, A., Bruck, C., Thiriart, C., Rosenberg, H., Debouck, C. & Rosenberg, M. (1988) Activation of human immunodeficiency virus type 1 by DNA damage in human cells. *Nature*, **333**, 78-81

Varghese, A.J. & Patrick, M.H. (1969) Cytosine derived heteroadduct formation in ultraviolet-irradiated DNA. *Nature*, **223**, 299-300

Varghese, A.J. & Wang, S.Y. (1967) Ultraviolet irradiation of DNA *in vitro* and *in vivo* produces a third thymine-derived product. *Science*, **156**, 955-957

van't Veer, L.J., Burgering, B.M.T., Versteeg, R., Boot, A.J.M., Ruiter, D.J., Osanto, S., Schrier, P.I. & Bos, J.L. (1989) N-*ras* Mutations in human cutaneous melanoma from sun-exposed body sites. *Mol. cell. Biol.*, **9**, 3114-3116

Verma, A.K., Lowe, N.J. & Boutwell, R.K. (1979) Induction of mouse epidermal ornithine decarboxylase activity and DNA synthesis by ultraviolet light. *Cancer Res.*, **39**, 1035-1040

Vermeer, M., Schmieder, G.J., Yoshikawa, T., van den Berg, J.-W., Metzman, M.S., Taylor, J.R. & Streilein, J.W. (1991) Effects of ultraviolet B light on cutaneous immune responses of humans with deeply pigmented skin. *J. invest. Dermatol.*, **97**, 729-734

Vitaliano, P.P. (1978) The use of logistic regression for modelling risk factors: with application to nonmelanoma skin cancer. *Am. J. Epidemiol.*, **108**, 402-414

Vitasa, B.C., Taylor, H.R., Strickland, P.T., Rosenthal, F.S., West, S., Abbey, H., Ng, S.K., Munoz, B. & Emmett, E.A. (1990) Association of nonmelanoma skin cancer and actinic keratosis with cumulative solar ultraviolet exposure in Maryland watermen. *Cancer*, **65**, 2811-2817

Vlček, D., Podstavkova, S., Miadokova, E., Adams, G.M.W. & Small, G.D. (1987) General characteristics, molecular and genetic analysis of two new UV-sensitive mutants of *Chlamydomonas reinhardtii*. *Mutat. Res.*, **183**, 169–175

Vogel, J., Cepeda, M., Tschachler, E., Napolitano, L.A. & Jay, G. (1992) UV activation of human immunodeficiency virus gene expression in transgenic mice. *J. Virol.*, **66**, 1–5

Vuillaume, M., Daya-Grosjean, L., Vincens, P., Pennetier, J.L., Tarroux, P., Baret, A., Calvayrac, R., Taieb, A. & Sarasin, A. (1992) Striking differences in cellular catalase activity between two DNA repair-deficient diseases: xeroderma pigmentosum and trichothiodystrophy. *Carcinogenesis*, **13**, 321–328

Walker, G.C. (1984) Mutagenesis and inducible responses to deoxyribonucleic acid damage in *Escherichia coli*. *Microbiol. Rev.*, **48**, 60–93

Walter, S.D., Marrett, L.D., From, L., Hertzman, C., Shannon, H.S. & Roy, P. (1990) The association of cutaneous malignant melanoma with the use of sunbeds and sunlamps. *Am. J. Epidemiol.*, **131**, 232–243

Walter, S.D., Marrett, L.D., Shannon, H.S., From, L. & Hertzman, C. (1992) The association of cutaneous malignant melanoma and fluorescent light exposure. *Am. J. Epidemiol.*, **135**, 749–762

Wang, S.-Y. & Varghese, A.J. (1967) Cytosine–thymine addition product from DNA irradiated with ultraviolet light. *Biochem. biophys. Res. Commun.*, **29**, 543–549

Waterhouse, J., Muir, C., Correa, P. & Powell, J., eds (1976) *Cancer Incidence in Five Continents, Vol. III* (IARC Scientific Publications No. 15), Lyon, IARC

Waterhouse, J., Muir, C., Shanmugaratnam, K. & Powell, J., eds (1982) *Cancer Incidence in Five Continents, Vol. IV* (IARC Scientific Publications No. 42), Lyon, IARC

Weast, R.C., ed. (1989) *CRC Handbook of Chemistry and Physics*, 70th ed., Boca Raton, FL, CRC Press, p. E-209

Webb, R.B. (1977) Lethal and mutagenic effects of near-ultraviolet radiation. *Photochem. Photobiol. Rev.*, **2**, 169–261

Webb, R.B. & Malina, M.M. (1970) Mutagenic effects of near ultraviolet and visible radiant energy on continuous cultures of *Escherichia coli*. *Photochem. Photobiol.*, **12**, 457–468

van Weelden, H. & van der Leun, J.C. (1986) Photorecovery by UVA. In: Urbach, F. & Gange, R.W., eds, *The Biological Effects of UVA Radiation*, New York, Praeger, pp. 147–155

van Weelden, H., de Gruijl, F.R. & van der Leun, J.C. (1986) Carcinogenesis by UVA, with an attempt to assess the carcinogenic risks of tanning with UVA and UVB. In: Urbach, F. & Gange, R.W., eds, *The Biological Effects of UVA Radiation*, New York, Praeger, pp. 137–146

van Weelden, H., de Gruijl, F.R., van der Putte, S.C.J., Toonstra, J. & van der Leun, J.C. (1988) The carcinogenic risks of modern tanning equipment: is UVA safer than UVB? *Arch. dermatol. Res.*, **280**, 300–307

van Weelden, H., van der Putte, S.C.J., Toonstra, J. & van der Leun, J.C. (1990a) UVA-Induced tumours in pigmented hairless mice and the carcinogenic risks of tanning with UVA. *Arch. dermatol. Res.*, **282**, 289–294

van Weelden, H., van der Putte, S.C.J., Toonstra, J. & van der Leun, J.C. (1990b) Ultraviolet B-induced tumors in pigmented hairless mice, with an unsuccessful attempt to induce cutaneous melanoma. *Photodermatol. Photoimmunol. Photomed.*, **7**, 68–72

Weinstock, M.A., Colditz, G.A., Willett, W.C., Stampfer, M.J., Bronstein, B.R., Mihm, M.C., Jr & Speizer, F.E. (1989) Nonfamilial cutaneous melanoma incidence in women associated with sun exposure before 20 years of age. *Pediatrics*, **84**, 199–204

Weinstock, M.A., Colditz, G.A., Willett, W.C., Stampfer, M.J., Bronstein, B.R., Mihm, M.C., Jr & Speizer, F.E. (1991a) Melanoma and the sun: the effect of swimsuits and a 'healthy' tan on the risk of nonfamilial malignant melanoma in women. *Am. J. Epidemiol.*, **134**, 462–470

Weinstock, M.A., Stryker, W.S., Stampfer, M.J., Lew, R.A., Willett, W.C. & Sober, A.J. (1991b) Sunlight and dysplastic nevus risk. Results of a clinic-based case–control study. *Cancer*, **67**, 1701–1706

Weiss, J., Garbe, C., Bertz, J., Biltz, H., Burg, G., Hennes, B., Jung, E.G., Kreysel, H.-W., Orfanos, C.E., Petzold, D., Schwermann, M., Stadler, R., Tilgen, W., Tronnier, H. & Völkers, W. (1990) Risk factors for the development of malignant melanoma in Germany (Ger.). *Hautartz*, **41**, 309–313

Wells, R.L. & Han, A. (1984) Action spectra for killing and mutation of Chinese hamster cells exposed to mid- and near-ultraviolet monochromatic light. *Mutat. Res.*, **129**, 251–258

Welsh, E.A. & Kripke, M.L. (1990) Murine Thy-1$^+$ dendritic epidermal cells induce immunologic tolerance *in vivo*. *J. Immunol.*, **144**, 883–891

West, S.K., Rosenthal, F.S., Bressler, N.M., Bressler, S.B., Muñoz, B., Fine, S.L. & Taylor, H.R. (1989) Exposure to sunlight and other risk factors for age-related macular degeneration. *Arch. ophthalmol.*, **107**, 875–879

Wheatcroft, R., Cox, B.S. & Haynes, R.H. (1975) Repair of UV-induced damage and survival in yeast. 1. Dimer excision. *Mutat. Res.*, **30**, 209–218

Wheeler, C.E., Jr (1975) Pathogenesis of recurrent herpes simplex infections. *J. invest. Dermatol.*, **65**, 341–346

Whelan, S.L., Parkin, D.M. & Masuyer, E., eds (1990) *Patterns of Cancer in Five Continents* (IARC Scientific Publications No. 102), Lyon, IARC, p. 29

Whitaker, C.J., Lee, W.R. & Downes, J.E. (1979) Squamous cell skin cancer in the north-west of England, 1967–69, and its relation to occupation. *Br. J. ind. Med.*, **36**, 43–51

WHO (1977) *Manual of the International Statistical Classification of Diseases, Injuries, and Causes of Death. International Classification of Diseases*, 1975 rev., Vol. 1, Geneva, p. 102

WHO (1979) *Ultraviolet Radiation* (Environmental Health Criteria 14), Geneva

Willis, I., Menter, J.M. & Whyte, H.J. (1981) The rapid induction of cancers in the hairless mouse utilizing the principle of photoaugmentation. *J. invest. Dermatol.*, **76**, 404–408

Willis, I., Menter, J.M., de Fabo, E., Margolin, R. & Fisher, M. (1986) Photoaugmentation of UVB effects by UVA. In: Urbach, F. & Gange, R.W., eds, *The Biological Effects of UVA Radiation*, New York, Praeger, pp. 187–201

Winkelmann, R.K., Baldes, E.J. & Zollman, P.E. (1960) Squamous cell tumors induced in hairless mice with ultraviolet light. *J. invest. Dermatol.*, **34**, 131–138

Winkelmann, R.K., Zollman, P.E. & Baldes, E.J. (1963) Squamous cell carcinoma produced by ultraviolet light in hairless mice. *J. invest. Dermatol.*, **40**, 217–224

Wiskemann, A., Sturm, E. & Klehr, N.W. (1986) Fluorescent lighting enhances chemically induced papilloma formation and increases susceptibility to tumor challenge in mice. *J. Cancer Res. clin. Oncol.*, **112**, 141–143

Withrow, T.J., Lugo, M.H. & Dempsey, M.J. (1980) Transformation of Balb 3T3 cells exposed to a germicidal UV lamp and a sunlamp. *Photochem. Photobiol.*, **31**, 135–141

Witkin, E.M. (1976) Ultraviolet mutagenesis and inducible DNA repair in *Escherichia coli*. *Bacteriol. Rev.*, **40**, 869–907

Wittenberg, S. (1986) Solar radiation and the eye: a review of knowledge relevant to eye care. *Am. J. Optom. physiol. Opt.*, **63**, 676–689

Wolf, C., Steiner, A. & Hönigsmann, H. (1988) Do oral carotenoids protect human skin against ultraviolet erythema, psoralen phototoxicity, and ultraviolet-induced DNA damage? *J. invest. Dermatol.*, **90**, 55–57

Wood, R.D., Skopek, T.R. & Hutchinson, F. (1984) Changes in DNA base sequence induced by targeted mutagenesis of lambda phage by ultraviolet light. *J. mol. Biol.*, **173**, 273–291

Wulff, D.L. & Fraenkel, G. (1961) On the nature of thymine photoproduct. *Biochim. biophys. Acta*, **51**, 332–339

Yang, X.-Y., Ronai, Z.A., Santella, R.M. & Weinstein, I.B. (1988) Effect of 8-methoxypsoralen and ultraviolet light A on EGF receptor (HER-1) expression. *Biochem. biophys. Res. Commun.*, **157**, 590–596

Yasumoto, S., Hayashi, Y. & Aurelian, L. (1987) Immunity to herpes simplex virus type 2. Suppression of virus-induced immune responses in ultraviolet B-irradiated mice. *J. Immunol.*, **139**, 2788–2793

Yoshikawa, J., Rae, V., Bruins-Slot, W., Van den Berg, J.-W., Taylor, J.R. & Streilein, J.W. (1990) Susceptibility to effects of UVB radiation on induction of contact hypersensitivity as a risk factor for skin cancer in humans. *J. invest. Dermatol.*, **95**, 530–536

Young, R.W. (1988) Solar radiation and age-related macular degeneration. *Surv. Ophthalmol.*, **32**, 252–269

Young, M. & Russell, W.T. (1926) *An Investigation into the Statistics of Cancer in Different Trades and Professions* (Medical Research Council Special Report No. 99), London, Her Majesty's Stationery Office

Young, A.R., Walker, S.L., Kinley, J.S., Plastow, S.R., Averbeck, D., Morlière, P. & Dubertret, L. (1990) Phototumorigenesis studies of 5-methoxypsoralen in bergamot oil: evaluation and modification of risk of human use in an albino mouse skin model. *J. Photochem. Photobiol. B. Biol.*, **7**, 231–250

Zanetti, R., Rosso, S., Faggiano, F., Roffino, R., Colonna, S. & Martina, G. (1988) A case–control study on cutaneous malignant melanoma in the province of Torino, Italy (Fr.). *Rev. Epidemiol. Santé publ.*, **36**, 309–317

Zelle, B., Reynolds, R.J., Kottenhagen, M.J., Schuite, A. & Lohman, P.H.M. (1980) The influence of the wavelength of ultraviolet radiation on survival, mutation induction and DNA repair in irradiated Chinese hamster cells. *Mutat. Res.*, **72**, 491–509

Zigman, S. & Vaughan, T. (1974) Near-ultraviolet light effects on the lenses and retinas of mice. *Invest. Ophthalmol.*, **13**, 462–465

Zigman, S., Yulo, T. & Schultz, J. (1974) Cataract induction in mice exposed to near UV light. *Ophthal. Res.*, **6**, 259–270

Zigman, S., Fowler, E. & Kraus, A.L. (1976) Black light induction of skin tumors in mice. *J. invest. Dermatol.*, **67**, 723–725

Zmudzka, B.Z. & Beer, J.Z. (1990) Activation of human immunodeficiency virus by ultraviolet radiation. *Photochem. Photobiol.*, **52**, 1153–1162

Zölzer, F. & Kiefer, J. (1983) Wavelength dependence of inactivation and mutagenesis in haploid yeast cells of different sensitivities. *Photochem. Photobiol.*, **37**, 39–48

Zölzer, F. & Kiefer, J. (1984) Wavelength dependence of inactivation and mutation induction to 6-thioguanine-resistance in V79 Chinese hamster fibroblasts. *Photochem. Photobiol.*, **40**, 49–53

Zölzer, F., Kiefer, J. & Rase, S. (1988) Inactivation and mutation induction to 6-thioguanine resistance in V79 hamster fibroblasts by simulated sunlight. *Photochem. Photobiol.*, **47**, 399–404

SUMMARY OF FINAL EVALUATIONS

Agent	Degree of evidence of carcinogenicity		Overall evaluation of carcinogenicity to humans
	Human	Animal	
Solar radiation	S	S	1
Broad-spectrum ultraviolet radiation		S	
Ultraviolet A radiation		S	2A
Ultraviolet B radiation		S	2A
Ultraviolet C radiation		S	2A
Fluorescent lighting	I		3
Sunlamps and sunbeds, use of	L		2A

S, sufficient evidence; L, limited evidence; I, inadequate evidence; for definitions of degrees of evidence and groupings of evaluations, see Preamble, pp. 32–35.

GLOSSARY OF TERMS

Actinic radiation: electromagnetic radiation capable of initiating photochemical reactions; UVB and UVC radiation (180–315 nm)

Albedo: that fraction of the radiation incident on a surface which is reflected back in all directions

Black light: primarily near-UV radiant energy in the 320–380 nm (or 400 nm) range

Effective irradiance: hypothetical irradiance of monochromatic radiation with a wavelength at which the action spectrum of the relevant photobiological effect is equal to unity (see also section 1.1)

Effective exposure dose: time integral of effective irradiance

Erythema: sunburn

Exposure dose: radiant exposure (J/m^2 unweighted) incident on biologically relevant surface

Fluence: radiant flux passing from all directions through a unit area in J/m^2 or J/cm^2; includes backscatter

Global irradiance: the irradiance of solar radiation at the Earth's surface

Global radiation: solar radiation at the Earth's surface comprising the sum of direct radiation from the sun and diffuse radiation from the sky

Minimal erythema dose (MED): the lowest radiant exposure of UVR that produces a threshold erythemal response 8–24 h after irradiation. There is no consensus on this response; a just perceptible reddening of the skin and erythema with sharp margins are both used as end-points.

Photoreactivation: the enzyme-mediated reversal of the biological effects of UVC or UVB radiation mediated by radiation of longer wavelength and associated with the reversion of cyclobutane-type pyrimidine dimers to monomeric pyrimidines

Radiant exposure: radiant energy delivered to a given area (J/m^2)

Radiant flux: rate of flow of radiant energy (in W)

Solar simulated radiation: radiation from an artificial source (e.g., an optically filtered xenon arc lamp) that approximates the terrestrial solar spectrum

Solar zenith angle: angle between the point in the sky directly overhead (the zenith) and the sun

Spectral distribution: relative intensity of radiation of different wavelengths present in a source emission spectrum

Spectral irradiance: surface density of the radiant flux that is incident on a unit surface area per unit wavelength (see Table 1)

UVA: electromagnetic radiation of wavelength 315–400 nm

UVB: electromagnetic radiation of wavelength 280–315 nm
UVC: electromagnetic radiation of wavelength 100–280 nm
UVR: electromagnetic radiation of wavelength 100–400 nm
Zenith angle: the angle between the point in the sky directly overhead (the zenith) and another point or object

APPENDIX 1. TOPICAL SUNSCREENS

1. General

Sunscreens are physical and chemical topical preparations which attenuate the transmission of solar radiation into the skin by absorption, reflection or scattering. Physical sunscreens (sunblocks), for example zinc oxide or titanium dioxide, function by reflecting and scattering and provide protection against a broad spectrum of UV and visible wavelengths. They are normally nontoxic and have few known adverse effects. Chemical sunscreens contain one or more colourless UV-absorbing ingredients which generally absorb UVB radiation more strongly than UVA. The application of any sunscreen thus normally changes the spectrum of radiation that reaches the target cells. General information is available on sunscreens that have been or are in use (Liem & Hilderink, 1979; Boger *et al.*, 1984; Murphy & Hawk, 1986; Pathak, 1986, 1987; Ramsay, 1989; Lowe & Shaath, 1990; Taylor *et al.*, 1990) and on procedures for testing them (Azizi *et al.*, 1987; Kaidbey & Gange, 1987; Urbach, 1989).

Although most sunscreens are designed to attenuate UVR, some contain additives such as bergamot oil (containing 5-methoxypsoralen; see IARC, 1986, 1987) to enhance pigmentation and photoprotection (Young *et al.*, 1991). The role of such preparations remains controversial.

The generally accepted parameter for evaluating the efficacy of sunscreen preparations is the sun protection factor (SPF), which is defined as the ratio of the least amount of UVR required to produce minimal erythema after application of a standard quantity of the sunscreen product film to the skin to that required to produce the same erythema without sunscreen application. The US Food and Drug Administration (1978) published recommendations for the testing of proprietary sunscreens. Many factors influence SPF values; particularly important are the spectral power distribution of the source used for SPF testing and a clear definition of the end-point used for assessment (see Urbach, 1989). Variations in these factors can lead to considerable differences in measured SPF values for the same product.

SPF values generally reflect the degree of protection against solar UVB radiation, but their protective capacity against UVA must also be defined. Several in-vivo and in-vitro methods have been proposed for defining protection against UVA, but there is no consensus on which is the most appropriate.

Correctly used, sunscreens are effective in preventing erythema. Little information is available, however, on their protective value against harmful immunological changes, photoageing or skin cancer or on their potential long-term adverse effects. The protective and adverse effects of sunscreen use are summarized below.

2. Protective effects

2.1 *Against DNA damage*

UVR inhibits normal (semi-conservative) DNA synthesis. Knowledge about the prevention of DNA damage is based on the results of studies of a small number of sunscreens. In a limited in-vitro study, two commercially available sunscreens (Spectraban, SPF 15.0 and Spectraban, SPF 5.6 [components unspecified]) were tested for their ability to protect against the inhibition of semi-conservative DNA synthesis or the induction of unscheduled DNA synthesis by UVB (300 nm) radiation (Arase & Jung, 1986). Protective factors were found to correlate with the stated SPF values of the sunscreens.

The ability of sunscreens to protect against UV-induced inhibition of DNA synthesis has also been tested in epidermal mouse skin. In a study of seven commercially available sunscreens [components unspecified], the calculated protection factors corresponded fairly well with the SPF values provided by the manufacturers (Walter, 1981). In a study of a single sunscreen (7.5% octyl methoxycinnamate, 4.5% benzophenone-3; SPF, 15), the induction of pyrimidine dimers in human skin *in situ* by a solar simulator (280–400 nm) was measured as a function of fluence (up to 10 times the MED), with or without application of the sunscreen. Dimer induction was reduced by 40-fold in sunscreen-treated skin (Freeman *et al.*, 1988).

2.2 *Against acute and chronic actinic damage*

Protection against erythema is well substantiated by extensive human experience; however, other cellular and metabolic activities may not be afforded the same degree of protection (Pearse & Marks, 1983). In a histological assessment of mouse skin damage, Kligman *et al.* (1982) found that sunscreens provided protection against the effects of chronic sunlamp irradiation. Furthermore, the application of sunscreens (SPF 6 or 15) allowed previously damaged dermis to be repaired despite continued irradiation (Kligman *et al.*, 1983). A UVB sunscreen (2-ethylhexyl 4'-methoxycinnamate, SPF 8) was shown to protect against biochemical changes induced in collagen by Westinghouse FS20 sunlamp irradiation of mouse skin over 12 weeks (Plastow *et al.*, 1988).

2.3 *Against immunological alterations*

Various investigators have examined the efficacy of sunscreens to inhibit photoimmunological reactions in the skin. Inhibition of the development of UV-induced suppression of contact hypersensitivity has been reported (Morison, 1984), but in other studies sunscreens have been ineffective in preventing immunosuppression (Gurish *et al.*, 1981; Hersey *et al.*, 1987; Fisher *et al.*, 1989; van Praag *et al.*, 1991), or mixed results have been obtained depending on the sunscreen used (Reeve *et al.*, 1991). [The Working Group concluded that no consistent relationship could be assumed between protection against photoimmunological events and erythema and other changes in the skin.]

2.4 *Against tumour formation*

Some sunscreens have been shown to protect mice against UV-induced skin tumour formation (Knox *et al.*, 1960; Kligman *et al.*, 1980; Wulf *et al.*, 1982; Gallagher *et al.*, 1984; Morison, 1984). Demonstration of effectiveness against skin tumour formation is, however, not required by regulatory bodies in evaluations of sunscreens. Sunscreen use may encourage people to have longer overall exposure to sunlight, because protection by the sunscreen reduces the effective irradiance. Kelfkens et al. (1991) observed that exposure of mice to a daily dose of UVB over a longer period gives a higher tumour yield than the same dose given over a shorter period. Accordingly, any assessment of the overall impact of sunscreens in reducing human skin cancer should take into account both the efficacy of sunscreens in reducing UV-induced damage to the skin and concomitant human behavioural changes with respect to time spent in the sun. In some case–control studies (e.g., Holman et al., 1986; Beitner et al., 1990), use of sunscreens has been associated with an increased risk for melanoma. This association is probably the result of confounding of sun exposure by skin type or amount of exposure, because individuals who easily get sunburned or expose themselves heavily (and who are at increased risk of skin cancer) may use sunscreens more frequently than other people.

3. Adverse effects

3.1 *Acute toxicity*

Acute toxic side-effects of specific sunscreen agents include contact irritation, allergic contact dermatitis, phototoxicity, photoallergy and staining of the skin (Schauder & Ippen, 1986; Pathak, 1987; Knobler et al., 1989).

3.2 *Chronic toxicity*

Relatively little information is available on the mutagenic and carcinogenic potential of sunscreen agents. This deficiency was reviewed in a report by the US National Cancer Institute (1989), which recommended the following six compounds for chronic testing in the US National Toxicology Program rodent test programme: cinoxate, 2-ethylhexyl 2-cyano-3,3-diphenyl-acrylate, 2-ethylhexyl *para*-methoxycinnamate, homosalate, methyl anthranilate and oxybenzone. The bases for selecting these compounds, together with extensive references, are given in the report. In short, neither epidemiological data nor long-term mammalian carcinogenicity studies are available on these compounds. The results of in-vitro testing were assessed as either negative or inconsistent among test systems or among batches of a compound (because of impurities). 2-Ethylhexyl *para*-methoxycinnamate was implicated as a potential tumour initiator in one study in which hairless mice were painted with the compound over a nine-week period and subsequently treated with the tumour promoter, croton oil (Gallagher et al., 1984). Subsequent work by Reeve et al. (1985), however, failed to confirm these results, and Forbes et al. (1989) found no evidence of tumour initiation by the compound in an initiation–promotion experiment in mice.

trans-Urocanic acid (an additive in some commercial sunscreen products) increased the yield of simulated solar UV-induced tumours in hairless mice (Reeve et al., 1989). The significance of this finding for human exposure has not been evaluated.

3.3 *Reduced vitamin D synthesis*

Vitamin D production is almost completely blocked in subjects who use UVB sunscreens (Matsuoka et al., 1987). This finding may be significant for elderly individuals, who are already at risk for vitamin D_3 deficiency (MacLaughlin & Holick, 1985), but its significance for clinical disease remains unknown (Fine, 1988).

4. References

Arase, S. & Jung, E.G. (1986) In vitro evaluation of the photoprotective efficacy of sunscreens against DNA damage by UVB. *Photodermatology*, 3, 56–59

Azizi, E., Modan, M., Kushelevsky, A.P. & Schewach-Millet, M. (1987) A more reliable index of sunscreen protection, based on life table analysis of individual sun protection factors. *Br. J. Dermatol.*, 116, 693–702

Beitner, H., Norell, S.E., Ringborg, U., Wennersten, G. & Mattson, B. (1990) Malignant melanoma: aetiological importance of individual pigmentation and sun exposure. *Br. J. Dermatol.*, 122, 43–51

Boger, J., Araujo, O.E. & Flowers, F. (1984) Sunscreens: efficacy, use and misuse. *South. med. J.*, 77, 1421–1427

Fine, R.M. (1988) Sunscreens and cutaneous vitamin D synthesis. *Int. J. Dermatol.*, 27, 300–301

Fisher, M.S., Menter, J.M. & Willis, I. (1989) Ultraviolet radiation-induced suppression of contact hypersensitivity in relation to Padimate O and oxybenzone. *J. invest. Dermatol.*, 92, 337–341

Forbes, P.D., Davies, R.E., Sambuco, C.P. & Urbach, F. (1989) Inhibition of ultraviolet radiation-induced skin tumors in hairless mice by topical application of the sunscreen 2-ethylhexyl-p-methoxycinnamate. *J. Toxicol. cutaneous ocul. Toxicol.*, 8, 209–226

Freeman, S.E., Ley, R.D. & Ley, K.D. (1988) Sunscreen protection against UV-induced pyrimidine dimers in DNA of human skin *in situ*. *Photodermatology*, 5, 243–247

Gallagher, C.H., Greenoak, G.E., Reeve, V.E., Canfield, P.J., Baker, R.S.U. & Bonin, A.M. (1984) Ultraviolet carcinogenesis in the hairless mouse skin—influence of the sunscreen 2-ethylhexyl-*p*-methoxycinnamate. *Aust. J. exp. Biol. med. Sci.*, 62, 577–588

Gurish, M.F., Roberts, L.K., Krueger, G.G. & Daynes, R.A. (1981) The effect of various sunscreen agents on skin damage and the induction of tumor susceptibility in mice subjected to ultraviolet irradiation. *J. invest. Dermatol.*, 76, 246–251

Hersey, P., MacDonald, M., Burns, C., Schibeci, S., Matthews, H. & Wilkinson, F.J. (1987) Analysis of the effect of a sunscreen agent on the suppression of natural killer cell activity induced in human subjects by radiation from solarium lamps. *J. invest. Dermatol.*, 88, 271–276

Holman, C.D.J., Armstrong, B.K. & Heenan, P.J. (1986) Relationship of cutaneous malignant melanoma to individual sunlight-exposure habits. *J. natl Cancer Inst.*, 76, 403–414

IARC (1986) *IARC Monographs on the Evaluation of the Carcinogenic Risk of Chemicals to Humans*, Vol. 40, *Some Naturally Occurring and Synthetic Food Components, Furocoumarins and Ultraviolet Radiation*, Lyon, pp. 327–347

IARC (1987) *IARC Monographs on the Evaluation of Carcinogenic Risks to Humans*, Suppl. 7, *Overall Evaluations of Carcinogenicity: An Updating of* IARC Monographs *Volumes 1 to 42*, Lyon, pp. 243–245

Kaidbey, K. & Gange, R.W. (1987) Comparison of methods for assessing photoprotection against ultraviolet A *in vivo*. *J. Am. Acad. Dermatol.*, 16, 346–353

Kelfkens, G., van Weelden, H., de Gruijl, F.R. & van der Leun, J.C. (1991) The influence of dose rate on ultraviolet tumorigenesis. *J. Photochem. Photobiol. B. Biol.*, 10, 41–50

Kligman, L.H., Akin, F.J. & Kligman, A.M. (1980) Sunscreens prevent ultraviolet photocarcinogenesis. *J. Am. Acad. Dermatol.*, 3, 30–35

Kligman, L.H., Akin, F.J. & Kligman, A.M. (1982) Prevention of ultraviolet damage to the dermis of hairless mice by sunscreens. *J. invest. Dermatol.*, 78, 181–189

Kligman, L.H., Akin, F.J. & Kligman, A.M. (1983) Sunscreens promote repair of ultraviolet radiation-induced dermal damage. *J. invest. Dermatol.*, **81**, 98–102

Knobler, E., Almeida, L., Ruzkowski, A.M., Held, J., Harber, L. & DeLeo, V. (1989) Photoallergy to benzophenone. *Arch. Dermatol.*, **125**, 801–804

Knox, J.M., Griffin, A.C. & Hakim, R.E. (1960) Protection from ultraviolet carcinogenesis. *J. invest. Dermatol.*, **34**, 51–58

Liem, D.H. & Hilderink, L.T.H. (1979) UV absorbers in sun cosmetics 1978. *Int. J. cosmet. Sci.*, **1**, 341–361

Lowe, N.J. & Shaath, N.A., eds (1990) *Sunscreens. Development, Evaluation and Regulatory Aspects*, New York, Marcel Dekker

MacLaughlin, J.A. & Holick, M.F. (1985) Aging decreases the capacity of human skin to produce vitamin D_3. *J. clin. Invest.*, **76**, 1536–1538

Matsuoka, L.Y., Ide, L., Wortsman J., McLaughlin, J.A. & Holick, M.F. (1987) Sunscreens suppress cutaneous vitamin D_3 synthesis. *J. clin. Endocrinol. Metab.*, **64**, 1165–1168

Morison, W.L. (1984) The effect of a sunscreen containing *para*-aminobenzoic acid on the systemic immunologic alterations induced in mice by exposure to UVB radiation. *J. invest. Dermatol.*, **83**, 405–408

Murphy, G.M. & Hawk, J.L.M. (1986) Sunscreens. *J. R. Soc. Med.*, **79**, 254–256

Pathak, M.A. (1986) Sunscreens. Topical and systemic approaches for the prevention of acute and chronic sun-induced skin reactions. *Dermatol. Clin.*, **4**, 321–334

Pathak, M.A. (1987) Sunscreens and their use in the preventive treatment of sunlight-induced skin damage. *J. Dermatol. surg. Oncol.*, **13**, 739–750

Pearse, A.D. & Marks, R. (1983) Response of human skin to ultraviolet radiation: dissociation of erythema and metabolic changes following sunscreen protection. *J. invest. Dermatol.*, **80**, 191–194

Plastow, S.R., Harrison, J.A. & Young, A.R. (1988) Early changes in dermal collagen of mice exposed to chronic UVB irradiation and the effects of a UVB sunscreen. *J. invest. Dermatol.*, **91**, 590–592

van Praag, M.C.G., Out-Luyting, C., Claas, F.H.J., Vermeer, B.-J. & Mommaas, A.M. (1991) Effect of topical sunscreens on the UV-radiation-induced suppression of the alloactivating capacity in human skin *in vivo*. *J. invest. Dermatol.*, **97**, 629–633

Ramsay, C.A. (1989) Ultraviolet A protective sunscreens. *Clin. Dermatol.*, **7**, 163–166

Reeve, V.E., Greenoak, G.E., Gallagher, C.H., Canfield, P.J. & Wilkinson, F.J. (1985) Effect of immunosuppressive agents and sunscreens on UV carcinogenesis in the hairless mouse. *Aust. J. exp. Biol. med. Sci.*, **63**, 655–665

Reeve, V.E., Greenoak, G.E., Canfield, P.J., Boehm-Wilcox, C. & Gallagher, C.H. (1989) Topical urocanic acid enhances UV-induced tumour yield and malignancy in the hairless mouse. *Photochem. Photobiol.*, **49**, 459–464

Reeve, V.E., Bosnic, M., Boehm-Wilcox, C. & Ley, R.D. (1991) Differential protection by two sunscreens from UV radiation-induced immunosuppression. *J. invest. Dermatol.*, **97**, 624–628

Schauder, S. & Ippen, H. (1986) Photoallergic and allergic contact dermatitis from dibenzoylmethanes. *Photodermatology*, **3**, 140–147

Taylor, C.R., Stern, R.S., Leyden, J.J. & Gilchrest, B.A. (1990) Photoaging/photodamage and photoprotection. *J. Am. Acad. Dermatol.*, **22**, 1–15

Urbach, F. (1989) Testing the efficacy of sunscreens: effect of choice of source and spectral power distribution of ultraviolet radiation, and choice of endpoint. *Photodermatology*, **6**, 177–181

US Food and Drug Administration (1978) Sunscreen drug products for over-the-counter human use. Establishment of a monograph; notice of proposed rulemaking. *Fed. Regist.*, **43**, 38206–38269

US National Cancer Institute (1989) *Sunscreens* (Class Study Report; Contract No. NO1-CP-71082 (7/89)), Rockville, MD, Tracor Technological Resources Inc.

Walter, J.F. (1981) Evaluation of seven sunscreens on hairless mouse skin. *Arch. Dermatol.*, **117**, 547–550

Wulf, H.C., Poulsen, T., Brodthagen, H. & Hou-Jensen, K. (1982) Sunscreens for delay of ultraviolet induction of skin tumors. *J. Am. Acad. Dermatol.*, **7**, 194–202

Young, A.R., Potten, C.S., Chadwick, C.A., Murphy, G.M., Hawk, J.L.M. & Cohen, A.J. (1991) Photoprotection and 5-MOP photochemoprotection from UVR-induced DNA damage in humans: the role of skin type. *J. invest. Dermatol.*, **97**, 942–948

CUMULATIVE CROSS INDEX TO *IARC MONOGRAPHS ON THE EVALUATION OF CARCINOGENIC RISKS TO HUMANS*

The volume, page and year are given. References to corrigenda are given in parentheses.

A

A-α-C	40, 245 (1986); *Suppl. 7*, 56 (1987)
Acetaldehyde	36, 101 (1985) (*corr. 42*, 263); *Suppl. 7*, 77 (1987)
Acetaldehyde formylmethylhydrazone (*see* Gyromitrin)	
Acetamide	7, 197 (1974); *Suppl. 7*, 389 (1987)
Acetaminophen (*see* Paracetamol)	
Acridine orange	16, 145 (1978); *Suppl. 7*, 56 (1987)
Acriflavinium chloride	13, 31 (1977); *Suppl. 7*, 56 (1987)
Acrolein	19, 479 (1979); 36, 133 (1985); *Suppl. 7*, 78 (1987)
Acrylamide	39, 41 (1986); *Suppl. 7*, 56 (1987)
Acrylic acid	19, 47 (1979); *Suppl. 7*, 56 (1987)
Acrylic fibres	19, 86 (1979); *Suppl. 7*, 56 (1987)
Acrylonitrile	19, 73 (1979); *Suppl. 7*, 79 (1987)
Acrylonitrile-butadiene-styrene copolymers	19, 91 (1979); *Suppl. 7*, 56 (1987)
Actinolite (*see* Asbestos)	
Actinomycins	10, 29 (1976) (*corr. 42*, 255); *Suppl. 7*, 80 (1987)
Adriamycin	10, 43 (1976); *Suppl. 7*, 82 (1987)
AF-2	31, 47 (1983); *Suppl. 7*, 56 (1987)
Aflatoxins	1, 145 (1972) (*corr. 42*, 251); 10, 51 (1976); *Suppl. 7*, 83 (1987)
Aflatoxin B_1 (*see* Aflatoxins)	
Aflatoxin B_2 (*see* Aflatoxins)	
Aflatoxin G_1 (*see* Aflatoxins)	
Aflatoxin G_2 (*see* Aflatoxins)	
Aflatoxin M_1 (*see* Aflatoxins)	
Agaritine	31, 63 (1983); *Suppl. 7*, 56 (1987)
Alcohol drinking	44 (1988)
Aldicarb	53, 93 (1991)
Aldrin	5, 25 (1974); *Suppl. 7*, 88 (1987)
Allyl chloride	36, 39 (1985); *Suppl. 7*, 56 (1987)
Allyl isothiocyanate	36, 55 (1985); *Suppl. 7*, 56 (1987)
Allyl isovalerate	36, 69 (1985); *Suppl. 7*, 56 (1987)
Aluminium production	34, 37 (1984); *Suppl. 7*, 89 (1987)
Amaranth	8, 41 (1975); *Suppl. 7*, 56 (1987)
5-Aminoacenaphthene	16, 243 (1978); *Suppl. 7*, 56 (1987)

2-Aminoanthraquinone	27, 191 (1982); *Suppl.* 7, 56 (1987)
para-Aminoazobenzene	8, 53 (1975); *Suppl.* 7, 390 (1987)
ortho-Aminoazotoluene	8, 61 (1975) (*corr.* 42, 254); *Suppl.* 7, 56 (1987)
para-Aminobenzoic acid	16, 249 (1978); *Suppl.* 7, 56 (1987)
4-Aminobiphenyl	1, 74 (1972) (*corr.* 42, 251); *Suppl.* 7, 91 (1987)
2-Amino-3,4-dimethylimidazo[4,5-*f*]quinoline (*see* MeIQ)	
2-Amino-3,8-dimethylimidazo[4,5-*f*]quinoxaline (*see* MeIQx)	
3-Amino-1,4-dimethyl-5*H*-pyrido[4,3-*b*]indole (*see* Trp-P-1)	
2-Aminodipyrido[1,2-*a*:3′,2′-*d*]imidazole (*see* Glu-P-2)	
1-Amino-2-methylanthraquinone	27, 199 (1982); *Suppl.* 7, 57 (1987)
2-Amino-3-methylimidazo[4,5-*f*]quinoline (*see* IQ)	
2-Amino-6-methyldipyrido[1,2-*a*:3′,2′-*d*]imidazole (*see* Glu-P-1)	
2-Amino-3-methyl-9*H*-pyrido[2,3-*b*]indole (*see* MeA-α-C)	
3-Amino-1-methyl-5*H*-pyrido[4,3-*b*]indole (*see* Trp-P-2)	
2-Amino-5-(5-nitro-2-furyl)-1,3,4-thiadiazole	7, 143 (1974); *Suppl.* 7, 57 (1987)
4-Amino-2-nitrophenol	16, 43 (1978); *Suppl.* 7, 57 (1987)
2-Amino-5-nitrothiazole	31, 71 (1983); *Suppl.* 7, 57 (1987)
2-Amino-9*H*-pyrido[2,3-*b*]indole (*see* A-α-C)	
11-Aminoundecanoic acid	39, 239 (1986); *Suppl.* 7, 57 (1987)
Amitrole	7, 31 (1974); 41, 293 (1986) (*corr.* 52, 513; *Suppl.* 7, 92 (1987)
Ammonium potassium selenide (*see* Selenium and selenium compounds)	
Amorphous silica (*see also* Silica)	42, 39 (1987); *Suppl.* 7, 341 (1987)
Amosite (*see* Asbestos)	
Ampicillin	50, 153 (1990)
Anabolic steroids (*see* Androgenic (anabolic) steroids)	
Anaesthetics, volatile	11, 285 (1976); *Suppl.* 7, 93 (1987)
Analgesic mixtures containing phenacetin (*see also* Phenacetin)	*Suppl.* 7, 310 (1987)
Androgenic (anabolic) steroids	*Suppl.* 7, 96 (1987)
Angelicin and some synthetic derivatives (*see also* Angelicins)	40, 291 (1986)
Angelicin plus ultraviolet radiation (*see also* Angelicin and some synthetic derivatives)	*Suppl.* 7, 57 (1987)
Angelicins	*Suppl.* 7, 57 (1987)
Aniline	4, 27 (1974) (*corr.* 42, 252); 27, 39 (1982); *Suppl.* 7, 99 (1987)
ortho-Anisidine	27, 63 (1982); *Suppl.* 7, 57 (1987)
para-Anisidine	27, 65 (1982); *Suppl.* 7, 57 (1987)
Anthanthrene	32, 95 (1983); *Suppl.* 7, 57 (1987)
Anthophyllite (*see* Asbestos)	
Anthracene	32, 105 (1983); *Suppl.* 7, 57 (1987)
Anthranilic acid	16, 265 (1978); *Suppl.* 7, 57 (1987)
Antimony trioxide	47, 291 (1989)
Antimony trisulfide	47, 291 (1989)
ANTU (*see* 1-Naphthylthiourea)	
Apholate	9, 31 (1975); *Suppl.* 7, 57 (1987)
Aramite®	5, 39 (1974); *Suppl.* 7, 57 (1987)
Areca nut (*see* Betel quid)	
Arsanilic acid (*see* Arsenic and arsenic compounds)	
Arsenic and arsenic compounds	1, 41 (1972); 2, 48 (1973); 23, 39 (1980); *Suppl.* 7, 100 (1987)

Arsenic pentoxide (see Arsenic and arsenic compounds)	
Arsenic sulfide (see Arsenic and arsenic compounds)	
Arsenic trioxide (see Arsenic and arsenic compounds)	
Arsine (see Arsenic and arsenic compounds)	
Asbestos	2, 17 (1973) (corr. 42, 252); 14 (1977) (corr. 42, 256); Suppl. 7, 106 (1987) (corr. 45, 283)
Atrazine	53, 441 (1991)
Attapulgite	42, 159 (1987); Suppl. 7, 117 (1987)
Auramine (technical-grade)	1, 69 (1972) (corr. 42, 251); Suppl. 7, 118 (1987)
Auramine, manufacture of (see also Auramine, technical-grade)	Suppl. 7, 118 (1987)
Aurothioglucose	13, 39 (1977); Suppl. 7, 57 (1987)
Azacitidine	26, 37 (1981); Suppl. 7, 57 (1987); 50, 47 (1990)
5-Azacytidine (see Azacitidine)	
Azaserine	10, 73 (1976) (corr. 42, 255); Suppl. 7, 57 (1987)
Azathioprine	26, 47 (1981); Suppl. 7, 119 (1987)
Aziridine	9, 37 (1975); Suppl. 7, 58 (1987)
2-(1-Aziridinyl)ethanol	9, 47 (1975); Suppl. 7, 58 (1987)
Aziridyl benzoquinone	9, 51 (1975); Suppl. 7, 58 (1987)
Azobenzene	8, 75 (1975); Suppl. 7, 58 (1987)

B

Barium chromate (see Chromium and chromium compounds)	
Basic chromic sulfate (see Chromium and chromium compounds)	
BCNU (see Bischloroethyl nitrosourea)	
Benz[a]acridine	32, 123 (1983); Suppl. 7, 58 (1987)
Benz[c]acridine	3, 241 (1973); 32, 129 (1983); Suppl. 7, 58 (1987)
Benzal chloride (see also α-Chlorinated toluenes)	29, 65 (1982); Suppl. 7, 148 (1987)
Benz[a]anthracene	3, 45 (1973); 32, 135 (1983); Suppl. 7, 58 (1987)
Benzene	7, 203 (1974) (corr. 42, 254); 29, 93, 391 (1982); Suppl. 7, 120 (1987)
Benzidine	1, 80 (1972); 29, 149, 391 (1982); Suppl. 7, 123 (1987)
Benzidine-based dyes	Suppl. 7, 125 (1987)
Benzo[b]fluoranthene	3, 69 (1973); 32, 147 (1983); Suppl. 7, 58 (1987)
Benzo[j]fluoranthene	3, 82 (1973); 32, 155 (1983); Suppl. 7, 58 (1987)
Benzo[k]fluoranthene	32, 163 (1983); Suppl. 7, 58 (1987)
Benzo[ghi]fluoranthene	32, 171 (1983); Suppl. 7, 58 (1987)
Benzo[a]fluorene	32, 177 (1983); Suppl. 7, 58 (1987)
Benzo[b]fluorene	32, 183 (1983); Suppl. 7, 58 (1987)
Benzo[c]fluorene	32, 189 (1983); Suppl. 7, 58 (1987)
Benzo[ghi]perylene	32, 195 (1983); Suppl. 7, 58 (1987)
Benzo[c]phenanthrene	32, 205 (1983); Suppl. 7, 58 (1987)

Benzo[a]pyrene	3, 91 (1973); 32, 211 (1983); Suppl. 7, 58 (1987)
Benzo[e]pyrene	3, 137 (1973); 32, 225 (1983); Suppl. 7, 58 (1987)
para-Benzoquinone dioxime	29, 185 (1982); Suppl. 7, 58 (1987)
Benzotrichloride (see also α-Chlorinated toluenes)	29, 73 (1982); Suppl. 7, 148 (1987)
Benzoyl chloride	29, 83 (1982) (corr. 42, 261); Suppl. 7, 126 (1987)
Benzoyl peroxide	36, 267 (1985); Suppl. 7, 58 (1987)
Benzyl acetate	40, 109 (1986); Suppl. 7, 58 (1987)
Benzyl chloride (see also α-Chlorinated toluenes)	11, 217 (1976) (corr. 42, 256); 29, 49 (1982); Suppl. 7, 148 (1987)
Benzyl violet 4B	16, 153 (1978); Suppl. 7, 58 (1987)
Bertrandite (see Beryllium and beryllium compounds)	
Beryllium and beryllium compounds	1, 17 (1972); 23, 143 (1980) (corr. 42, 260); Suppl. 7, 127 (1987)
Beryllium acetate (see Beryllium and beryllium compounds)	
Beryllium acetate, basic (see Beryllium and beryllium compounds)	
Beryllium–aluminium alloy (see Beryllium and beryllium compounds)	
Beryllium carbonate (see Beryllium and beryllium compounds)	
Beryllium chloride (see Beryllium and beryllium compounds)	
Beryllium–copper alloy (see Beryllium and beryllium compounds)	
Beryllium–copper–cobalt alloy (see Beryllium and beryllium compounds)	
Beryllium fluoride (see Beryllium and beryllium compounds)	
Beryllium hydroxide (see Beryllium and beryllium compounds)	
Beryllium–nickel alloy (see Beryllium and beryllium compounds)	
Beryllium oxide (see Beryllium and beryllium compounds)	
Beryllium phosphate (see Beryllium and beryllium compounds)	
Beryllium silicate (see Beryllium and beryllium compounds)	
Beryllium sulfate (see Beryllium and beryllium compounds)	
Beryl ore (see Beryllium and beryllium compounds)	
Betel quid	37, 141 (1985); Suppl. 7, 128 (1987)
Betel-quid chewing (see Betel quid)	
BHA (see Butylated hydroxyanisole)	
BHT (see Butylated hydroxytoluene)	
Bis(1-aziridinyl)morpholinophosphine sulfide	9, 55 (1975); Suppl. 7, 58 (1987)
Bis(2-chloroethyl)ether	9, 117 (1975); Suppl. 7, 58 (1987)
N,N-Bis(2-chloroethyl)-2-naphthylamine	4, 119 (1974) (corr. 42, 253); Suppl. 7, 130 (1987)
Bischloroethyl nitrosourea (see also Chloroethyl nitrosoureas)	26, 79 (1981); Suppl. 7, 150 (1987)
1,2-Bis(chloromethoxy)ethane	15, 31 (1977); Suppl. 7, 58 (1987)
1,4-Bis(chloromethoxymethyl)benzene	15, 37 (1977); Suppl. 7, 58 (1987)
Bis(chloromethyl)ether	4, 231 (1974) (corr. 42, 253); Suppl. 7, 131 (1987)
Bis(2-chloro-1-methylethyl)ether	41, 149 (1986); Suppl. 7, 59 (1987)
Bis(2,3-epoxycyclopentyl)ether	47, 231 (1989)
Bisphenol A diglycidyl ether (see Glycidyl ethers)	
Bisulfites (see Sulfur dioxide and some sulfites, bisulfites and metabisulfites)	
Bitumens	35, 39 (1985); Suppl. 7, 133 (1987)
Bleomycins	26, 97 (1981); Suppl. 7, 134 (1987)
Blue VRS	16, 163 (1978); Suppl. 7, 59 (1987)
Boot and shoe manufacture and repair	25, 249 (1981); Suppl. 7, 232 (1987)

Bracken fern	*40*, 47 (1986); *Suppl. 7*, 135 (1987)
Brilliant Blue FCF, disodium salt	*16*, 171 (1978) (*corr. 42*, 257); *Suppl. 7*, 59 (1987)
Bromochloroacetonitrile (*see* Halogenated acetonitriles)	
Bromodichloromethane	*52*, 179 (1991)
Bromoethane	*52*, 299 (1991)
Bromoform	*52*, 213 (1991)
1,3-Butadiene	*39*, 155 (1986) (*corr. 42*, 264); *Suppl. 7*, 136 (1987); *54*, 237 (1992)
1,4-Butanediol dimethanesulfonate	*4*, 247 (1974); *Suppl. 7*, 137 (1987)
n-Butyl acrylate	*39*, 67 (1986); *Suppl. 7*, 59 (1987)
Butylated hydroxyanisole	*40*, 123 (1986); *Suppl. 7*, 59 (1987)
Butylated hydroxytoluene	*40*, 161 (1986); *Suppl. 7*, 59 (1987)
Butyl benzyl phthalate	*29*, 193 (1982) (*corr. 42*, 261); *Suppl. 7*, 59 (1987)
β-Butyrolactone	*11*, 225 (1976); *Suppl. 7*, 59 (1987)
γ-Butyrolactone	*11*, 231 (1976); *Suppl. 7*, 59 (1987)

C

Cabinet-making (*see* Furniture and cabinet-making)	
Cadmium acetate (*see* Cadmium and cadmium compounds)	
Cadmium and cadmium compounds	*2*, 74 (1973); *11*, 39 (1976) (*corr. 42*, 255); *Suppl. 7*, 139 (1987)
Cadmium chloride (*see* Cadmium and cadmium compounds)	
Cadmium oxide (*see* Cadmium and cadmium compounds)	
Cadmium sulfate (*see* Cadmium and cadmium compounds)	
Cadmium sulfide (*see* Cadmium and cadmium compounds)	
Caffeine	*51*, 291 (1991)
Calcium arsenate (*see* Arsenic and arsenic compounds)	
Calcium chromate (*see* Chromium and chromium compounds)	
Calcium cyclamate (*see* Cyclamates)	
Calcium saccharin (*see* Saccharin)	
Cantharidin	*10*, 79 (1976); *Suppl. 7*, 59 (1987)
Caprolactam	*19*, 115 (1979) (*corr. 42*, 258); *39*, 247 (1986) (*corr. 42*, 264); *Suppl. 7*, 390 (1987)
Captafol	*53*, 353 (1991)
Captan	*30*, 295 (1983); *Suppl. 7*, 59 (1987)
Carbaryl	*12*, 37 (1976); *Suppl. 7*, 59 (1987)
Carbazole	*32*, 239 (1983); *Suppl. 7*, 59 (1987)
3-Carbethoxypsoralen	*40*, 317 (1986); *Suppl. 7*, 59 (1987)
Carbon blacks	*3*, 22 (1973); *33*, 35 (1984); *Suppl. 7*, 142 (1987)
Carbon tetrachloride	*1*, 53 (1972); *20*, 371 (1979); *Suppl. 7*, 143 (1987)
Carmoisine	*8*, 83 (1975); *Suppl. 7*, 59 (1987)
Carpentry and joinery	*25*, 139 (1981); *Suppl. 7*, 378 (1987)
Carrageenan	*10*, 181 (1976) (*corr. 42*, 255); *31*, 79 (1983); *Suppl. 7*, 59 (1987)
Catechol	*15*, 155 (1977); *Suppl. 7*, 59 (1987)
CCNU (*see* 1-(2-Chloroethyl)-3-cyclohexyl-1-nitrosourea)	

Ceramic fibres (*see* Man-made mineral fibres)
Chemotherapy, combined, including alkylating agents (*see* MOPP and other combined chemotherapy including alkylating agents)

Chlorambucil	9, 125 (1975); 26, 115 (1981); *Suppl.* 7, 144 (1987)
Chloramphenicol	10, 85 (1976); *Suppl.* 7, 145 (1987); 50, 169 (1990)
Chlorendic acid	48, 45 (1990)
Chlordane (*see also* Chlordane/Heptachlor)	20, 45 (1979) (*corr.* 42, 258)
Chlordane/Heptachlor	*Suppl.* 7, 146 (1987); 53, 115 (1991)
Chlordecone	20, 67 (1979); *Suppl.* 7, 59 (1987)
Chlordimeform	30, 61 (1983); *Suppl.* 7, 59 (1987)
Chlorinated dibenzodioxins (other than TCDD)	15, 41 (1977); *Suppl.* 7, 59 (1987)
Chlorinated drinking-water	52, 45 (1991)
Chlorinated paraffins	48, 55 (1990)
α-Chlorinated toluenes	*Suppl.* 7, 148 (1987)
Chlormadinone acetate (*see also* Progestins; Combined oral contraceptives)	6, 149 (1974); 21, 365 (1979)
Chlornaphazine (*see* N,N-Bis(2-chloroethyl)-2-naphthylamine)	
Chloroacetonitrile (*see* Halogenated acetonitriles)	
Chlorobenzilate	5, 75 (1974); 30, 73 (1983); *Suppl.* 7, 60 (1987)
Chlorodibromomethane	52, 243 (1991)
Chlorodifluoromethane	41, 237 (1986) (*corr.* 51, 483); *Suppl.* 7, 149 (1987)
Chloroethane	52, 315 (1991)
1-(2-Chloroethyl)-3-cyclohexyl-1-nitrosourea (*see also* Chloroethyl nitrosoureas)	26, 137 (1981) (*corr.* 42, 260); *Suppl.* 7, 150 (1987)
1-(2-Chloroethyl)-3-(4-methylcyclohexyl)-1-nitrosourea (*see also* Chloroethyl nitrosoureas)	*Suppl.* 7, 150 (1987)
Chloroethyl nitrosoureas	*Suppl.* 7, 150 (1987)
Chlorofluoromethane	41, 229 (1986); *Suppl.* 7, 60 (1987)
Chloroform	1, 61 (1972); 20, 401 (1979); *Suppl.* 7, 152 (1987)
Chloromethyl methyl ether (technical-grade) (*see also* Bis(chloromethyl)ether)	4, 239 (1974); *Suppl.* 7, 131 (1987)
(4-Chloro-2-methylphenoxy)acetic acid (*see* MCPA)	
Chlorophenols	*Suppl.* 7, 154 (1987)
Chlorophenols (occupational exposures to)	41, 319 (1986)
Chlorophenoxy herbicides	*Suppl.* 7, 156 (1987)
Chlorophenoxy herbicides (occupational exposures to)	41, 357 (1986)
4-Chloro-*ortho*-phenylenediamine	27, 81 (1982); *Suppl.* 7, 60 (1987)
4-Chloro-*meta*-phenylenediamine	27, 82 (1982); *Suppl.* 7, 60 (1987)
Chloroprene	19, 131 (1979); *Suppl.* 7, 160 (1987)
Chloropropham	12, 55 (1976); *Suppl.* 7, 60 (1987)
Chloroquine	13, 47 (1977); *Suppl.* 7, 60 (1987)
Chlorothalonil	30, 319 (1983); *Suppl.* 7, 60 (1987)
para-Chloro-*ortho*-toluidine and its strong acid salts (*see also* Chlordimeform)	16, 277 (1978); 30, 65 (1983); *Suppl.* 7, 60 (1987); 48, 123 (1990)
Chlorotrianisene (*see also* Nonsteroidal oestrogens)	21, 139 (1979)
2-Chloro-1,1,1-trifluoroethane	41, 253 (1986); *Suppl.* 7, 60 (1987)
Chlorozotocin	50, 65 (1990)

Cholesterol	10, 99 (1976); 31, 95 (1983); Suppl. 7, 161 (1987)
Chromic acetate (see Chromium and chromium compounds)	
Chromic chloride (see Chromium and chromium compounds)	
Chromic oxide (see Chromium and chromium compounds)	
Chromic phosphate (see Chromium and chromium compounds)	
Chromite ore (see Chromium and chromium compounds)	
Chromium and chromium compounds	2, 100 (1973); 23, 205 (1980); Suppl. 7, 165 (1987); 49, 49 (1990) (corr. 51, 483)
Chromium carbonyl (see Chromium and chromium compounds)	
Chromium potassium sulfate (see Chromium and chromium compounds)	
Chromium sulfate (see Chromium and chromium compounds)	
Chromium trioxide (see Chromium and chromium compounds)	
Chrysazin (see Dantron)	
Chrysene	3, 159 (1973); 32, 247 (1983); Suppl. 7, 60 (1987)
Chrysoidine	8, 91 (1975); Suppl. 7, 169 (1987)
Chrysotile (see Asbestos)	
Ciclosporin	50, 77 (1990)
CI Disperse Yellow 3	8, 97 (1975); Suppl. 7, 60 (1987)
Cimetidine	50, 235 (1990)
Cinnamyl anthranilate	16, 287 (1978); 31, 133 (1983); Suppl. 7, 60 (1987)
Cisplatin	26, 151 (1981); Suppl. 7, 170 (1987)
Citrinin	40, 67 (1986); Suppl. 7, 60 (1987)
Citrus Red No. 2	8, 101 (1975) (corr. 42, 254); Suppl. 7, 60 (1987)
Clofibrate	24, 39 (1980); Suppl. 7, 171 (1987)
Clomiphene citrate	21, 551 (1979); Suppl. 7, 172 (1987)
Coal gasification	34, 65 (1984); Suppl. 7, 173 (1987)
Coal-tar pitches (see also Coal-tars)	35, 83 (1985); Suppl. 7, 174 (1987)
Coal-tars	35, 83 (1985); Suppl. 7, 175 (1987)
Cobalt[III] acetate (see Cobalt and cobalt compounds)	
Cobalt-aluminium-chromium spinel (see Cobalt and cobalt compounds)	
Cobalt and cobalt compounds	52, 363 (1991)
Cobalt[II] chloride (see Cobalt and cobalt compounds)	
Cobalt-chromium alloy (see Chromium and chromium compounds)	
Cobalt-chromium-molybdenum alloys (see Cobalt and cobalt compounds)	
Cobalt metal powder (see Cobalt and cobalt compounds)	
Cobalt naphthenate (see Cobalt and cobalt compounds)	
Cobalt[II] oxide (see Cobalt and cobalt compounds)	
Cobalt[II,III] oxide (see Cobalt and cobalt compounds)	
Cobalt[II] sulfide (see Cobalt and cobalt compounds)	
Coffee	51, 41 (1991) (corr. 52, 513)
Coke production	34, 101 (1984); Suppl. 7, 176 (1987)
Combined oral contraceptives (see also Oestrogens, progestins and combinations)	Suppl. 7, 297 (1987)
Conjugated oestrogens (see also Steroidal oestrogens)	21, 147 (1979)

Contraceptives, oral (*see* Combined oral contraceptives;
   Sequential oral contraceptives)
Copper 8-hydroxyquinoline   15, 103 (1977); *Suppl.* 7, 61 (1987)
Coronene   32, 263 (1983); *Suppl.* 7, 61 (1987)
Coumarin   10, 113 (1976); *Suppl.* 7, 61 (1987)
Creosotes (*see also* Coal-tars)   35, 83 (1985); *Suppl.* 7, 177 (1987)
meta-Cresidine   27, 91 (1982); *Suppl.* 7, 61 (1987)
para-Cresidine   27, 92 (1982); *Suppl.* 7, 61 (1987)
Crocidolite (*see* Asbestos)
Crude oil   45, 119 (1989)
Crystalline silica (*see also* Silica)   42, 39 (1987); *Suppl.* 7, 341 (1987)
Cycasin   1, 157 (1972) (*corr.* 42, 251); 10,
                                                               121 (1976); *Suppl.* 7, 61 (1987)
Cyclamates   22, 55 (1980); *Suppl.* 7, 178 (1987)
Cyclamic acid (*see* Cyclamates)
Cyclochlorotine   10, 139 (1976); *Suppl.* 7, 61 (1987)
Cyclohexanone   47, 157 (1989)
Cyclohexylamine (*see* Cyclamates)
Cyclopenta[*cd*]pyrene   32, 269 (1983); *Suppl.* 7, 61 (1987)
Cyclopropane (*see* Anaesthetics, volatile)
Cyclophosphamide   9, 135 (1975); 26, 165 (1981);
                                                               Suppl. 7, 182 (1987)

D

2,4-D (*see also* Chlorophenoxy herbicides; Chlorophenoxy   15, 111 (1977)
   herbicides, occupational exposures to)
Dacarbazine   26, 203 (1981); *Suppl.* 7, 184 (1987)
Dantron   50, 265 (1990)
D & C Red No. 9   8, 107 (1975); *Suppl.* 7, 61 (1987)
Dapsone   24, 59 (1980); *Suppl.* 7, 185 (1987)
Daunomycin   10, 145 (1976); *Suppl.* 7, 61 (1987)
DDD (*see* DDT)
DDE (*see* DDT)
DDT   5, 83 (1974) (*corr.* 42, 253);
                                                               Suppl. 7, 186 (1987); 53, 179 (1991)
Decabromodiphenyl oxide   48, 73 (1990)
Deltamethrin   53, 251 (1991)
Diacetylaminoazotoluene   8, 113 (1975); *Suppl.* 7, 61 (1987)
N,N'-Diacetylbenzidine   16, 293 (1978); *Suppl.* 7, 61 (1987)
Dichlorvos   53, 267 (1991)
Diallate   12, 69 (1976); 30, 235 (1983);
                                                               Suppl. 7, 61 (1987)
2,4-Diaminoanisole   16, 51 (1978); 27, 103 (1982);
                                                               Suppl. 7, 61 (1987)
4,4'-Diaminodiphenyl ether   16, 301 (1978); 29, 203 (1982);
                                                               Suppl. 7, 61 (1987)
1,2-Diamino-4-nitrobenzene   16, 63 (1978); *Suppl.* 7, 61 (1987)
1,4-Diamino-2-nitrobenzene   16, 73 (1978); *Suppl.* 7, 61 (1987)
2,6-Diamino-3-(phenylazo)pyridine (*see* Phenazopyridine
   hydrochloride)
2,4-Diaminotoluene (*see also* Toluene diisocyanates)   16, 83 (1978); *Suppl.* 7, 61 (1987)

2,5-Diaminotoluene (*see also* Toluene diisocyanates)	*16*, 97 (1978); *Suppl. 7*, 61 (1987)
ortho-Dianisidine (*see* 3,3'-Dimethoxybenzidine)	
Diazepam	*13*, 57 (1977); *Suppl. 7*, 189 (1987)
Diazomethane	*7*, 223 (1974); *Suppl. 7*, 61 (1987)
Dibenz[*a,h*]acridine	*3*, 247 (1973); *32*, 277 (1983); *Suppl. 7*, 61 (1987)
Dibenz[*a,j*]acridine	*3*, 254 (1973); *32*, 283 (1983); *Suppl. 7*, 61 (1987)
Dibenz[*a,c*]anthracene	*32*, 289 (1983) (*corr. 42*, 262); *Suppl. 7*, 61 (1987)
Dibenz[*a,h*]anthracene	*3*, 178 (1973) (*corr. 43*, 261); *32*, 299 (1983); *Suppl. 7*, 61 (1987)
Dibenz[*a,j*]anthracene	*32*, 309 (1983); *Suppl. 7*, 61 (1987)
7*H*-Dibenzo[*c,g*]carbazole	*3*, 260 (1973); *32*, 315 (1983); *Suppl. 7*, 61 (1987)
Dibenzodioxins, chlorinated (other than TCDD) [*see* Chlorinated dibenzodioxins (other than TCDD)]	
Dibenzo[*a,e*]fluoranthene	*32*, 321 (1983); *Suppl. 7*, 61 (1987)
Dibenzo[*h,rst*]pentaphene	*3*, 197 (1973); *Suppl. 7*, 62 (1987)
Dibenzo[*a,e*]pyrene	*3*, 201 (1973); *32*, 327 (1983); *Suppl. 7*, 62 (1987)
Dibenzo[*a,h*]pyrene	*3*, 207 (1973); *32*, 331 (1983); *Suppl. 7*, 62 (1987)
Dibenzo[*a,i*]pyrene	*3*, 215 (1973); *32*, 337 (1983); *Suppl. 7*, 62 (1987)
Dibenzo[*a,l*]pyrene	*3*, 224 (1973); *32*, 343 (1983); *Suppl. 7*, 62 (1987)
Dibromoacetonitrile (*see* Halogenated acetonitriles)	
1,2-Dibromo-3-chloropropane	*15*, 139 (1977); *20*, 83 (1979); *Suppl. 7*, 191 (1987)
Dichloroacetonitrile (*see* Halogenated acetonitriles)	
Dichloroacetylene	*39*, 369 (1986); *Suppl. 7*, 62 (1987)
ortho-Dichlorobenzene	*7*, 231 (1974); *29*, 213 (1982); *Suppl. 7*, 192 (1987)
para-Dichlorobenzene	*7*, 231 (1974); *29*, 215 (1982); *Suppl. 7*, 192 (1987)
3,3'-Dichlorobenzidine	*4*, 49 (1974); *29*, 239 (1982); *Suppl. 7*, 193 (1987)
trans-1,4-Dichlorobutene	*15*, 149 (1977); *Suppl. 7*, 62 (1987)
3,3'-Dichloro-4,4'-diaminodiphenyl ether	*16*, 309 (1978); *Suppl. 7*, 62 (1987)
1,2-Dichloroethane	*20*, 429 (1979); *Suppl. 7*, 62 (1987)
Dichloromethane	*20*, 449 (1979); *41*, 43 (1986); *Suppl. 7*, 194 (1987)
2,4-Dichlorophenol (*see* Chlorophenols; Chlorophenols, occupational exposures to)	
(2,4-Dichlorophenoxy)acetic acid (*see* 2,4-D)	
2,6-Dichloro-*para*-phenylenediamine	*39*, 325 (1986); *Suppl. 7*, 62 (1987)
1,2-Dichloropropane	*41*, 131 (1986); *Suppl. 7*, 62 (1987)
1,3-Dichloropropene (technical-grade)	*41*, 113 (1986); *Suppl. 7*, 195 (1987)
Dichlorvos	*20*, 97 (1979); *Suppl. 7*, 62 (1987); *53*, 267 (1991)
Dicofol	*30*, 87 (1983); *Suppl. 7*, 62 (1987)

Dicyclohexylamine (*see* Cyclamates)
Dieldrin 5, 125 (1974); *Suppl.* 7, 196 (1987)
Dienoestrol (*see also* Nonsteroidal oestrogens) 21, 161 (1979)
Diepoxybutane 11, 115 (1976) (*corr.* 42, 255); *Suppl.* 7, 62 (1987)

Diesel and gasoline engine exhausts 46, 41 (1989)
Diesel fuels 45, 219 (1989) (*corr.* 47, 505)
Diethyl ether (*see* Anaesthetics, volatile)
Di(2-ethylhexyl)adipate 29, 257 (1982); *Suppl.* 7, 62 (1987)
Di(2-ethylhexyl)phthalate 29, 269 (1982) (*corr.* 42, 261); *Suppl.* 7, 62 (1987)

1,2-Diethylhydrazine 4, 153 (1974); *Suppl.* 7, 62 (1987)
Diethylstilboestrol 6, 55 (1974); 21, 173 (1979) (*corr.* 42, 259); *Suppl.* 7, 273 (1987)

Diethylstilboestrol dipropionate (*see* Diethylstilboestrol)
Diethyl sulfate 4, 277 (1974); *Suppl.* 7, 198 (1987); 54, 213 (1992)

Diglycidyl resorcinol ether 11, 125 (1976); 36, 181 (1985); *Suppl.* 7, 62 (1987)

Dihydrosafrole 1, 170 (1972); 10, 233 (1976); *Suppl.* 7, 62 (1987)

1,8-Dihydroxyanthraquinone (*see* Dantron)
Dihydroxybenzenes (*see* Catechol; Hydroquinone; Resorcinol)
Dihydroxymethylfuratrizine 24, 77 (1980); *Suppl.* 7, 62 (1987)
Diisopropyl sulfate 54, 229 (1992)
Dimethisterone (*see also* Progestins; Sequential oral contraceptives) 6, 167 (1974); 21, 377 (1979)
Dimethoxane 15, 177 (1977); *Suppl.* 7, 62 (1987)
3,3'-Dimethoxybenzidine 4, 41 (1974); *Suppl.* 7, 198 (1987)
3,3'-Dimethoxybenzidine-4,4'-diisocyanate 39, 279 (1986); *Suppl.* 7, 62 (1987)
para-Dimethylaminoazobenzene 8, 125 (1975); *Suppl.* 7, 62 (1987)
para-Dimethylaminoazobenzenediazo sodium sulfonate 8, 147 (1975); *Suppl.* 7, 62 (1987)
trans-2-[(Dimethylamino)methylimino]-5-[2-(5-nitro-2-furyl)-vinyl]-1,3,4-oxadiazole 7, 147 (1974) (*corr.* 42, 253); *Suppl.* 7, 62 (1987)
4,4'-Dimethylangelicin plus ultraviolet radiation (*see also* Angelicin and some synthetic derivatives) *Suppl.* 7, 57 (1987)
4,5'-Dimethylangelicin plus ultraviolet radiation (*see also* Angelicin and some synthetic derivatives) *Suppl.* 7, 57 (1987)
Dimethylarsinic acid (*see* Arsenic and arsenic compounds)
3,3'-Dimethylbenzidine 1, 87 (1972); *Suppl.* 7, 62 (1987)
Dimethylcarbamoyl chloride 12, 77 (1976); *Suppl.* 7, 199 (1987)
Dimethylformamide 47, 171 (1989)
1,1-Dimethylhydrazine 4, 137 (1974); *Suppl.* 7, 62 (1987)
1,2-Dimethylhydrazine 4, 145 (1974) (*corr.* 42, 253); *Suppl.* 7, 62 (1987)

Dimethyl hydrogen phosphite 48, 85 (1990)
1,4-Dimethylphenanthrene 32, 349 (1983); *Suppl.* 7, 62 (1987)
Dimethyl sulfate 4, 271 (1974); *Suppl.* 7, 200 (1987)
3,7-Dinitrofluoranthene 46, 189 (1989)
3,9-Dinitrofluoranthene 46, 195 (1989)
1,3-Dinitropyrene 46, 201 (1989)
1,6-Dinitropyrene 46, 215 (1989)

1,8-Dinitropyrene	*33*, 171 (1984); *Suppl. 7*, 63 (1987); *46*, 231 (1989)
Dinitrosopentamethylenetetramine	*11*, 241 (1976); *Suppl. 7*, 63 (1987)
1,4-Dioxane	*11*, 247 (1976); *Suppl. 7*, 201 (1987)
2,4'-Diphenyldiamine	*16*, 313 (1978); *Suppl. 7*, 63 (1987)
Direct Black 38 (*see also* Benzidine-based dyes)	*29*, 295 (1982) (*corr. 42*, 261)
Direct Blue 6 (*see also* Benzidine-based dyes)	*29*, 311 (1982)
Direct Brown 95 (*see also* Benzidine-based dyes)	*29*, 321 (1982)
Disperse Blue 1	*48*, 139 (1990)
Disperse Yellow 3	*48*, 149 (1990)
Disulfiram	*12*, 85 (1976); *Suppl. 7*, 63 (1987)
Dithranol	*13*, 75 (1977); *Suppl. 7*, 63 (1987)
Divinyl ether (*see* Anaesthetics, volatile)	
Dulcin	*12*, 97 (1976); *Suppl. 7*, 63 (1987)

E

Endrin	*5*, 157 (1974); *Suppl. 7*, 63 (1987)
Enflurane (*see* Anaesthetics, volatile)	
Eosin	*15*, 183 (1977); *Suppl. 7*, 63 (1987)
Epichlorohydrin	*11*, 131 (1976) (*corr. 42*, 256); *Suppl. 7*, 202 (1987)
1,2-Epoxybutane	*47*, 217 (1989)
1-Epoxyethyl-3,4-epoxycyclohexane	*11*, 141 (1976); *Suppl. 7*, 63 (1987)
3,4-Epoxy-6-methylcyclohexylmethyl-3,4-epoxy-6-methyl-cyclohexane carboxylate	*11*, 147 (1976); *Suppl. 7*, 63 (1987)
cis-9,10-Epoxystearic acid	*11*, 153 (1976); *Suppl. 7*, 63 (1987)
Erionite	*42*, 225 (1987); *Suppl. 7*, 203 (1987)
Ethinyloestradiol (*see also* Steroidal oestrogens)	*6*, 77 (1974); *21*, 233 (1979)
Ethionamide	*13*, 83 (1977); *Suppl. 7*, 63 (1987)
Ethyl acrylate	*19*, 57 (1979); *39*, 81 (1986); *Suppl. 7*, 63 (1987)
Ethylene	*19*, 157 (1979); *Suppl. 7*, 63 (1987)
Ethylene dibromide	*15*, 195 (1977); *Suppl. 7*, 204 (1987)
Ethylene oxide	*11*, 157 (1976); *36*, 189 (1985) (*corr. 42*, 263); *Suppl. 7*, 205 (1987)
Ethylene sulfide	*11*, 257 (1976); *Suppl. 7*, 63 (1987)
Ethylene thiourea	*7*, 45 (1974); *Suppl. 7*, 207 (1987)
Ethyl methanesulfonate	*7*, 245 (1974); *Suppl. 7*, 63 (1987)
N-Ethyl-*N*-nitrosourea	*1*, 135 (1972); *17*, 191 (1978); *Suppl. 7*, 63 (1987)
Ethyl selenac (*see also* Selenium and selenium compounds)	*12*, 107 (1976); *Suppl. 7*, 63 (1987)
Ethyl tellurac	*12*, 115 (1976); *Suppl. 7*, 63 (1987)
Ethynodiol diacetate (*see also* Progestins; Combined oral contraceptives)	*6*, 173 (1974); *21*, 387 (1979)
Eugenol	*36*, 75 (1985); *Suppl. 7*, 63 (1987)
Evans blue	*8*, 151 (1975); *Suppl. 7*, 63 (1987)

F

Fast Green FCF	*16*, 187 (1978); *Suppl. 7*, 63 (1987)
Fenvalerate	*53*, 309 (1991)

Ferbam	12, 121 (1976) (corr. 42, 256); Suppl. 7, 63 (1987)
Ferric oxide	1, 29 (1972); Suppl. 7, 216 (1987)
Ferrochromium (see Chromium and chromium compounds)	
Fluometuron	30, 245 (1983); Suppl. 7, 63 (1987)
Fluoranthene	32, 355 (1983); Suppl. 7, 63 (1987)
Fluorene	32, 365 (1983); Suppl. 7, 63 (1987)
Fluorides (inorganic, used in drinking-water)	27, 237 (1982); Suppl. 7, 208 (1987)
5-Fluorouracil	26, 217 (1981); Suppl. 7, 210 (1987)
Fluorspar (see Fluorides)	
Fluosilicic acid (see Fluorides)	
Fluroxene (see Anaesthetics, volatile)	
Formaldehyde	29, 345 (1982); Suppl. 7, 211 (1987)
2-(2-Formylhydrazino)-4-(5-nitro-2-furyl)thiazole	7, 151 (1974) (corr. 42, 253); Suppl. 7, 63 (1987)
Frusemide (see Furosemide)	
Fuel oils (heating oils)	45, 239 (1989) (corr. 47, 505)
Furazolidone	31, 141 (1983); Suppl. 7, 63 (1987)
Furniture and cabinet-making	25, 99 (1981); Suppl. 7, 380 (1987)
Furosemide	50, 277 (1990)
2-(2-Furyl)-3-(5-nitro-2-furyl)acrylamide (see AF-2)	
Fusarenon-X	11, 169 (1976); 31, 153 (1983); Suppl. 7, 64 (1987)

G

Gasoline	45, 159 (1989) (corr. 47, 505)
Gasoline engine exhaust (see Diesel and gasoline engine exhausts)	
Glass fibres (see Man-made mineral fibres)	
Glasswool (see Man-made mineral fibres)	
Glass filaments (see Man-made mineral fibres)	
Glu-P-1	40, 223 (1986); Suppl. 7, 64 (1987)
Glu-P-2	40, 235 (1986); Suppl. 7, 64 (1987)
L-Glutamic acid, 5-[2-(4-hydroxymethyl)phenylhydrazide] (see Agaritine)	
Glycidaldehyde	11, 175 (1976); Suppl. 7, 64 (1987)
Glycidyl ethers	47, 237 (1989)
Glycidyl oleate	11, 183 (1976); Suppl. 7, 64 (1987)
Glycidyl stearate	11, 187 (1976); Suppl. 7, 64 (1987)
Griseofulvin	10, 153 (1976); Suppl. 7, 391 (1987)
Guinea Green B	16, 199 (1978); Suppl. 7, 64 (1987)
Gyromitrin	31, 163 (1983); Suppl. 7, 391 (1987)

H

Haematite	1, 29 (1972); Suppl. 7, 216 (1987)
Haematite and ferric oxide	Suppl. 7, 216 (1987)
Haematite mining, underground, with exposure to radon	1, 29 (1972); Suppl. 7, 216 (1987)
Hair dyes, epidemiology of	16, 29 (1978); 27, 307 (1982)
Halogenated acetonitriles	52, 269 (1991)
Halothane (see Anaesthetics, volatile)	
α-HCH (see Hexachlorocyclohexanes)	

β-HCH (*see* Hexachlorocyclohexanes)
γ-HCH (*see* Hexachlorocyclohexanes)
Heating oils (*see* Fuel oils)
Heptachlor (*see also* Chlordane/Heptachlor) 5, 173 (1974); 20, 129 (1979)
Hexachlorobenzene 20, 155 (1979); *Suppl. 7*, 219 (1987)
Hexachlorobutadiene 20, 179 (1979); *Suppl. 7*; 64 (1987)
Hexachlorocyclohexanes 5, 47 (1974); 20, 195 (1979) (*corr.* 42, 258); *Suppl. 7*, 220 (1987)

Hexachlorocyclohexane, technical-grade (*see* Hexachlorocyclohexanes)
Hexachloroethane 20, 467 (1979); *Suppl. 7*, 64 (1987)
Hexachlorophene 20, 241 (1979); *Suppl. 7*, 64 (1987)
Hexamethylphosphoramide 15, 211 (1977); *Suppl. 7*, 64 (1987)
Hexoestrol (*see* Nonsteroidal oestrogens)
Hycanthone mesylate 13, 91 (1977); *Suppl. 7*, 64 (1987)
Hydralazine 24, 85 (1980); *Suppl. 7*, 222 (1987)
Hydrazine 4, 127 (1974); *Suppl. 7*, 223 (1987)
Hydrochloric acid 54, 189 (1992)
Hydrochlorothiazide 50, 293 (1990)
Hydrogen peroxide 36, 285 (1985); *Suppl. 7*, 64 (1987)
Hydroquinone 15, 155 (1977); *Suppl. 7*, 64 (1987)
4-Hydroxyazobenzene 8, 157 (1975); *Suppl. 7*, 64 (1987)
17α-Hydroxyprogesterone caproate (*see also* Progestins) 21, 399 (1979) (*corr.* 42, 259)
8-Hydroxyquinoline 13, 101 (1977); *Suppl. 7*, 64 (1987)
8-Hydroxysenkirkine 10, 265 (1976); *Suppl. 7*, 64 (1987)
Hypochlorite salts 52, 159 (1991)

I

Indeno[1,2,3-*cd*]pyrene 3, 229 (1973); 32, 373 (1983); *Suppl. 7*, 64 (1987)

Inorganic acids (*see* Sulfuric acid and other strong inorganic acids, occupational exposures to mists and vapours from)
Insecticides, occupational exposures in spraying and application of 53, 45 (1991)
IQ 40, 261 (1986); *Suppl. 7*, 64 (1987)
Iron and steel founding 34, 133 (1984); *Suppl. 7*, 224 (1987)
Iron-dextran complex 2, 161 (1973); *Suppl. 7*, 226 (1987)
Iron-dextrin complex 2, 161 (1973) (*corr.* 42, 252); *Suppl. 7*, 64 (1987)

Iron oxide (*see* Ferric oxide)
Iron oxide, saccharated (*see* Saccharated iron oxide)
Iron sorbitol–citric acid complex 2, 161 (1973); *Suppl. 7*, 64 (1987)
Isatidine 10, 269 (1976); *Suppl. 7*, 65 (1987)
Isoflurane (*see* Anaesthetics, volatile)
Isoniazid (*see* Isonicotinic acid hydrazide)
Isonicotinic acid hydrazide 4, 159 (1974); *Suppl. 7*, 227 (1987)
Isophosphamide 26, 237 (1981); *Suppl. 7*, 65 (1987)
Isopropyl alcohol 15, 223 (1977); *Suppl. 7*, 229 (1987)
Isopropyl alcohol manufacture (strong-acid process) *Suppl. 7*, 229 (1987)
    (*see also* Isopropyl alcohol; Sulfuric acid and other strong inorganic acids, occupational exposures to mists and vapours from)
Isopropyl oils 15, 223 (1977); *Suppl. 7*, 229 (1987)

Isosafrole   *1*, 169 (1972); *10*, 232 (1976);
                                                                Suppl. 7, 65 (1987)

J

Jacobine   *10*, 275 (1976); Suppl. 7, 65 (1987)
Jet fuel   *45*, 203 (1989)
Joinery (see Carpentry and joinery)

K

Kaempferol   31, 171 (1983); Suppl. 7, 65 (1987)
Kepone (see Chlordecone)

L

Lasiocarpine   *10*, 281 (1976); Suppl. 7, 65 (1987)
Lauroyl peroxide   *36*, 315 (1985); Suppl. 7, 65 (1987)
Lead acetate (see Lead and lead compounds)
Lead and lead compounds   *1*, 40 (1972) (corr. *42*, 251); *2*, 52,
                                                                150 (1973); *12*, 131 (1976);
                                                                23, 40, 208, 209, 325 (1980);
                                                                Suppl. 7, 230 (1987)
Lead arsenate (see Arsenic and arsenic compounds)
Lead carbonate (see Lead and lead compounds)
Lead chloride (see Lead and lead compounds)
Lead chromate (see Chromium and chromium compounds)
Lead chromate oxide (see Chromium and chromium compounds)
Lead naphthenate (see Lead and lead compounds)
Lead nitrate (see Lead and lead compounds)
Lead oxide (see Lead and lead compounds)
Lead phosphate (see Lead and lead compounds)
Lead subacetate (see Lead and lead compounds)
Lead tetroxide (see Lead and lead compounds)
Leather goods manufacture   *25*, 279 (1981); Suppl. 7, 235 (1987)
Leather industries   *25*, 199 (1981); Suppl. 7, 232 (1987)
Leather tanning and processing   *25*, 201 (1981); Suppl. 7, 236 (1987)
Ledate (see also Lead and lead compounds)   *12*, 131 (1976)
Light Green SF   *16*, 209 (1978); Suppl. 7, 65 (1987)
Lindane (see Hexachlorocyclohexanes)
The lumber and sawmill industries (including logging)   *25*, 49 (1981); Suppl. 7, 383 (1987)
Luteoskyrin   *10*, 163 (1976); Suppl. 7, 65 (1987)
Lynoestrenol (see also Progestins; Combined oral contraceptives) *21*, 407 (1979)

M

Magenta   *4*, 57 (1974) (corr. *42*, 252);
                                                                Suppl. 7, 238 (1987)
Magenta, manufacture of (see also Magenta)   Suppl. 7, 238 (1987)
Malathion   *30*, 103 (1983); Suppl. 7, 65 (1987)
Maleic hydrazide   *4*, 173 (1974) (corr. *42*, 253);
                                                                Suppl. 7, 65 (1987)

Malonaldehyde	36, 163 (1985); *Suppl. 7*, 65 (1987)
Maneb	12, 137 (1976); *Suppl. 7*, 65 (1987)
Man-made mineral fibres	43, 39 (1988)
Mannomustine	9, 157 (1975); *Suppl. 7*, 65 (1987)
Mate	51, 273 (1991)
MCPA (*see also* Chlorophenoxy herbicides; Chlorophenoxy herbicides, occupational exposures to)	30, 255 (1983)
MeA-α-C	40, 253 (1986); *Suppl. 7*, 65 (1987)
Medphalan	9, 168 (1975); *Suppl. 7*, 65 (1987)
Medroxyprogesterone acetate	6, 157 (1974); 21, 417 (1979) (*corr. 42*, 259); *Suppl. 7*, 289 (1987)
Megestrol acetate (*see also* Progestins; Combined oral contraceptives)	
MeIQ	40, 275 (1986); *Suppl. 7*, 65 (1987)
MeIQx	40, 283 (1986); *Suppl. 7*, 65 (1987)
Melamine	39, 333 (1986); *Suppl. 7*, 65 (1987)
Melphalan	9, 167 (1975); *Suppl. 7*, 239 (1987)
6-Mercaptopurine	26, 249 (1981); *Suppl. 7*, 240 (1987)
Merphalan	9, 169 (1975); *Suppl. 7*, 65 (1987)
Mestranol (*see also* Steroidal oestrogens)	6, 87 (1974); 21, 257 (1979) (*corr. 42*, 259)
Metabisulfites (*see* Sulfur dioxide and some sulfites, bisulfites and metabisulfites)	
Methanearsonic acid, disodium salt (*see* Arsenic and arsenic compounds)	
Methanearsonic acid, monosodium salt (*see* Arsenic and arsenic compounds)	
Methotrexate	26, 267 (1981); *Suppl. 7*, 241 (1987)
Methoxsalen (*see* 8-Methoxypsoralen)	
Methoxychlor	5, 193 (1974); 20, 259 (1979); *Suppl. 7*, 66 (1987)
Methoxyflurane (*see* Anaesthetics, volatile)	
5-Methoxypsoralen	40, 327 (1986); *Suppl. 7*, 242 (1987)
8-Methoxypsoralen (*see also* 8-Methoxypsoralen plus ultraviolet radiation)	24, 101 (1980)
8-Methoxypsoralen plus ultraviolet radiation	*Suppl. 7*, 243 (1987)
Methyl acrylate	19, 52 (1979); 39, 99 (1986); *Suppl. 7*, 66 (1987)
5-Methylangelicin plus ultraviolet radiation (*see also* Angelicin and some synthetic derivatives)	*Suppl. 7*, 57 (1987)
2-Methylaziridine	9, 61 (1975); *Suppl. 7*, 66 (1987)
Methylazoxymethanol acetate	1, 164 (1972); 10, 131 (1976); *Suppl. 7*, 66 (1987)
Methyl bromide	41, 187 (1986) (*corr. 45*, 283); *Suppl. 7*, 245 (1987)
Methyl carbamate	12, 151 (1976); *Suppl. 7*, 66 (1987)
Methyl-CCNU [*see* 1-(2-Chloroethyl)-3-(4-methylcyclohexyl)-1-nitrosourea]	
Methyl chloride	41, 161 (1986); *Suppl. 7*, 246 (1987)
1-, 2-, 3-, 4-, 5- and 6-Methylchrysenes	32, 379 (1983); *Suppl. 7*, 66 (1987)
N-Methyl-N,4-dinitrosoaniline	1, 141 (1972); *Suppl. 7*, 66 (1987)

4,4'-Methylene bis(2-chloroaniline)	4, 65 (1974) (corr. 42, 252); Suppl. 7, 246 (1987)
4,4'-Methylene bis(N,N-dimethyl)benzenamine	27, 119 (1982); Suppl. 7, 66 (1987)
4,4'-Methylene bis(2-methylaniline)	4, 73 (1974); Suppl. 7, 248 (1987)
4,4'-Methylenedianiline	4, 79 (1974) (corr. 42, 252); 39, 347 (1986); Suppl. 7, 66 (1987)
4,4'-Methylenediphenyl diisocyanate	19, 314 (1979); Suppl. 7, 66 (1987)
2-Methylfluoranthene	32, 399 (1983); Suppl. 7, 66 (1987)
3-Methylfluoranthene	32, 399 (1983); Suppl. 7, 66 (1987)
Methylglyoxal	51, 443 (1991)
Methyl iodide	15, 245 (1977); 41, 213 (1986); Suppl. 7, 66 (1987)
Methyl methacrylate	19, 187 (1979); Suppl. 7, 66 (1987)
Methyl methanesulfonate	7, 253 (1974); Suppl. 7, 66 (1987)
2-Methyl-1-nitroanthraquinone	27, 205 (1982); Suppl. 7, 66 (1987)
N-Methyl-N'-nitro-N-nitrosoguanidine	4, 183 (1974); Suppl. 7, 248 (1987)
3-Methylnitrosaminopropionaldehyde [see 3-(N-Nitrosomethylamino)-propionaldehyde]	
3-Methylnitrosaminopropionitrile [see 3-(N-Nitrosomethylamino)-propionitrile]	
4-(Methylnitrosamino)-4-(3-pyridyl)-1-butanal [see 4-(N-Nitrosomethyl-amino)-4-(3-pyridyl)-1-butanal]	
4-(Methylnitrosamino)-1-(3-pyridyl)-1-butanone [see 4-(N-Nitrosomethyl-amino)-1-(3-pyridyl)-1-butanone]	
N-Methyl-N-nitrosourea	1, 125 (1972); 17, 227 (1978); Suppl. 7, 66 (1987)
N-Methyl-N-nitrosourethane	4, 211 (1974); Suppl. 7, 66 (1987)
Methyl parathion	30, 131 (1983); Suppl. 7, 392 (1987)
1-Methylphenanthrene	32, 405 (1983); Suppl. 7, 66 (1987)
7-Methylpyrido[3,4-c]psoralen	40, 349 (1986); Suppl. 7, 71 (1987)
Methyl red	8, 161 (1975); Suppl. 7, 66 (1987)
Methyl selenac (see also Selenium and selenium compounds)	12, 161 (1976); Suppl. 7, 66 (1987)
Methylthiouracil	7, 53 (1974); Suppl. 7, 66 (1987)
Metronidazole	13, 113 (1977); Suppl. 7, 250 (1987)
Mineral oils	3, 30 (1973); 33, 87 (1984) (corr. 42, 262); Suppl. 7, 252 (1987)
Mirex	5, 203 (1974); 20, 283 (1979) (corr. 42, 258); Suppl. 7, 66 (1987)
Mitomycin C	10, 171 (1976); Suppl. 7, 67 (1987)
MNNG [see N-Methyl-N'-nitro-N-nitrosoguanidine]	
MOCA [see 4,4'-Methylene bis(2-chloroaniline)]	
Modacrylic fibres	19, 86 (1979); Suppl. 7, 67 (1987)
Monocrotaline	10, 291 (1976); Suppl. 7, 67 (1987)
Monuron	12, 167 (1976); Suppl. 7, 67 (1987); 53, 467 (1991)
MOPP and other combined chemotherapy including alkylating agents	Suppl. 7, 254 (1987)
Morpholine	47, 199 (1989)
5-(Morpholinomethyl)-3-[(5-nitrofurfurylidene)amino]-2-oxazolidinone	7, 161 (1974); Suppl. 7, 67 (1987)
Mustard gas	9, 181 (1975) (corr. 42, 254); Suppl. 7, 259 (1987)

Myleran (see 1,4-Butanediol dimethanesulfonate)

N

Nafenopin	24, 125 (1980); Suppl. 7, 67 (1987)
1,5-Naphthalenediamine	27, 127 (1982); Suppl. 7, 67 (1987)
1,5-Naphthalene diisocyanate	19, 311 (1979); Suppl. 7, 67 (1987)
1-Naphthylamine	4, 87 (1974) (corr. 42, 253); Suppl. 7, 260 (1987)
2-Naphthylamine	4, 97 (1974); Suppl. 7, 261 (1987)
1-Naphthylthiourea	30, 347 (1983); Suppl. 7, 263 (1987)
Nickel acetate (see Nickel and nickel compounds)	
Nickel ammonium sulfate (see Nickel and nickel compounds)	
Nickel and nickel compounds	2, 126 (1973) (corr. 42, 252); 11, 75 (1976); Suppl. 7, 264 (1987) (corr. 45, 283); 49, 257 (1990)
Nickel carbonate (see Nickel and nickel compounds)	
Nickel carbonyl (see Nickel and nickel compounds)	
Nickel chloride (see Nickel and nickel compounds)	
Nickel-gallium alloy (see Nickel and nickel compounds)	
Nickel hydroxide (see Nickel and nickel compounds)	
Nickelocene (see Nickel and nickel compounds)	
Nickel oxide (see Nickel and nickel compounds)	
Nickel subsulfide (see Nickel and nickel compounds)	
Nickel sulfate (see Nickel and nickel compounds)	
Niridazole	13, 123 (1977); Suppl. 7, 67 (1987)
Nithiazide	31, 179 (1983); Suppl. 7, 67 (1987)
Nitrilotriacetic acid and its salts	48, 181 (1990)
5-Nitroacenaphthene	16, 319 (1978); Suppl. 7, 67 (1987)
5-Nitro-ortho-anisidine	27, 133 (1982); Suppl. 7, 67 (1987)
9-Nitroanthracene	33, 179 (1984); Suppl. 7, 67 (1987)
7-Nitrobenz[a]anthracene	46, 247 (1989)
6-Nitrobenzo[a]pyrene	33, 187 (1984); Suppl. 7, 67 (1987); 46, 255 (1989)
4-Nitrobiphenyl	4, 113 (1974); Suppl. 7, 67 (1987)
6-Nitrochrysene	33, 195 (1984); Suppl. 7, 67 (1987); 46, 267 (1989)
Nitrofen (technical-grade)	30, 271 (1983); Suppl. 7, 67 (1987)
3-Nitrofluoranthene	33, 201 (1984); Suppl. 7, 67 (1987)
2-Nitrofluorene	46, 277 (1989)
Nitrofural	7, 171 (1974); Suppl. 7, 67 (1987); 50, 195 (1990)
5-Nitro-2-furaldehyde semicarbazone (see Nitrofural)	
Nitrofurantoin	50, 211 (1990)
Nitrofurazone (see Nitrofural)	
1-[(5-Nitrofurfurylidene)amino]-2-imidazolidinone	7, 181 (1974); Suppl. 7, 67 (1987)
N-[4-(5-Nitro-2-furyl)-2-thiazolyl]acetamide	1, 181 (1972); 7, 185 (1974); Suppl. 7, 67 (1987)
Nitrogen mustard	9, 193 (1975); Suppl. 7, 269 (1987)
Nitrogen mustard N-oxide	9, 209 (1975); Suppl. 7, 67 (1987)
1-Nitronaphthalene	46, 291 (1989)
2-Nitronaphthalene	46, 303 (1989)

3-Nitroperylene	46, 313 (1989)
2-Nitropropane	29, 331 (1982); *Suppl.* 7, 67 (1987)
1-Nitropyrene	33, 209 (1984); *Suppl.* 7, 67 (1987); 46, 321 (1989)
2-Nitropyrene	46, 359 (1989)
4-Nitropyrene	46, 367 (1989)
N-Nitrosatable drugs	24, 297 (1980) *(corr. 42, 260)*
N-Nitrosatable pesticides	30, 359 (1983)
N'-Nitrosoanabasine	37, 225 (1985); *Suppl.* 7, 67 (1987)
N'-Nitrosoanatabine	37, 233 (1985); *Suppl.* 7, 67 (1987)
N-Nitrosodi-n-butylamine	4, 197 (1974); 17, 51 (1978); *Suppl.* 7, 67 (1987)
N-Nitrosodiethanolamine	17, 77 (1978); *Suppl.* 7, 67 (1987)
N-Nitrosodiethylamine	1, 107 (1972) *(corr. 42, 251)*; 17, 83 (1978) *(corr. 42, 257)*; *Suppl.* 7, 67 (1987)
N-Nitrosodimethylamine	1, 95 (1972); 17, 125 (1978) *(corr. 42, 257)*; *Suppl.* 7, 67 (1987)
N-Nitrosodiphenylamine	27, 213 (1982); *Suppl.* 7, 67 (1987)
para-Nitrosodiphenylamine	27, 227 (1982) *(corr. 42, 261)*; *Suppl.* 7, 68 (1987)
N-Nitrosodi-n-propylamine	17, 177 (1978); *Suppl.* 7, 68 (1987)
N-Nitroso-N-ethylurea (see N-Ethyl-N-nitrosourea)	
N-Nitrosofolic acid	17, 217 (1978); *Suppl.* 7, 68 (1987)
N-Nitrosoguvacine	37, 263 (1985); *Suppl.* 7, 68 (1987)
N-Nitrosoguvacoline	37, 263 (1985); *Suppl.* 7, 68 (1987)
N-Nitrosohydroxyproline	17, 304 (1978); *Suppl.* 7, 68 (1987)
3-(N-Nitrosomethylamino)propionaldehyde	37, 263 (1985); *Suppl.* 7, 68 (1987)
3-(N-Nitrosomethylamino)propionitrile	37, 263 (1985); *Suppl.* 7, 68 (1987)
4-(N-Nitrosomethylamino)-4-(3-pyridyl)-1-butanal	37, 205 (1985); *Suppl.* 7, 68 (1987)
4-(N-Nitrosomethylamino)-1-(3-pyridyl)-1-butanone	37, 209 (1985); *Suppl.* 7, 68 (1987)
N-Nitrosomethylethylamine	17, 221 (1978); *Suppl.* 7, 68 (1987)
N-Nitroso-N-methylurea (see N-Methyl-N-nitrosourea)	
N-Nitroso-N-methylurethane (see N-Methyl-N-methylurethane)	
N-Nitrosomethylvinylamine	17, 257 (1978); *Suppl.* 7, 68 (1987)
N-Nitrosomorpholine	17, 263 (1978); *Suppl.* 7, 68 (1987)
N'-Nitrosonornicotine	17, 281 (1978); 37, 241 (1985); *Suppl.* 7, 68 (1987)
N-Nitrosopiperidine	17, 287 (1978); *Suppl.* 7, 68 (1987)
N-Nitrosoproline	17, 303 (1978); *Suppl.* 7, 68 (1987)
N-Nitrosopyrrolidine	17, 313 (1978); *Suppl.* 7, 68 (1987)
N-Nitrososarcosine	17, 327 (1978); *Suppl.* 7, 68 (1987)
Nitrosoureas, chloroethyl (see Chloroethyl nitrosoureas)	
5-Nitro-*ortho*-toluidine	48, 169 (1990)
Nitrous oxide (see Anaesthetics, volatile)	
Nitrovin	31, 185 (1983); *Suppl.* 7, 68 (1987)
NNA [see 4-(N-Nitrosomethylamino)-4-(3-pyridyl)-1-butanal]	
NNK [see 4-(N-Nitrosomethylamino)-1-(3-pyridyl)-1-butanone]	
Nonsteroidal oestrogens (see also Oestrogens, progestins and combinations)	*Suppl.* 7, 272 (1987)
Norethisterone (see also Progestins; Combined oral contraceptives)	6, 179 (1974); 21, 461 (1979)

Norethynodrel (see also Progestins; Combined oral contraceptives)	6, 191 (1974); 21, 461 (1979) (corr. 42, 259)
Norgestrel (see also Progestins, Combined oral contraceptives)	6, 201 (1974); 21, 479 (1979)
Nylon 6	19, 120 (1979); Suppl. 7, 68 (1987)

O

Ochratoxin A	10, 191 (1976); 31, 191 (1983) (corr. 42, 262); Suppl. 7, 271 (1987)
Oestradiol-17β (see also Steroidal oestrogens)	6, 99 (1974); 21, 279 (1979)
Oestradiol 3-benzoate (see Oestradiol-17β)	
Oestradiol dipropionate (see Oestradiol-17β)	
Oestradiol mustard	9, 217 (1975)
Oestradiol-17β-valerate (see Oestradiol-17β)	
Oestriol (see also Steroidal oestrogens)	6, 117 (1974); 21, 327 (1979)
Oestrogen–progestin combinations (see Oestrogens, progestins and combinations)	
Oestrogen–progestin replacement therapy (see also Oestrogens, progestins and combinations)	Suppl. 7, 308 (1987)
Oestrogen replacement therapy (see also Oestrogens, progestins and combinations)	Suppl. 7, 280 (1987)
Oestrogens (see Oestrogens, progestins and combinations)	
Oestrogens, conjugated (see Conjugated oestrogens)	
Oestrogens, nonsteroidal (see Nonsteroidal oestrogens)	
Oestrogens, progestins and combinations	6 (1974); 21 (1979); Suppl. 7, 272 (1987)
Oestrogens, steroidal (see Steroidal oestrogens)	
Oestrone (see also Steroidal oestrogens)	6, 123 (1974); 21, 343 (1979) (corr. 42, 259)
Oestrone benzoate (see Oestrone)	
Oil Orange SS	8, 165 (1975); Suppl. 7, 69 (1987)
Oral contraceptives, combined (see Combined oral contraceptives)	
Oral contraceptives, investigational (see Combined oral contraceptives)	
Oral contraceptives, sequential (see Sequential oral contraceptives)	
Orange I	8, 173 (1975); Suppl. 7, 69 (1987)
Orange G	8, 181 (1975); Suppl. 7, 69 (1987)
Organolead compounds (see also Lead and lead compounds)	Suppl. 7, 230 (1987)
Oxazepam	13, 58 (1977); Suppl. 7, 69 (1987)
Oxymetholone [see also Androgenic (anabolic) steroids]	13, 131 (1977)
Oxyphenbutazone	13, 185 (1977); Suppl. 7, 69 (1987)

P

Paint manufacture and painting (occupational exposures in)	47, 329 (1989)
Panfuran S (see also Dihydroxymethylfuratrizine)	24, 77 (1980); Suppl. 7, 69 (1987)
Paper manufacture (see Pulp and paper manufacture)	
Paracetamol	50, 307 (1990)
Parasorbic acid	10, 199 (1976) (corr. 42, 255); Suppl. 7, 69 (1987)
Parathion	30, 153 (1983); Suppl. 7, 69 (1987)
Patulin	10, 205 (1976); 40, 83 (1986); Suppl. 7, 69 (1987)

Penicillic acid	10, 211 (1976); *Suppl. 7*, 69 (1987)
Pentachloroethane	41, 99 (1986); *Suppl. 7*, 69 (1987)
Pentachloronitrobenzene (*see* Quintozene)	
Pentachlorophenol (*see also* Chlorophenols; Chlorophenols, occupational exposures to)	20, 303 (1979); 53, 371 (1991)
Permethrin	53, 329 (1991)
Perylene	32, 411 (1983); *Suppl. 7*, 69 (1987)
Petasitenine	31, 207 (1983); *Suppl. 7*, 69 (1987)
Petasites japonicus (*see* Pyrrolizidine alkaloids)	
Petroleum refining (occupational exposures in)	45, 39 (1989)
Some petroleum solvents	47, 43 (1989)
Phenacetin	13, 141 (1977); 24, 135 (1980); *Suppl. 7*, 310 (1987)
Phenanthrene	32, 419 (1983); *Suppl. 7*, 69 (1987)
Phenazopyridine hydrochloride	8, 117 (1975); 24, 163 (1980) (*corr.* 42, 260); *Suppl. 7*, 312 (1987)
Phenelzine sulfate	24, 175 (1980); *Suppl. 7*, 312 (1987)
Phenicarbazide	12, 177 (1976); *Suppl. 7*, 70 (1987)
Phenobarbital	13, 157 (1977); *Suppl. 7*, 313 (1987)
Phenol	47, 263 (1989) (*corr.* 50, 385)
Phenoxyacetic acid herbicides (*see* Chlorophenoxy herbicides)	
Phenoxybenzamine hydrochloride	9, 223 (1975); 24, 185 (1980); *Suppl. 7*, 70 (1987)
Phenylbutazone	13, 183 (1977); *Suppl. 7*, 316 (1987)
meta-Phenylenediamine	16, 111 (1978); *Suppl. 7*, 70 (1987)
para-Phenylenediamine	16, 125 (1978); *Suppl. 7*, 70 (1987)
Phenyl glycidyl ether (*see* Glycidyl ethers)	
N-Phenyl-2-naphthylamine	16, 325 (1978) (*corr.* 42, 257); *Suppl. 7*, 318 (1987)
ortho-Phenylphenol	30, 329 (1983); *Suppl. 7*, 70 (1987)
Phenytoin	13, 201 (1977); *Suppl. 7*, 319 (1987)
Picloram	53, 481 (1991)
Piperazine oestrone sulfate (*see* Conjugated oestrogens)	
Piperonyl butoxide	30, 183 (1983); *Suppl. 7*, 70 (1987)
Pitches, coal-tar (*see* Coal-tar pitches)	
Polyacrylic acid	19, 62 (1979); *Suppl. 7*, 70 (1987)
Polybrominated biphenyls	18, 107 (1978); 41, 261 (1986); *Suppl. 7*, 321 (1987)
Polychlorinated biphenyls	7, 261 (1974); 18, 43 (1978) (*corr.* 42, 258); *Suppl. 7*, 322 (1987)
Polychlorinated camphenes (*see* Toxaphene)	
Polychloroprene	19, 141 (1979); *Suppl. 7*, 70 (1987)
Polyethylene	19, 164 (1979); *Suppl. 7*, 70 (1987)
Polymethylene polyphenyl isocyanate	19, 314 (1979); *Suppl. 7*, 70 (1987)
Polymethyl methacrylate	19, 195 (1979); *Suppl. 7*, 70 (1987)
Polyoestradiol phosphate (*see* Oestradiol-17β)	
Polypropylene	19, 218 (1979); *Suppl. 7*, 70 (1987)
Polystyrene	19, 245 (1979); *Suppl. 7*, 70 (1987)
Polytetrafluoroethylene	19, 288 (1979); *Suppl. 7*, 70 (1987)
Polyurethane foams	19, 320 (1979); *Suppl. 7*, 70 (1987)
Polyvinyl acetate	19, 346 (1979); *Suppl. 7*, 70 (1987)
Polyvinyl alcohol	19, 351 (1979); *Suppl. 7*, 70 (1987)

Polyvinyl chloride	7, 306 (1974); *19*, 402 (1979); *Suppl. 7*, 70 (1987)
Polyvinyl pyrrolidone	*19*, 463 (1979); *Suppl. 7*, 70 (1987)
Ponceau MX	*8*, 189 (1975); *Suppl. 7*, 70 (1987)
Ponceau 3R	*8*, 199 (1975); *Suppl. 7*, 70 (1987)
Ponceau SX	*8*, 207 (1975); *Suppl. 7*, 70 (1987)
Potassium arsenate (*see* Arsenic and arsenic compounds)	
Potassium arsenite (*see* Arsenic and arsenic compounds)	
Potassium bis(2-hydroxyethyl)dithiocarbamate	*12*, 183 (1976); *Suppl. 7*, 70 (1987)
Potassium bromate	*40*, 207 (1986); *Suppl. 7*, 70 (1987)
Potassium chromate (*see* Chromium and chromium compounds)	
Potassium dichromate (*see* Chromium and chromium compounds)	
Prednimustine	*50*, 115 (1990)
Prednisone	*26*, 293 (1981); *Suppl. 7*, 326 (1987)
Procarbazine hydrochloride	*26*, 311 (1981); *Suppl. 7*, 327 (1987)
Proflavine salts	*24*, 195 (1980); *Suppl. 7*, 70 (1987)
Progesterone (*see also* Progestins; Combined oral contraceptives)	*6*, 135 (1974); *21*, 491 (1979) (*corr. 42*, 259)
Progestins (*see also* Oestrogens, progestins and combinations)	*Suppl. 7*, 289 (1987)
Pronetalol hydrochloride	*13*, 227 (1977) (*corr. 42*, 256); *Suppl. 7*, 70 (1987)
1,3-Propane sultone	*4*, 253 (1974) (*corr. 42*, 253); *Suppl. 7*, 70 (1987)
Propham	*12*, 189 (1976); *Suppl. 7*, 70 (1987)
β-Propiolactone	*4*, 259 (1974) (*corr. 42*, 253); *Suppl. 7*, 70 (1987)
n-Propyl carbamate	*12*, 201 (1976); *Suppl. 7*, 70 (1987)
Propylene	*19*, 213 (1979); *Suppl. 7*, 71 (1987)
Propylene oxide	*11*, 191 (1976); *36*, 227 (1985) (*corr. 42*, 263); *Suppl. 7*, 328 (1987)
Propylthiouracil	7, 67 (1974); *Suppl. 7*, 329 (1987)
Ptaquiloside (*see also* Bracken fern)	*40*, 55 (1986); *Suppl. 7*, 71 (1987)
Pulp and paper manufacture	*25*, 157 (1981); *Suppl. 7*, 385 (1987)
Pyrene	*32*, 431 (1983); *Suppl. 7*, 71 (1987)
Pyrido[3,4-*c*]psoralen	*40*, 349 (1986); *Suppl. 7*, 71 (1987)
Pyrimethamine	*13*, 233 (1977); *Suppl. 7*, 71 (1987)
Pyrrolizidine alkaloids (*see* Hydroxysenkirkine; Isatidine; Jacobine; Lasiocarpine; Monocrotaline; Retrorsine; Riddelliine; Seneciphylline; Senkirkine)	

Q

Quercetin (*see also* Bracken fern)	*31*, 213 (1983); *Suppl. 7*, 71 (1987)
para-Quinone	*15*, 255 (1977); *Suppl. 7*, 71 (1987)
Quintozene	*5*, 211 (1974); *Suppl. 7*, 71 (1987)

R

Radon	*43*, 173 (1988) (*corr. 45*, 283)
Reserpine	*10*, 217 (1976); *24*, 211 (1980) (*corr. 42*, 260); *Suppl. 7*, 330 (1987)
Resorcinol	*15*, 155 (1977); *Suppl. 7*, 71 (1987)

Retrorsine	*10*, 303 (1976); *Suppl. 7*, 71 (1987)
Rhodamine B	*16*, 221 (1978); *Suppl. 7*, 71 (1987)
Rhodamine 6G	*16*, 233 (1978); *Suppl. 7*, 71 (1987)
Riddelliine	*10*, 313 (1976); *Suppl. 7*, 71 (1987)
Rifampicin	*24*, 243 (1980); *Suppl. 7*, 71 (1987)
Rockwool (*see* Man-made mineral fibres)	
The rubber industry	*28* (1982) (*corr. 42*, 261); *Suppl. 7*, 332 (1987)
Rugulosin	*40*, 99 (1986); *Suppl. 7*, 71 (1987)

S

Saccharated iron oxide	*2*, 161 (1973); *Suppl. 7*, 71 (1987)
Saccharin	*22*, 111 (1980) (*corr. 42*, 259); *Suppl. 7*, 334 (1987)
Safrole	*1*, 169 (1972); *10*, 231 (1976); *Suppl. 7*, 71 (1987)
The sawmill industry (including logging) [*see* The lumber and sawmill industry (including logging)]	
Scarlet Red	*8*, 217 (1975); *Suppl. 7*, 71 (1987)
Selenium and selenium compounds	*9*, 245 (1975) (*corr. 42*, 255); *Suppl. 7*, 71 (1987)
Selenium dioxide (*see* Selenium and selenium compounds)	
Selenium oxide (*see* Selenium and selenium compounds)	
Semicarbazide hydrochloride	*12*, 209 (1976) (*corr. 42*, 256); *Suppl. 7*, 71 (1987)
Senecio jacobaea L. (*see* Pyrrolizidine alkaloids)	
Senecio longilobus (*see* Pyrrolizidine alkaloids)	
Seneciphylline	*10*, 319, 335 (1976); *Suppl. 7*, 71 (1987)
Senkirkine	*10*, 327 (1976); *31*, 231 (1983); *Suppl. 7*, 71 (1987)
Sepiolite	*42*, 175 (1987); *Suppl. 7*, 71 (1987)
Sequential oral contraceptives (*see also* Oestrogens, progestins and combinations)	*Suppl. 7*, 296 (1987)
Shale-oils	*35*, 161 (1985); *Suppl. 7*, 339 (1987)
Shikimic acid (*see also* Bracken fern)	*40*, 55 (1986); *Suppl. 7*, 71 (1987)
Shoe manufacture and repair (*see* Boot and shoe manufacture and repair)	
Silica (*see also* Amorphous silica; Crystalline silica)	*42*, 39 (1987)
Simazine	*53*, 495 (1991)
Slagwool (*see* Man-made mineral fibres)	
Sodium arsenate (*see* Arsenic and arsenic compounds)	
Sodium arsenite (*see* Arsenic and arsenic compounds)	
Sodium cacodylate (*see* Arsenic and arsenic compounds)	
Sodium chlorite	*52*, 145 (1991)
Sodium chromate (*see* Chromium and chromium compounds)	
Sodium cyclamate (*see* Cyclamates)	
Sodium dichromate (*see* Chromium and chromium compounds)	
Sodium diethyldithiocarbamate	*12*, 217 (1976); *Suppl. 7*, 71 (1987)
Sodium equilin sulfate (*see* Conjugated oestrogens)	
Sodium fluoride (*see* Fluorides)	

Sodium monofluorophosphate (*see* Fluorides)
Sodium oestrone sulfate (*see* Conjugated oestrogens)
Sodium *ortho*-phenylphenate (*see also ortho*-Phenylphenol)   30, 329 (1983); *Suppl. 7*, 392 (1987)
Sodium saccharin (*see* Saccharin)
Sodium selenate (*see* Selenium and selenium compounds)
Sodium selenite (*see* Selenium and selenium compounds)
Sodium silicofluoride (*see* Fluorides)
Solar radiation   55 (1992)
Soots   3, 22 (1973); 35, 219 (1985); *Suppl. 7*, 343 (1987)

Spironolactone   24, 259 (1980); *Suppl. 7*, 344 (1987)
Stannous fluoride (*see* Fluorides)
Steel founding (*see* Iron and steel founding)
Sterigmatocystin   1, 175 (1972); 10, 245 (1976); *Suppl. 7*, 72 (1987)

Steroidal oestrogens (*see also* Oestrogens, progestins and   *Suppl. 7*, 280 (1987)
  combinations)
Streptozotocin   4, 221 (1974); 17, 337 (1978); *Suppl. 7*, 72 (1987)

Strobane® (*see* Terpene polychlorinates)
Strontium chromate (*see* Chromium and chromium compounds)
Styrene   19, 231 (1979) (*corr.* 42, 258); *Suppl. 7*, 345 (1987)

Styrene-acrylonitrile copolymers   19, 97 (1979); *Suppl. 7*, 72 (1987)
Styrene-butadiene copolymers   19, 252 (1979); *Suppl. 7*, 72 (1987)
Styrene oxide   11, 201 (1976); 19, 275 (1979); 36, 245 (1985); *Suppl. 7*, 72 (1987)

Succinic anhydride   15, 265 (1977); *Suppl. 7*, 72 (1987)
Sudan I   8, 225 (1975); *Suppl. 7*, 72 (1987)
Sudan II   8, 233 (1975); *Suppl. 7*, 72 (1987)
Sudan III   8, 241 (1975); *Suppl. 7*, 72 (1987)
Sudan Brown RR   8, 249 (1975); *Suppl. 7*, 72 (1987)
Sudan Red 7B   8, 253 (1975); *Suppl. 7*, 72 (1987)
Sulfafurazole   24, 275 (1980); *Suppl. 7*, 347 (1987)
Sulfallate   30, 283 (1983); *Suppl. 7*, 72 (1987)
Sulfamethoxazole   24, 285 (1980); *Suppl. 7*, 348 (1987)
Sulfites (*see* Sulfur dioxide and some sulfites, bisulfites and metabisulfites)
Sulfur dioxide and some sulfites, bisulfites and metabisulfites   54, 131 (1992)
Sulfur mustard (*see* Mustard gas)
Sulfuric acid and other strong inorganic acids, occupational exposures   54, 41 (1992)
  to mists and vapours from
Sulfur trioxide   54, 121 (1992)
Sulphisoxazole (*see* Sulfafurazole)
Sunset Yellow FCF   8, 257 (1975); *Suppl. 7*, 72 (1987)
Symphytine   31, 239 (1983); *Suppl. 7*, 72 (1987)

T

2,4,5-T (*see also* Chlorophenoxy herbicides; Chlorophenoxy   15, 273 (1977)
  herbicides, occupational exposures to)
Talc   42, 185 (1987); *Suppl. 7*, 349 (1987)

Tannic acid	10, 253 (1976) (corr. 42, 255); Suppl. 7, 72 (1987)
Tannins (see also Tannic acid)	10, 254 (1976); Suppl. 7, 72 (1987)
TCDD (see 2,3,7,8-Tetrachlorodibenzo-para-dioxin)	
TDE (see DDT)	
Tea	51, 207 (1991)
Terpene polychlorinates	5, 219 (1974); Suppl. 7, 72 (1987)
Testosterone (see also Androgenic (anabolic) steroids)	6, 209 (1974); 21, 519 (1979)
Testosterone oenanthate (see Testosterone)	
Testosterone propionate (see Testosterone)	
2,2′,5,5′-Tetrachlorobenzidine	27, 141 (1982); Suppl. 7, 72 (1987)
2,3,7,8-Tetrachlorodibenzo-para-dioxin	15, 41 (1977); Suppl. 7, 350 (1987)
1,1,1,2-Tetrachloroethane	41, 87 (1986); Suppl. 7, 72 (1987)
1,1,2,2-Tetrachloroethane	20, 477 (1979); Suppl. 7, 354 (1987)
Tetrachloroethylene	20, 491 (1979); Suppl. 7, 355 (1987)
2,3,4,6-Tetrachlorophenol (see Chlorophenols; Chlorophenols, occupational exposures to)	
Tetrachlorvinphos	30, 197 (1983); Suppl. 7, 72 (1987)
Tetraethyllead (see Lead and lead compounds)	
Tetrafluoroethylene	19, 285 (1979); Suppl. 7, 72 (1987)
Tetrakis(hydroxymethyl) phosphonium salts	48, 95 (1990)
Tetramethyllead (see Lead and lead compounds)	
Textile manufacturing industry, exposures in	48, 215 (1990) (corr. 51, 483)
Theobromine	51, 421 (1991)
Theophylline	51, 391 (1991)
Thioacetamide	7, 77 (1974); Suppl. 7, 72 (1987)
4,4′-Thiodianiline	16, 343 (1978); 27, 147 (1982); Suppl. 7, 72 (1987)
Thiotepa	9, 85 (1975); Suppl. 7, 368 (1987); 50, 123 (1990)
Thiouracil	7, 85 (1974); Suppl. 7, 72 (1987)
Thiourea	7, 95 (1974); Suppl. 7, 72 (1987)
Thiram	12, 225 (1976); Suppl. 7, 72 (1987); 53, 403 (1991)
Titanium dioxide	47, 307 (1989)
Tobacco habits other than smoking (see Tobacco products, smokeless)	
Tobacco products, smokeless	37 (1985) (corr. 42, 263; 52, 513); Suppl. 7, 357 (1987)
Tobacco smoke	38 (1986) (corr. 42, 263); Suppl. 7, 357 (1987)
Tobacco smoking (see Tobacco smoke)	
ortho-Tolidine (see 3,3′-Dimethylbenzidine)	
2,4-Toluene diisocyanate (see also Toluene diisocyanates)	19, 303 (1979); 39, 287 (1986)
2,6-Toluene diisocyanate (see also Toluene diisocyanates)	19, 303 (1979); 39, 289 (1986)
Toluene	47, 79 (1989)
Toluene diisocyanates	39, 287 (1986) (corr. 42, 264); Suppl. 7, 72 (1987)
Toluenes, α-chlorinated (see α-Chlorinated toluenes)	
ortho-Toluenesulfonamide (see Saccharin)	
ortho-Toluidine	16, 349 (1978); 27, 155 (1982); Suppl. 7, 362 (1987)

Toxaphene	20, 327 (1979); *Suppl.* 7, 72 (1987)
Tremolite (*see* Asbestos)	
Treosulfan	26, 341 (1981); *Suppl.* 7, 363 (1987)
Triaziquone [*see* Tris(aziridinyl)-*para*-benzoquinone]	
Trichlorfon	30, 207 (1983); *Suppl.* 7, 73 (1987)
Trichlormethine	9, 229 (1975); *Suppl.* 7, 73 (1987); 50, 143 (1990)
Trichloroacetonitrile (*see* Halogenated acetonitriles)	
1,1,1-Trichloroethane	20, 515 (1979); *Suppl.* 7, 73 (1987)
1,1,2-Trichloroethane	20, 533 (1979); *Suppl.* 7, 73 (1987); 52, 337 (1991)
Trichloroethylene	11, 263 (1976); 20, 545 (1979); *Suppl.* 7, 364 (1987)
2,4,5-Trichlorophenol (*see also* Chlorophenols; Chlorophenols occupational exposures to)	20, 349 (1979)
2,4,6-Trichlorophenol (*see also* Chlorophenols; Chlorophenols, occupational exposures to)	20, 349 (1979)
(2,4,5-Trichlorophenoxy)acetic acid (*see* 2,4,5-T)	
Trichlorotriethylamine hydrochloride (*see* Trichlormethine)	
T$_2$-Trichothecene	31, 265 (1983); *Suppl.* 7, 73 (1987)
Triethylene glycol diglycidyl ether	11, 209 (1976); *Suppl.* 7, 73 (1987)
Trifluralin	53, 515 (1991)
4,4',6-Trimethylangelicin plus ultraviolet radiation (*see also* Angelicin and some synthetic derivatives)	*Suppl.* 7, 57 (1987)
2,4,5-Trimethylaniline	27, 177 (1982); *Suppl.* 7, 73 (1987)
2,4,6-Trimethylaniline	27, 178 (1982); *Suppl.* 7, 73 (1987)
4,5',8-Trimethylpsoralen	40, 357 (1986); *Suppl.* 7, 366 (1987)
Trimustine hydrochloride (*see* Trichlormethine)	
Triphenylene	32, 447 (1983); *Suppl.* 7, 73 (1987)
Tris(aziridinyl)-*para*-benzoquinone	9, 67 (1975); *Suppl.* 7, 367 (1987)
Tris(1-aziridinyl)phosphine oxide	9, 75 (1975); *Suppl.* 7, 73 (1987)
Tris(1-aziridinyl)phosphine sulphide (*see* Thiotepa)	
2,4,6-Tris(1-aziridinyl)-*s*-triazine	9, 95 (1975); *Suppl.* 7, 73 (1987)
Tris(2-chloroethyl) phosphate	48, 109 (1990)
1,2,3-Tris(chloromethoxy)propane	15, 301 (1977); *Suppl.* 7, 73 (1987)
Tris(2,3-dibromopropyl)phosphate	20, 575 (1979); *Suppl.* 7, 369 (1987)
Tris(2-methyl-1-aziridinyl)phosphine oxide	9, 107 (1975); *Suppl.* 7, 73 (1987)
Trp-P-1	31, 247 (1983); *Suppl.* 7, 73 (1987)
Trp-P-2	31, 255 (1983); *Suppl.* 7, 73 (1987)
Trypan blue	8, 267 (1975); *Suppl.* 7, 73 (1987)
Tussilago farfara L. (*see* Pyrrolizidine alkaloids)	

U

Ultraviolet radiation	40, 379 (1986); 55 (1992)
Underground haematite mining with exposure to radon	1, 29 (1972); *Suppl.* 7, 216 (1987)
Uracil mustard	9, 235 (1975); *Suppl.* 7, 370 (1987)
Urethane	7, 111 (1974); *Suppl.* 7, 73 (1987)

V

Vat Yellow 4	48, 161 (1990)

Vinblastine sulfate	26, 349 (1981) (corr. 42, 261); Suppl. 7, 371 (1987)
Vincristine sulfate	26, 365 (1981); Suppl. 7, 372 (1987)
Vinyl acetate	19, 341 (1979); 39, 113 (1986); Suppl. 7, 73 (1987)
Vinyl bromide	19, 367 (1979); 39, 133 (1986); Suppl. 7, 73 (1987)
Vinyl chloride	7, 291 (1974); 19, 377 (1979) (corr. 42, 258); Suppl. 7, 373 (1987)
Vinyl chloride–vinyl acetate copolymers	7, 311 (1976); 19, 412 (1979) (corr. 42, 258); Suppl. 7, 73 (1987)
4-Vinylcyclohexene	11, 277 (1976); 39, 181 (1986); Suppl. 7, 73 (1987)
Vinyl fluoride	39, 147 (1986); Suppl. 7, 73 (1987)
Vinylidene chloride	19, 439 (1979); 39, 195 (1986); Suppl. 7, 376 (1987)
Vinylidene chloride–vinyl chloride copolymers	19, 448 (1979) (corr. 42, 258); Suppl. 7, 73 (1987)
Vinylidene fluoride	39, 227 (1986); Suppl. 7, 73 (1987)
N-Vinyl-2-pyrrolidone	19, 461 (1979); Suppl. 7, 73 (1987)

W

Welding	49, 447 (1990) (corr. 52, 513)
Wollastonite	42, 145 (1987); Suppl. 7, 377 (1987)
Wood industries	25 (1981); Suppl. 7, 378 (1987)

X

Xylene	47, 125 (1989)
2,4-Xylidine	16, 367 (1978); Suppl. 7, 74 (1987)
2,5-Xylidine	16, 377 (1978); Suppl. 7, 74 (1987)

Y

Yellow AB	8, 279 (1975); Suppl. 7, 74 (1987)
Yellow OB	8, 287 (1975); Suppl. 7, 74 (1987)

Z

Zearalenone	31, 279 (1983); Suppl. 7, 74 (1987)
Zectran	12, 237 (1976); Suppl. 7, 74 (1987)
Zinc beryllium silicate (see Beryllium and beryllium compounds)	
Zinc chromate (see Chromium and chromium compounds)	
Zinc chromate hydroxide (see Chromium and chromium compounds)	
Zinc potassium chromate (see Chromium and chromium compounds)	
Zinc yellow (see Chromium and chromium compounds)	
Zineb	12, 245 (1976); Suppl. 7, 74 (1987)
Ziram	12, 259 (1976); Suppl. 7, 74 (1987); 53, 423 (1991)

PUBLICATIONS OF THE INTERNATIONAL AGENCY FOR RESEARCH ON CANCER

Scientific Publications Series

(Available from Oxford University Press through local bookshops)

No. 1 Liver Cancer
1971; 176 pages (*out of print*)

No. 2 Oncogenesis and Herpesviruses
Edited by P.M. Biggs, G. de-Thé and L.N. Payne
1972; 515 pages (*out of print*)

No. 3 N-Nitroso Compounds: Analysis and Formation
Edited by P. Bogovski, R. Preussman and E.A. Walker
1972; 140 pages (*out of print*)

No. 4 Transplacental Carcinogenesis
Edited by L. Tomatis and U. Mohr
1973; 181 pages (*out of print*)

No. 5/6 Pathology of Tumours in Laboratory Animals, Volume 1, Tumours of the Rat
Edited by V.S. Turusov
1973/1976; 533 pages (*out of print*)

No. 7 Host Environment Interactions in the Etiology of Cancer in Man
Edited by R. Doll and I. Vodopija
1973; 464 pages (*out of print*)

No. 8 Biological Effects of Asbestos
Edited by P. Bogovski, J.C. Gilson, V. Timbrell and J.C. Wagner
1973; 346 pages (*out of print*)

No. 9 N-Nitroso Compounds in the Environment
Edited by P. Bogovski and E.A. Walker
1974; 243 pages (*out of print*)

No. 10 Chemical Carcinogenesis Essays
Edited by R. Montesano and L. Tomatis
1974; 230 pages (*out of print*)

No. 11 Oncogenesis and Herpesviruses II
Edited by G. de-Thé, M.A. Epstein and H. zur Hausen
1975; Part I: 511 pages
Part II: 403 pages (*out of print*)

No. 12 Screening Tests in Chemical Carcinogenesis
Edited by R. Montesano, H. Bartsch and L. Tomatis
1976; 666 pages (*out of print*)

No. 13 Environmental Pollution and Carcinogenic Risks
Edited by C. Rosenfeld and W. Davis
1975; 441 pages (*out of print*)

No. 14 Environmental N-Nitroso Compounds. Analysis and Formation
Edited by E.A. Walker, P. Bogovski and L. Griciute
1976; 512 pages (*out of print*)

No. 15 Cancer Incidence in Five Continents, Volume III
Edited by J.A.H. Waterhouse, C. Muir, P. Correa and J. Powell
1976; 584 pages (*out of print*)

No. 16 Air Pollution and Cancer in Man
Edited by U. Mohr, D. Schmähl and L. Tomatis
1977; 328 pages (*out of print*)

No. 17 Directory of On-going Research in Cancer Epidemiology 1977
Edited by C.S. Muir and G. Wagner
1977; 599 pages (*out of print*)

No. 18 Environmental Carcinogens. Selected Methods of Analysis. Volume 1: Analysis of Volatile Nitrosamines in Food
Editor-in-Chief: H. Egan
1978; 212 pages (*out of print*)

No. 19 Environmental Aspects of N-Nitroso Compounds
Edited by E.A. Walker, M. Castegnaro, L. Griciute and R.E. Lyle
1978; 561 pages (*out of print*)

No. 20 Nasopharyngeal Carcinoma: Etiology and Control
Edited by G. de-Thé and Y. Ito
1978; 606 pages (*out of print*)

No. 21 Cancer Registration and its Techniques
Edited by R. MacLennan, C. Muir, R. Steinitz and A. Winkler
1978; 235 pages (*out of print*)

No. 22 Environmental Carcinogens. Selected Methods of Analysis. Volume 2: Methods for the Measurement of Vinyl Chloride in Poly(vinyl chloride), Air, Water and Foodstuffs
Editor-in-Chief: H. Egan
1978; 142 pages (*out of print*)

No. 23 Pathology of Tumours in Laboratory Animals. Volume II: Tumours of the Mouse
Editor-in-Chief: V.S. Turusov
1979; 669 pages (*out of print*)

No. 24 Oncogenesis and Herpesviruses III
Edited by G. de-Thé, W. Henle and F. Rapp
1978; Part I: 580 pages, Part II: 512 pages (*out of print*)

Prices, valid for November 1992, are subject to change without notice

List of IARC Publications

No. 25 Carcinogenic Risk. Strategies for Intervention
Edited by W. Davis and C. Rosenfeld
1979; 280 pages (*out of print*)

No. 26 Directory of On-going Research in Cancer Epidemiology 1978
Edited by C.S. Muir and G. Wagner
1978; 550 pages (*out of print*)

No. 27 Molecular and Cellular Aspects of Carcinogen Screening Tests
Edited by R. Montesano, H. Bartsch and L. Tomatis
1980; 372 pages £29.00

No. 28 Directory of On-going Research in Cancer Epidemiology 1979
Edited by C.S. Muir and G. Wagner
1979; 672 pages (*out of print*)

No. 29 Environmental Carcinogens. Selected Methods of Analysis. Volume 3: Analysis of Polycyclic Aromatic Hydrocarbons in Environmental Samples
Editor-in-Chief: H. Egan
1979; 240 pages (*out of print*)

No. 30 Biological Effects of Mineral Fibres
Editor-in-Chief: J.C. Wagner
1980; Volume 1: 494 pages Volume 2: 513 pages (*out of print*)

No. 31 N-Nitroso Compounds: Analysis, Formation and Occurrence
Edited by E.A. Walker, L. Griciute, M. Castegnaro and M. Börzsönyi
1980; 835 pages (*out of print*)

No. 32 Statistical Methods in Cancer Research. Volume 1. The Analysis of Case-control Studies
By N.E. Breslow and N.E. Day
1980; 338 pages £25.00

No. 33 Handling Chemical Carcinogens in the Laboratory
Edited by R. Montesano *et al.*
1979; 32 pages (*out of print*)

No. 34 Pathology of Tumours in Laboratory Animals. Volume III. Tumours of the Hamster
Editor-in-Chief: V.S. Turusov
1982; 461 pages (*out of print*)

No. 35 Directory of On-going Research in Cancer Epidemiology 1980
Edited by C.S. Muir and G. Wagner
1980; 660 pages (*out of print*)

No. 36 Cancer Mortality by Occupation and Social Class 1851-1971
Edited by W.P.D. Logan
1982; 253 pages (*out of print*)

No. 37 Laboratory Decontamination and Destruction of Aflatoxins B_1, B_2, G_1, G_2 in Laboratory Wastes
Edited by M. Castegnaro *et al.*
1980; 56 pages (*out of print*)

No. 38 Directory of On-going Research in Cancer Epidemiology 1981
Edited by C.S. Muir and G. Wagner
1981; 696 pages (*out of print*)

No. 39 Host Factors in Human Carcinogenesis
Edited by H. Bartsch and B. Armstrong
1982; 583 pages (*out of print*)

No. 40 Environmental Carcinogens. Selected Methods of Analysis. Volume 4: Some Aromatic Amines and Azo Dyes in the General and Industrial Environment
Edited by L. Fishbein, M. Castegnaro, I.K. O'Neill and H. Bartsch
1981; 347 pages (*out of print*)

No. 41 N-Nitroso Compounds: Occurrence and Biological Effects
Edited by H. Bartsch, I.K. O'Neill, M. Castegnaro and M. Okada
1982; 755 pages £50.00

No. 42 Cancer Incidence in Five Continents, Volume IV
Edited by J. Waterhouse, C. Muir, K. Shanmugaratnam and J. Powell
1982; 811 pages (*out of print*)

No. 43 Laboratory Decontamination and Destruction of Carcinogens in Laboratory Wastes: Some N-Nitrosamines
Edited by M. Castegnaro *et al.*
1982; 73 pages £7.50

No. 44 Environmental Carcinogens. Selected Methods of Analysis. Volume 5: Some Mycotoxins
Edited by L. Stoloff, M. Castegnaro, P. Scott, I.K. O'Neill and H. Bartsch
1983; 455 pages £32.50

No. 45 Environmental Carcinogens. Selected Methods of Analysis. Volume 6: N-Nitroso Compounds
Edited by R. Preussmann, I.K. O'Neill, G. Eisenbrand, B. Spiegelhalder and H. Bartsch
1983; 508 pages £32.50

No. 46 Directory of On-going Research in Cancer Epidemiology 1982
Edited by C.S. Muir and G. Wagner
1982; 722 pages (*out of print*)

No. 47 Cancer Incidence in Singapore 1968-1977
Edited by K. Shanmugaratnam, H.P. Lee and N.E. Day
1983; 171 pages (*out of print*)

No. 48 Cancer Incidence in the USSR (2nd Revised Edition)
Edited by N.P. Napalkov, G.F. Tserkovny, V.M. Merabishvili, D.M. Parkin, M. Smans and C.S. Muir
1983; 75 pages (*out of print*)

No. 49 Laboratory Decontamination and Destruction of Carcinogens in Laboratory Wastes: Some Polycyclic Aromatic Hydrocarbons
Edited by M. Castegnaro *et al.*
1983; 87 pages (*out of print*)

No. 50 Directory of On-going Research in Cancer Epidemiology 1983
Edited by C.S. Muir and G. Wagner
1983; 731 pages (*out of print*)

No. 51 Modulators of Experimental Carcinogenesis
Edited by V. Turusov and R. Montesano
1983; 307 pages (*out of print*)

List of IARC Publications

No. 52 Second Cancers in Relation to Radiation Treatment for Cervical Cancer: Results of a Cancer Registry Collaboration
Edited by N.E. Day and J.C. Boice, Jr
1984; 207 pages (*out of print*)

No. 53 Nickel in the Human Environment
Editor-in-Chief: F.W. Sunderman, Jr
1984; 529 pages (*out of print*)

No. 54 Laboratory Decontamination and Destruction of Carcinogens in Laboratory Wastes: Some Hydrazines
Edited by M. Castegnaro et al.
1983; 87 pages (*out of print*)

No. 55 Laboratory Decontamination and Destruction of Carcinogens in Laboratory Wastes: Some N-Nitrosamides
Edited by M. Castegnaro et al.
1984; 66 pages (*out of print*)

No. 56 Models, Mechanisms and Etiology of Tumour Promotion
Edited by M. Börzsönyi, N.E. Day, K. Lapis and H. Yamasaki
1984; 532 pages (*out of print*)

No. 57 N-Nitroso Compounds: Occurrence, Biological Effects and Relevance to Human Cancer
Edited by I.K. O'Neill, R.C. von Borstel, C.T. Miller, J. Long and H. Bartsch
1984; 1013 pages (*out of print*)

No. 58 Age-related Factors in Carcinogenesis
Edited by A. Likhachev, V. Anisimov and R. Montesano
1985; 288 pages (*out of print*)

No. 59 Monitoring Human Exposure to Carcinogenic and Mutagenic Agents
Edited by A. Berlin, M. Draper, K. Hemminki and H. Vainio
1984; 457 pages (*out of print*)

No. 60 Burkitt's Lymphoma: A Human Cancer Model
Edited by G. Lenoir, G. O'Conor and C.L.M. Olweny
1985; 484 pages (*out of print*)

No. 61 Laboratory Decontamination and Destruction of Carcinogens in Laboratory Wastes: Some Haloethers
Edited by M. Castegnaro et al.
1985; 55 pages (*out of print*)

No. 62 Directory of On-going Research in Cancer Epidemiology 1984
Edited by C.S. Muir and G. Wagner
1984; 717 pages (*out of print*)

No. 63 Virus-associated Cancers in Africa
Edited by A.O. Williams, G.T. O'Conor, G.B. de-Thé and C.A. Johnson
1984; 773 pages (*out of print*)

No. 64 Laboratory Decontamination and Destruction of Carcinogens in Laboratory Wastes: Some Aromatic Amines and 4-Nitrobiphenyl
Edited by M. Castegnaro et al.
1985; 84 pages (*out of print*)

No. 65 Interpretation of Negative Epidemiological Evidence for Carcinogenicity
Edited by N.J. Wald and R. Doll
1985; 232 pages (*out of print*)

No. 66 The Role of the Registry in Cancer Control
Edited by D.M. Parkin, G. Wagner and C.S. Muir
1985; 152 pages £10.00

No. 67 Transformation Assay of Established Cell Lines: Mechanisms and Application
Edited by T. Kakunaga and H. Yamasaki
1985; 225 pages (*out of print*)

No. 68 Environmental Carcinogens. Selected Methods of Analysis. Volume 7. Some Volatile Halogenated Hydrocarbons
Edited by L. Fishbein and I.K. O'Neill
1985; 479 pages (*out of print*)

No. 69 Directory of On-going Research in Cancer Epidemiology 1985
Edited by C.S. Muir and G. Wagner
1985; 745 pages (*out of print*)

No. 70 The Role of Cyclic Nucleic Acid Adducts in Carcinogenesis and Mutagenesis
Edited by B. Singer and H. Bartsch
1986; 467 pages (*out of print*)

No. 71 Environmental Carcinogens. Selected Methods of Analysis. Volume 8: Some Metals: As, Be, Cd, Cr, Ni, Pb, Se Zn
Edited by I.K. O'Neill, P. Schuller and L. Fishbein
1986; 485 pages (*out of print*)

No. 72 Atlas of Cancer in Scotland, 1975–1980. Incidence and Epidemiological Perspective
Edited by I. Kemp, P. Boyle, M. Smans and C.S. Muir
1985; 285 pages (*out of print*)

No. 73 Laboratory Decontamination and Destruction of Carcinogens in Laboratory Wastes: Some Antineoplastic Agents
Edited by M. Castegnaro et al.
1985; 163 pages £12.50

No. 74 Tobacco: A Major International Health Hazard
Edited by D. Zaridze and R. Peto
1986; 324 pages £22.50

No. 75 Cancer Occurrence in Developing Countries
Edited by D.M. Parkin
1986; 339 pages £22.50

No. 76 Screening for Cancer of the Uterine Cervix
Edited by M. Hakama, A.B. Miller and N.E. Day
1986; 315 pages £30.00

List of IARC Publications

No. 77 Hexachlorobenzene: Proceedings of an International Symposium
Edited by C.R. Morris and J.R.P. Cabral
1986; 668 pages (*out of print*)

No. 78 Carcinogenicity of Alkylating Cytostatic Drugs
Edited by D. Schmähl and J.M. Kaldor
1986; 337 pages (*out of print*)

No. 79 Statistical Methods in Cancer Research. Volume III: The Design and Analysis of Long-term Animal Experiments
By J.J. Gart, D. Krewski, P.N. Lee, R.E. Tarone and J. Wahrendorf
1986; 213 pages £22.00

No. 80 Directory of On-going Research in Cancer Epidemiology 1986
Edited by C.S. Muir and G. Wagner
1986; 805 pages (*out of print*)

No. 81 Environmental Carcinogens: Methods of Analysis and Exposure Measurement. Volume 9: Passive Smoking
Edited by I.K. O'Neill, K.D. Brunnemann, B. Dodet and D. Hoffmann
1987; 383 pages £35.00

No. 82 Statistical Methods in Cancer Research. Volume II: The Design and Analysis of Cohort Studies
By N.E. Breslow and N.E. Day
1987; 404 pages £35.00

No. 83 Long-term and Short-term Assays for Carcinogens: A Critical Appraisal
Edited by R. Montesano, H. Bartsch, H. Vainio, J. Wilbourn and H. Yamasaki
1986; 575 pages £35.00

No. 84 The Relevance of *N*-Nitroso Compounds to Human Cancer: Exposure and Mechanisms
Edited by H. Bartsch, I.K. O'Neill and R. Schulte-Hermann
1987; 671 pages (*out of print*)

No. 85 Environmental Carcinogens: Methods of Analysis and Exposure Measurement. Volume 10: Benzene and Alkylated Benzenes
Edited by L. Fishbein and I.K. O'Neill
1988; 327 pages £40.00

No. 86 Directory of On-going Research in Cancer Epidemiology 1987
Edited by D.M. Parkin and J. Wahrendorf
1987; 676 pages (*out of print*)

No. 87 International Incidence of Childhood Cancer
Edited by D.M. Parkin, C.A. Stiller, C.A. Bieber, G.J. Draper, B. Terracini and J.L. Young
1988; 401 pages £35.00

No. 88 Cancer Incidence in Five Continents Volume V
Edited by C. Muir, J. Waterhouse, T. Mack, J. Powell and S. Whelan
1987; 1004 pages £55.00

No. 89 Method for Detecting DNA Damaging Agents in Humans: Applications in Cancer Epidemiology and Prevention
Edited by H. Bartsch, K. Hemminki and I.K. O'Neill
1988; 518 pages £50.00

No. 90 Non-occupational Exposure to Mineral Fibres
Edited by J. Bignon, J. Peto and R. Saracci
1989; 500 pages £50.00

No. 91 Trends in Cancer Incidence in Singapore 1968–1982
Edited by H.P. Lee, N.E. Day and K. Shanmugaratnam
1988; 160 pages (*out of print*)

No. 92 Cell Differentiation, Genes and Cancer
Edited by T. Kakunaga, T. Sugimura, L. Tomatis and H. Yamasaki
1988; 204 pages £27.50

No. 93 Directory of On-going Research in Cancer Epidemiology 1988
Edited by M. Coleman and J. Wahrendorf
1988; 662 pages (*out of print*)

No. 94 Human Papillomavirus and Cervical Cancer
Edited by N. Muñoz, F.X. Bosch and O.M. Jensen
1989; 154 pages £22.50

No. 95 Cancer Registration: Principles and Methods
Edited by O.M. Jensen, D.M. Parkin, R. MacLennan, C.S. Muir and R. Skeet
1991; 288 pages £28.00

No. 96 Perinatal and Multigeneration Carcinogenesis
Edited by N.P. Napalkov, J.M. Rice, L. Tomatis and H. Yamasaki
1989; 436 pages £50.00

No. 97 Occupational Exposure to Silica and Cancer Risk
Edited by L. Simonato, A.C. Fletcher, R. Saracci and T. Thomas
1990; 124 pages £22.50

No. 98 Cancer Incidence in Jewish Migrants to Israel, 1961–1981
Edited by R. Steinitz, D.M. Parkin, J.L. Young, C.A. Bieber and L. Katz
1989; 320 pages £35.00

No. 99 Pathology of Tumours in Laboratory Animals, Second Edition, Volume 1, Tumours of the Rat
Edited by V.S. Turusov and U. Mohr
740 pages £85.00

No. 100 Cancer: Causes, Occurrence and Control
Editor-in-Chief L. Tomatis
1990; 352 pages £24.00

No. 101 Directory of On-going Research in Cancer Epidemiology 1989/90
Edited by M. Coleman and J. Wahrendorf
1989; 818 pages £36.00

List of IARC Publications

No. 102 **Patterns of Cancer in Five Continents**
Edited by S.L. Whelan and D.M. Parkin
1990; 162 pages £25.00

No. 103 **Evaluating Effectiveness of Primary Prevention of Cancer**
Edited by M. Hakama, V. Beral, J.W. Cullen and D.M. Parkin
1990; 250 pages £32.00

No. 104 **Complex Mixtures and Cancer Risk**
Edited by H. Vainio, M. Sorsa and A.J. McMichael
1990; 442 pages £38.00

No. 105 **Relevance to Human Cancer of N-Nitroso Compounds, Tobacco Smoke and Mycotoxins**
Edited by I.K. O'Neill, J. Chen and H. Bartsch
1991; 614 pages £70.00

No. 106 **Atlas of Cancer Incidence in the Former German Democratic Republic** Edited by W.H. Mehnert, M. Smans, C.S. Muir, M. Möhner & D. Schön
1992; 384 pages £55.00

No. 107 **Atlas of Cancer Mortality in the European Economic Community**
Edited by M. Smans, C.S. Muir and P. Boyle
Publ. due 1992; 280 pages £35.00

No. 108 **Environmental Carcinogens: Methods of Analysis and Exposure Measurement. Volume 11: Polychlorinated Dioxins and Dibenzofurans**
Edited by C. Rappe, H.R. Buser, B. Dodet and I.K. O'Neill
1991; 426 pages £45.00

No. 109 **Environmental Carcinogens: Methods of Analysis and Exposure Measurement. Volume 12: Indoor Air Contaminants**
Edited by B. Seifert, B. Dodet and I.K. O'Neill
Publ. due 1992; approx. 400 pages

No. 110 **Directory of On-going Research in Cancer Epidemiology 1991**
Edited by M. Coleman and J. Wahrendorf
1991; 753 pages £38.00

No. 111 **Pathology of Tumours in Laboratory Animals, Second Edition, Volume 2, Tumours of the Mouse**
Edited by V.S. Turusov and U. Mohr
Publ. due 1993; approx. 500 pages

No. 112 **Autopsy in Epidemiology and Medical Research**
Edited by E. Riboli and M. Delendi
1991; 288 pages £25.00

No. 113 **Laboratory Decontamination and Destruction of Carcinogens in Laboratory Wastes: Some Mycotoxins**
Edited by M. Castegnaro, J. Barek, J.-M. Frémy, M. Lafontaine, M. Miraglia, E.B. Sansone and G.M. Telling
1991; 64 pages £11.00

No. 114 **Laboratory Decontamination and Destruction of Carcinogens in Laboratory Wastes: Some Polycyclic Heterocyclic Hydrocarbons**
Edited by M. Castegnaro, J. Barek, J. Jacob, U. Kirso, M. Lafontaine, E.B. Sansone, G.M. Telling and T. Vu Duc
1991; 50 pages £8.00

No. 115 **Mycotoxins, Endemic Nephropathy and Urinary Tract Tumours**
Edited by M. Castegnaro, R. Plestina, G. Dirheimer, I.N. Chernozemsky and H Bartsch
1991; 340 pages £45.00

No. 116 **Mechanisms of Carcinogenesis in Risk Identification**
Edited by H. Vainio, P.N. Magee, D.B. McGregor & A.J. McMichael
1992; 616 pages £65.00

No. 117 **Directory of On-going Research in Cancer Epidemiology 1992**
Edited by M. Coleman, J. Wahrendorf & E. Démaret
1992; 773 pages £42.00

No. 118 **Cadmium in the Human Environment: Toxicity and Carcinogenicity**
Edited by G.F. Nordberg, R.F.M. Herber & L. Alessio
Publ. due 1992; approx. 450 pages

No. 119 **The Epidemiology of Cervical Cancer and Human Papillomavirus**
Edited by N. Muñoz, F.X. Bosch, K.V. Shah & A. Meheus
1992; 288 pages £28.00

No. 120 **Cancer Incidence in Five Continents, Volume VI**
Edited by D.M. Parkin, C.S. Muir, S.L. Whelan, Y.T. Gao, J. Ferlay & J. Powell
1992; 1050 pages £120.00

No. 122 **International Classification of Rodent Tumours. Part 1. The Rat**
Editor-in-Chief: U. Möhr
1992/93, 10 fascicles, approx. 600 pages, £120.00

List of IARC Publications

IARC MONOGRAPHS ON THE EVALUATION OF CARCINOGENIC RISKS TO HUMANS

(Available from booksellers through the network of WHO Sales Agents)

Volume 1 Some Inorganic Substances, Chlorinated Hydrocarbons, Aromatic Amines, N-Nitroso Compounds, and Natural Products
1972; 184 pages (*out of print*)

Volume 2 Some Inorganic and Organometallic Compounds
1973; 181 pages (*out of print*)

Volume 3 Certain Polycyclic Aromatic Hydrocarbons and Heterocyclic Compounds
1973; 271 pages (*out of print*)

Volume 4 Some Aromatic Amines, Hydrazine and Related Substances, N-Nitroso Compounds and Miscellaneous Alkylating Agents
1974; 286 pages Sw. fr. 18.

Volume 5 Some Organochlorine Pesticides
1974; 241 pages (*out of print*)

Volume 6 Sex Hormones
1974; 243 pages (*out of print*)

Volume 7 Some Anti-Thyroid and Related Substances, Nitrofurans and Industrial Chemicals
1974; 326 pages (*out of print*)

Volume 8 Some Aromatic Azo Compounds
1975; 357 pages Sw. fr. 36.

Volume 9 Some Aziridines, N-, S- and O-Mustards and Selenium
1975; 268 pages Sw.fr. 27.

Volume 10 Some Naturally Occurring Substances
1976; 353 pages (*out of print*)

Volume 11 Cadmium, Nickel, Some Epoxides, Miscellaneous Industrial Chemicals and General Considerations on Volatile Anaesthetics
1976; 306 pages (*out of print*)

Volume 12 Some Carbamates, Thiocarbamates and Carbazides
1976; 282 pages Sw. fr. 34.-

Volume 13 Some Miscellaneous Pharmaceutical Substances
1977; 255 pages Sw. fr. 30.

Volume 14 Asbestos
1977; 106 pages (*out of print*)

Volume 15 Some Fumigants, The Herbicides 2,4-D and 2,4,5-T, Chlorinated Dibenzodioxins and Miscellaneous Industrial Chemicals
1977; 354 pages Sw. fr. 50.

Volume 16 Some Aromatic Amines and Related Nitro Compounds - Hair Dyes, Colouring Agents and Miscellaneous Industrial Chemicals
1978; 400 pages Sw. fr. 50.

Volume 17 Some N-Nitroso Compounds
1978; 365 pages Sw. fr. 50.

Volume 18 Polychlorinated Biphenyls and Polybrominated Biphenyls
1978; 140 pages Sw. fr. 20.

Volume 19 Some Monomers, Plastics and Synthetic Elastomers, and Acrolein
1979; 513 pages (*out of print*)

Volume 20 Some Halogenated Hydrocarbons
1979; 609 pages (*out of print*)

Volume 21 Sex Hormones (II)
1979; 583 pages Sw. fr. 60.

Volume 22 Some Non-Nutritive Sweetening Agents
1980; 208 pages Sw. fr. 25.

Volume 23 Some Metals and Metallic Compounds
1980; 438 pages (*out of print*)

Volume 24 Some Pharmaceutical Drugs
1980; 337 pages Sw. fr. 40.

Volume 25 Wood, Leather and Some Associated Industries
1981; 412 pages Sw. fr. 60

Volume 26 Some Antineoplastic and Immunosuppressive Agents
1981; 411 pages Sw. fr. 62.

Volume 27 Some Aromatic Amines, Anthraquinones and Nitroso Compounds, and Inorganic Fluorides Used in Drinking Water and Dental Preparations
1982; 341 pages Sw. fr. 40.

Volume 28 The Rubber Industry
1982; 486 pages Sw. fr. 70.

Volume 29 Some Industrial Chemicals and Dyestuffs
1982; 416 pages Sw. fr. 60.

Volume 30 Miscellaneous Pesticides
1983; 424 pages Sw. fr. 60.

Volume 31 Some Food Additives, Feed Additives and Naturally Occurring Substances
1983; 314 pages Sw. fr. 60

Volume 32 Polynuclear Aromatic Compounds, Part 1: Chemical, Environmental and Experimental Data
1983; 477 pages Sw. fr. 60.

Volume 33 Polynuclear Aromatic Compounds, Part 2: Carbon Blacks, Mineral Oils and Some Nitroarenes
1984; 245 pages Sw. fr. 50.

Volume 34 Polynuclear Aromatic Compounds, Part 3: Industrial Exposures in Aluminium Production, Coal Gasification, Coke Production, and Iron and Steel Founding
1984; 219 pages Sw. fr. 48.

Volume 35 Polynuclear Aromatic Compounds, Part 4: Bitumens, Coal-tars and Derived Products, Shale-oils and Soots
1985; 271 pages Sw. fr. 70.

List of IARC Publications

Volume 36 **Allyl Compounds, Aldehydes, Epoxides and Peroxides**
1985; 369 pages Sw. fr. 70.

Volume 37 **Tobacco Habits Other than Smoking: Betel-quid and Areca-nut Chewing; and some Related Nitrosamines**
1985; 291 pages Sw. fr. 70.

Volume 38 **Tobacco Smoking**
1986; 421 pages Sw. fr. 75.

Volume 39 **Some Chemicals Used in Plastics and Elastomers**
1986; 403 pages Sw. fr. 60.

Volume 40 **Some Naturally Occurring and Synthetic Food Components, Furocoumarins and Ultraviolet Radiation**
1986; 444 pages Sw. fr. 65.

Volume 41 **Some Halogenated Hydrocarbons and Pesticide Exposures**
1986; 434 pages Sw. fr. 65.

Volume 42 **Silica and Some Silicates**
1987; 289 pages Sw. fr. 65.

Volume 43 **Man-Made Mineral Fibres and Radon**
1988; 300 pages Sw. fr. 65.

Volume 44 **Alcohol Drinking**
1988; 416 pages Sw. fr. 65.

Volume 45 **Occupational Exposures in Petroleum Refining; Crude Oil and Major Petroleum Fuels**
1989; 322 pages Sw. fr. 65.

Volume 46 **Diesel and Gasoline Engine Exhausts and Some Nitroarenes**
1989; 458 pages Sw. fr. 65.

Volume 47 **Some Organic Solvents, Resin Monomers and Related Compounds, Pigments and Occupational Exposures in Paint Manufacture and Painting**
1989; 536 pages Sw. fr. 85.

Volume 48 **Some Flame Retardants and Textile Chemicals, and Exposures in the Textile Manufacturing Industry**
1990; 345 pages Sw. fr. 65.

Volume 49 **Chromium, Nickel and Welding**
1990; 677 pages Sw. fr. 95.-

Volume 50 **Pharmaceutical Drugs**
1990; 415 pages Sw. fr. 65.-

Volume 51 **Coffee, Tea, Mate, Methylxanthines and Methylglyoxal**
1991; 513 pages Sw. fr. 80.-

Volume 52 **Chlorinated Drinking-water; Chlorination By-products; Some Other Halogenated Compounds; Cobalt and Cobalt Compounds**
1991; 544 pages Sw. fr. 80.-

Volume 53 **Occupational Exposures in Insecticide Application and some Pesticides**
1991; 612 pages Sw. fr. 95.-

Volume 54 **Occupational Exposures to Mists and Vapours from Strong Inorganic Acids; and Other Industrial Chemicals**
1992; 336 pages Sw. fr. 65.-

Volume 55 **Solar and Ultraviolet Radiation**
1992; 316 pages Sw. fr. 65.-

Supplement No. 1
Chemicals and Industrial Processes Associated with Cancer in Humans (IARC Monographs, Volumes 1 to 20)
1979; 71 pages (*out of print*)

Supplement No. 2
Long-term and Short-term Screening Assays for Carcinogens: A Critical Appraisal
1980; 426 pages Sw. fr. 40.-

Supplement No. 3
Cross Index of Synonyms and Trade Names in Volumes 1 to 26
1982; 199 pages (*out of print*)

Supplement No. 4
Chemicals, Industrial Processes and Industries Associated with Cancer in Humans (IARC Monographs, Volumes 1 to 29)
1982; 292 pages (*out of print*)

Supplement No. 5
Cross Index of Synonyms and Trade Names in Volumes 1 to 36
1985; 259 pages (*out of print*)

Supplement No. 6
Genetic and Related Effects: An Updating of Selected IARC Monographs from Volumes 1 to 42
1987; 729 pages Sw. fr. 80.-

Supplement No. 7
Overall Evaluations of Carcinogenicity: An Updating of IARC Monographs Volumes 1-42
1987; 440 pages Sw. fr. 65.-

Supplement No. 8
Cross Index of Synonyms and Trade Names in Volumes 1 to 46
1990; 346 pages Sw. fr. 60.-

List of IARC Publications

IARC TECHNICAL REPORTS*

No. 1 Cancer in Costa Rica
Edited by R. Sierra,
R. Barrantes, G. Muñoz Leiva, D.M. Parkin, C.A. Bieber and
N. Muñoz Calero
1988; 124 pages Sw. fr. 30.-

No. 2 SEARCH: A Computer Package to Assist the Statistical Analysis of Case-control Studies
Edited by G.J. Macfarlane,
P. Boyle and P. Maisonneuve
1991; 80 pages (out of print)

No. 3 Cancer Registration in the European Economic Community
Edited by M.P. Coleman and
E. Démaret
1988; 188 pages Sw. fr. 30.-

No. 4 Diet, Hormones and Cancer: Methodological Issues for Prospective Studies
Edited by E. Riboli and
R. Saracci
1988; 156 pages Sw. fr. 30.-

No. 5 Cancer in the Philippines
Edited by A.V. Laudico,
D. Esteban and D.M. Parkin
1989; 186 pages Sw. fr. 30.-

No. 6 La genèse du Centre International de Recherche sur le Cancer
Par R. Sohier et A.G.B. Sutherland
1990; 104 pages Sw. fr. 30.-

No. 7 Epidémiologie du cancer dans les pays de langue latine
1990; 310 pages Sw. fr. 30.-

No. 8 Comparative Study of Anti-smoking Legislation in Countries of the European Economic Community
Edited by A. Sasco, P. Dalla Vorgia and P. Van der Elst
1990; 82 pages Sw. fr. 30.-

No. 9 Epidémiologie du cancer dans les pays de langue latine
1991; 346 pages Sw. fr. 30.-

No. 11 Nitroso Compounds: Biological Mechanisms, Exposures and Cancer Etiology
Edited by I.K. O'Neill & H. Bartsch
1992; 149 pages Sw. fr. 30.-

No. 12 Epidémiologie du cancer dans les pays de langue latine
1992; 375 pages Sw. fr. 30.-

DIRECTORY OF AGENTS BEING TESTED FOR CARCINOGENICITY (Until Vol. 13 Information Bulletin on the Survey of Chemicals Being Tested for Carcinogenicity)*

No. 8 Edited by M.-J. Ghess,
H. Bartsch and L. Tomatis
1979; 604 pages Sw. fr. 40.-

No. 9 Edited by M.-J. Ghess,
J.D. Wilbourn, H. Bartsch and
L. Tomatis
1981; 294 pages Sw. fr. 41.-

No. 10 Edited by M.-J. Ghess,
J.D. Wilbourn and H. Bartsch
1982; 362 pages Sw. fr. 42.-

No. 11 Edited by M.-J. Ghess,
J.D. Wilbourn, H. Vainio and
H. Bartsch
1984; 362 pages Sw. fr. 50.-

No. 12 Edited by M.-J. Ghess,
J.D. Wilbourn, A. Tossavainen and
H. Vainio
1986; 385 pages Sw. fr. 50.-

No. 13 Edited by M.-J. Ghess,
J.D. Wilbourn and A. Aitio 1988;
404 pages Sw. fr. 43.-

No. 14 Edited by M.-J. Ghess,
J.D. Wilbourn and H. Vainio
1990; 370 pages Sw. fr. 45.-

No. 15 Edited by M.-J. Ghess, J.D. Wilbourn and H. Vainio
1992; 318 pages Sw. fr. 45.-

NON-SERIAL PUBLICATIONS †

Alcool et Cancer
By A. Tuyns (in French only)
1978; 42 pages Fr. fr. 35.-

Cancer Morbidity and Causes of Death Among Danish Brewery Workers
By O.M. Jensen
1980; 143 pages Fr. fr. 75.-

Directory of Computer Systems Used in Cancer Registries
By H.R. Menck and D.M. Parkin
1986; 236 pages Fr. fr. 50.-

---

* Available from booksellers through the network of WHO Sales agents.

† Available directly from IARC

www.ingramcontent.com/pod-product-compliance
Ingram Content Group UK Ltd.
Pitfield, Milton Keynes, MK11 3LW, UK
UKHW051258180426
11947UKWH00020B/1788